Springer Finance

The Springer Finance series, launched in 1998, is addressed to students, academic researchers and practitioners working on increasingly technical approaches to the analysis of financial markets. It covers mathematical and computational finance broadly, reaching into foreign exchange, term structure, risk measure and management, portfolio theory, equity derivatives, energy finance and commodities, financial economics.

All titles in this series are peer-reviewed to the usual standards of mathematics and its applications.

Giulia Di Nunno · Yuliya Mishura ·
Anton Yurchenko-Tytarenko

Volterra Volatility Models

Option Pricing and Hedging

Springer

Giulia Di Nunno
Department of Mathematics
University of Oslo
Oslo, Norway

Anton Yurchenko-Tytarenko
Statkraft Energi AS
Oslo, Norway

Yuliya Mishura
Department of Probability Theory, Statistics and Actuarial Mathematics
Taras Shevchenko National University of Kyiv
Kyiv, Ukraine

ISSN 1616-0533 ISSN 2195-0687 (electronic)
Springer Finance
ISBN 978-3-032-26575-3 ISBN 978-3-032-26576-0 (eBook)
https://doi.org/10.1007/978-3-032-26576-0

This work was supported by Norges Forskningsråd (274410).

This Springer imprint is published by the registered company Springer Nature Switzerland AG
The registered company address is: Gewerbestrasse 11, 6330 Cham, Switzerland

To Maria and Christian
Giulia Di Nunno

To my best friends
Yuliya Mishura

To Hanna, Maryna and Vitaly
Anton Yurchenko-Tytarenko

Preface

Stochastic volatility models have become a key part of modern quantitative finance, especially in pricing derivatives, managing risk and designing hedging strategies. However, as financial markets evolve and we learn more from the data, our models need to keep up: methods that once seemed good enough often have to be updated to reflect things that have only recently come into focus. This book is our contribution to the continuous pursuit for better ways to describe how volatility behaves. It comes out of our long-term research project focused on building a volatility model that tackles some of the long-standing challenges in the field.

In our approach, we kept in mind several very simple requirements.

- To begin with, we want to build a framework that is flexible enough to capture a wide range of empirical features observed in the market. These include stylized facts such as implied volatility smiles and skews, long memory, roughness. Ideally, the model should be able to reproduce as many of these patterns as possible.
- Next, volatility must remain positive. This demand is very natural from the modeling perspective: after all, the very concept of *volatility* usually refers to a positive quantity. In addition, if volatility is allowed to reach zero, then essential mathematical tools, such as change-of-measure techniques, may break down. This can create serious issues when pricing derivatives or constructing hedging strategies.
- Finally, the model must allow for efficient numerical implementation. Indeed, many of the more statistically accurate models—especially those involving fractional Brownian motion—tend to be non-Markovian, which makes simulation and optimization very difficult. We want our approach to be computationally usable.

With these considerations in mind, we propose a model based on a certain class of stochastic integro-differential equations. The solution to such an equations turns out to be constrained—or "*sandwiched*"—between two deterministic bounds. Such a feature is enforced by an explosive drift term that "pushes" the paths away from the boundaries, like a repulsive magnetic force.

This sandwiched framework turns out to be surprisingly versatile. It works with a wide class of driving noises—essentially any process with Hölder-continuous paths—which opens the door to modeling a variety of empirically observed behaviors. On the numerical side, we develop simulation schemes that preserve the bounds of the process—they can be used in Monte Carlo pricing and hedging algorithms. All of this culminates in a stochastic volatility model that we have named the *Sandwiched Volterra Volatility (SVV) model*. Our intention is to accompany our reader through the development of this model step by step.

- Chapter 1 begins with a broad overview of stylized facts about volatility, along with a brief survey of existing continuous-time models. Our focus is on continuous models with no jumps. Our goal is to set the stage for the rest of the book by highlighting several modeling challenges that will be explored in more depth in the following chapters.
- Chapter 2 takes a closer look at the Cox-Ingersoll-Ross (CIR) process, the main component of the Heston stochastic volatility model. We focus on the behavior of its square root and explain how the structure of the drift term naturally keeps the process away from zero. This idea of a "repelling drift" later serves as the main motivation behind our sandwiched process construction.
- Chapter 3 introduces the sandwiched process itself—a stochastic equation with an explosive drift term that forces the solution to remain within fixed bounds.
- Chapter 4 applies this framework to extend classical processes like CIR by introducing memory into the driving noise.
- Chapter 5 turns to numerical implementation. We develop a drift-implicit simulation scheme that preserves the sandwiching property of the process.
- Chapter 6 moves into more advanced territory, studying sandwiched processes driven by Gaussian Volterra noise. We analyze their Malliavin differentiability and introduce Markovian approximations that will be a foundation of our hedging algorithms.
- Chapter 7 presents the full SVV model, bringing together the elements from the earlier chapters into a unified stochastic volatility framework. We demonstrate that the SVV model is analytically tractable and capable of generating implied volatility surfaces with realistic smiles and skews.
- Chapter 8 focuses on numerical pricing under the SVV model. We propose a Malliavin-duality-based algorithm that improves convergence of Monte Carlo pricing algorithms for options with discontinuous payoffs.
- Chapter 9 concludes with a discussion on hedging within the SVV framework, with particular attention to quadratic hedging strategies and their practical computation.
- The Appendix collects several useful results from stochastic analysis that are referenced in the main text.

The chapters ahead reflect the full journey of our work and we have tried to keep the overall flow as clear and focused as we could. In order to keep the exposition self-contained, we adopt a bottom-up approach. We start with technical results in stochastic calculus and move to the practice-oriented setting in the second part of the book. Readers who are primarily interested in the financial applications can skip the more theoretical Chaps. 2–6 and start directly at Chap. 7 that introduces the volatility modeling framework.

We hope this book will be useful to a wide audience: to researchers interested in stochastic analysis, practitioners looking for robust models, or students trying to make sense of the many interlaced components of modern volatility modeling. As proper of mathematics, this model which is born within finance has universal nature and we may expect future applications beyond the original context.

Oslo, Norway Giulia Di Nunno
Kyiv, Ukraine Yuliya Mishura
Oslo, Norway Anton Yurchenko-Tytarenko
February 2026

Acknowledgements We are truly grateful to all the colleagues and friends who supported us during the work on this book. The constant exchange of ideas, the questions during presentations at the many scientific events, the ongoing collaborations, and the many insightful discussions played a big role in shaping our thinking on the subject. Many have played a direct and indirect role in this, helping us mature this work. In particular, we would like to thank (in alphabetic ordering) Elisa Alòs, Saul Jacka, Kestutis Kubilius, Maasaki Fukasawa, Dilip Madan, Andrey Pilipenko, Kostiantyn Ralchenko, Aasmund Sande, Dennis Schroers, Martin Schweizer, Knut Sølna, and Lorenzo Torricelli. The list of people who have helped us is, in truth, much longer, and anyone not mentioned here is no less appreciated or important.

This book was written with support from the Research Council of Norway within the frame of the ToppForsk project no. 274410, *STORM: Stochastics for Time-Space Risk Models*. Yuliya Mishura also warmly thanks the Swedish Foundation for Strategic Research (grant no. UKR24-0004), Mälardalen University in Sweden and University of Padova for offering a calm and welcoming place to work during a very difficult time for Ukraine.

Competing Interests The authors have no competing interests to declare that are relevant to the content of this manuscript.

Contents

Acronyms

CBOE	Chicago Board Options Exchange
CEV	Constant elasticity of variance model
CIR	Cox-Ingersoll-Ross process
CKLS	Chan–Karolyi–Longstaff–Sanders process
ROU	Reflected Ornstein-Uhlenbeck process
SDE	Stochastic differential equation
SPX	The Standard and Poor's 500 Index
SVV	Sandwiched Volterra Volatility model
u.c.p	Uniform convergence in probability
VIX	The CBOE Volatility Index

Chapter 1
Volatility: Data, Models, and the Road Ahead

Abstract In this chapter, we provide a broad overview of continuous stochastic volatility models, tracing how the field has developed and identifying the key empirical patterns that have shaped its trajectory. We pay particular attention to fractional and rough volatility approaches: while not advocating for either long memory or roughness, we explain what motivates these ideas and walk through some of the most influential models. We also briefly discuss the challenges of VIX modeling and recent progress on the SPX–VIX joint calibration puzzle. Our aim is to highlight the progress made so far and to point toward several modeling challenges that the following chapters explore in more detail.

1.1 Introduction

At its core, finance revolves around the interplay between expected return and risk, and one of the standard measures for the latter is *volatility*. Informally speaking, this term refers to the degree of variability of an asset S over time which is traditionally quantified as

$$\sqrt{\frac{1}{T}\sum_{t=1}^{T}\left(\log\frac{S(t)}{S(t-1)}\right)^2}. \tag{1.1}$$

The use of logarithmic returns in (1.1) has a natural justification that can be traced back to P. Samuelson: since the prices are usually strictly positive, it is reasonable to model them as the exponential of some random variable. For example, if one assumes that the returns follow

$$\frac{S(t)}{S(t-1)} = e^{\xi_t}, \quad t = 1, \dots, T,$$

where ξ_t are i.i.d. normal random variables with mean zero and variance σ^2, the value (1.1) becomes a strongly consistent estimator for the *standard deviation* parameter σ.

G. Di Nunno et al., *Volterra Volatility Models*, Springer Finance,
https://doi.org/10.1007/978-3-032-26576-0_1

However, the metric in (1.1) is not as straightforward and innocent as it might initially appear. For example, calculating (1.1) over two different time periods can yield *drastically* different results. To some extent, this variability is not entirely surprising: markets naturally alternate between quiet phases with little price movement and turbulent periods with extreme fluctuations (especially during crises). Furthermore, what is also interesting is that such variability cannot always be fully explained by shifts in the underlying "fundamental" economic factors [60, 72]. The key takeaway from this observation is as follows:

market volatility changes over time in an unpredictable manner.

One prominent approach to capturing this behavior is to *model volatility itself as a stochastic process*. This idea, dating back to Clark's early discrete-time model [56], has proven remarkably successful and it has since developed into a rich, sophisticated, and very active area of research.

The first objective of this chapter is to provide an overview of this *stochastic volatility* framework. Differently from earlier surveys by Skiadopoulous [199], Shephard and Andersen [197] and Duong and Swanson [92], we put a particular emphasis on more recent advances in fractional and rough volatility models. Our intention here is to establish the groundwork for the rest of the book: namely, to illustrate the benefits of these approaches while also highlighting some of their limitations—the limitations that we seek to address through our own models and techniques in the subsequent chapters. This chapter is based on substantially extended and reworked material from [83].

Before proceeding to the main part of our historical detour, let us make two important disclaimers.

- First, beyond its more direct applications in areas like mean-variance portfolio theory originating from the paper [161] by Markowitz, the concept of volatility has a crucial role in another significant domain within finance: *no-arbitrage pricing theory* conceptualized by F. Black, M. Scholes, and R. Merton in [29, 166]. This theory offers a distinctive perspective on volatility and has been the major driving force behind volatility modeling for the past four decades. Accordingly, in this book we primarily examine volatility through the lens of *option pricing*.
- Second, this book targets *continuous stochastic volatility models in continuous time*. To keep the discussion focused, we have decided to impose some limitations on our presentation and not to cover several important—but distinct—modeling frameworks. For example, we do not address various discrete-time models such as ARCH, GARCH or EARCH developed by Engle [101], Bollerslev [33] and Nelson [174]. Also, we do not cover jump-diffusion models including the approaches of Duffie et al. [90] or Barndorff-Nielsen and Shephard [18]. While we acknowledge that jumps add an important layer of realism to financial modeling, doing justice to these models would require a substantial shift in emphasis that lies outside the scope of this book.

That said, readers with a specific interest in discrete-time models are referred to [35, 111]; for jump models, see the specialized books by Barndorff-Nielsen and Shephard [20], Gatheral [122], Rachev et al. [179] or Tankov and Cont [202]. For broader perspectives on financial modeling, we refer our readers to the overviews provided by Shiryaev [198] and Mariani and Florescu [160].

We have made an effort to keep the presentation as accessible as possible, and we hope that readers with a basic background in stochastic analysis will find the material in this chapter approachable, interesting, and useful.

1.2 Empirical Data and Stylized Facts

1.2.1 The Dawn of Times: Early Market Models

The first attempts to model financial markets using probabilistic methods can be traced back to the French economist Jules Regnault[1]: as early as 1863 [180], he proposed modeling stock prices with, what we call now, *symmetric random walks*. However, the foundation for using probabilistic modeling in finance is commonly attributed to another French economist, Louis Bachelier with his dissertation titled "*Théorie de la spéculation*" [14], 1900. There he considered market modeling and even derivative pricing in a very comprehensive manner. Of course, the tools of stochastic calculus had not yet been developed, which means that the dissertation might lack some mathematical rigor. Nonetheless, when translated into modern mathematical language—an excellent work in that regard was done by M. Davis and A. Etheridge in [15]—it becomes clear that Bachelier essentially modeled stock price dynamics using a prototype concept of a standard Brownian motion, five years before A. Einstein and his famous paper [96].

Unfortunately, Bachelier's ideas did not receive immediate recognition and remained largely overlooked for decades. Only more than fifty years after the publication of the original dissertation, statistician Jimmy Savage happened upon the work by chance and brought it to the attention of several researchers in economics. One of those researchers turned out to be Paul Samuelson who described Savage's finding in the foreword to the English translation of Bachelier's dissertation [15] as follows:

> ...Discovery or rediscovery of Louis Bachelier's 1900 Sorbonne thesis [...] initially involved a dozen or so postcards sent out from Yale by the late Jimmie Savage, a pioneer in bringing back into fashion statistical use of Bayesian probabilities. In paraphrase, the postcard's message said, approximately, 'Do any of you economist guys know about a 1914 French book on the theory of speculation by some French professor named Bachelier?'
>
> Apparently I was the only fish to respond to Savage's cast. The good MIT mathematical library did not possess Savage's 1914 reference. But it did have something better, namely Bachelier's original thesis itself.
>
> I rapidly spread the news of the Bachelier gem among early finance theorists. And when our MIT PhD Paul Cootner edited his collection of worthy finance papers, on my suggestion he included an English version of Bachelier's 1900 French text...

[1] For more details about this lesser known and somewhat underestimated figure and his contributions to finance, the reader is referred to the paper [146].

Samuelson observed that Bachelier's model allowed prices to take negative values, which usually does not happen in real-life stock prices. Therefore, he proposed[2] a simple, but very useful modification of Bachelier's original approach: instead of modeling the prices themselves, Brownian dynamics should be used to model the *logarithms of prices*. After a small adjustment with a linear trend, Samuelson's model took the form of a *geometric* (or of an *economic relative* as originally named by Samuelson himself [191]) *Brownian motion*:

$$S(t) = S(0) \exp\left\{\left(\mu - \frac{\sigma^2}{2}\right)t + \sigma W(t)\right\}, \quad \mu \in \mathbb{R}, \quad S(0), \ \sigma > 0, \tag{1.2}$$

or equivalently written as a stochastic differential equation (SDE),

$$dS(t) = \mu S(t)dt + \sigma S(t)dW(t). \tag{1.3}$$

Within the model (1.2)–(1.3), the term *volatility* acquires a very specific and concrete meaning: the parameter σ. This terminology is not only intuitive but also aligns well with the empirical metric (1.1) mentioned above: if $0 = t_0 < t_1 < ... < t_n = T$ is a partition of the interval $[0, T]$, then

$$\frac{1}{T}\sum_{k=1}^{n}\left(\log\frac{S(t_{k+1})}{S(t_k)}\right)^2 \to \sigma^2$$

in L^2 as the diameter of the partition $\max_k |t_{k+1} - t_k| \to 0$ (see, e.g., [182]). Due to its simplicity and tractability, the log-normal process (1.2)–(1.3), together with the corresponding notion of volatility, subsequently became a mainstream choice for stock price models for the next few decades. Even now, despite being well informed about multiple arguments against the geometric Brownian motion, practitioners still use this model as a benchmark or a reliable "first approximation".

In addition to market price modeling, Bachelier also considered the problem of option pricing and eventually derived an expression that can be called a precursor of the now famous *Black-Scholes formula*. Of course, Bachelier's derivation did not rely on the no-arbitrage principle and had several conceptual gaps—as it is often the case with pioneering works. The correction of those shortcomings became the subject of a number of studies in the 1960s, among which one can mention [36, 200, 203]. Samuelson himself also contributed heavily to this topic, see, e.g., [192] or his paper [193] co-authored with Robert Merton, where they suggested to consider a warrant/option payoff as a function of the price of the underlying asset. One could argue that these works were just a few steps away from the breakthrough made by Black, Scholes, and Merton just few years later.

[2] Samuelson himself acknowledged (see, e.g., his comments in the foreword to [15]) that the same idea was independently expressed by an astronomer M. Osborne in [177].

1.2.2 *Implied Volatility and the Smile Phenomenon*

Until the late 1960s, options markets remained relatively illiquid. A major reason for this was the absence of a consistent and widely accepted pricing methodology, and many serious investors viewed options as speculative instruments—closer to gambling than to legitimate financial assets. Somewhat ironically, even Robert Merton, right in his seminal 1973 paper [166], wrote the following:

> Because options are specialized and relatively unimportant financial securities, the amount of time and space devoted to the development of a pricing theory might be questioned...

However, in 1968, a demand for that type of contract suddenly emerged from the Chicago Board of Trade. The organization observed a significant decline in commodity futures trading on its exchange and, in response, sought to introduce new instruments to attract investors. They settled on options, and the Chicago Board Options Exchange (CBOE) officially started its operations in 1973. That same year saw the publication of two groundbreaking papers: "*The pricing of options and corporate liabilities*" [29] by Fischer Black and Myron Scholes and "*Theory of rational option pricing*" [166] by Robert Merton. The ideas expressed therein revolutionized mathematical finance and enjoyed overwhelming empirical success: Stephen Ross, for instance, claimed in 1987 [188] that

> When judged by its ability to explain the empirical data, option pricing theory is the most successful theory not only in finance, but in all of economics.

As a side note, the publication of the Black and Scholes paper was far from a smooth process: in 1987 [31], Black recalled that the manuscript was rejected first by the *Journal of Political Economy* and then by the *Review of Economics and Statistics*. The paper was published only after Eugene Fama and Merton Miller urged the *Journal of Political Economy* to reconsider its initial decision. In the meanwhile, Robert Merton showed a great deal of academic integrity by delaying the publication of his own article so that Black and Scholes would have precedence.

The main result of Black, Scholes and Merton can be formulated as follows: if a stock follows the model (1.2)–(1.3), then, under some assumptions, the discounted *no-arbitrage price* of a standard European call option $C^{\text{B-S}} = C^{\text{B-S}}(t, S)$ evolves as a function of the current time t and price S, and must satisfy a partial differential equation, now known as the *Black-Scholes formula*:

$$\frac{\partial C^{\text{B-S}}}{\partial t} + \frac{1}{2}\sigma^2 S^2 \frac{\partial^2 C^{\text{B-S}}}{\partial S^2} + rS\frac{\partial C^{\text{B-S}}}{\partial S} - rC^{\text{B-S}} = 0 \tag{1.4}$$

with a boundary condition[3]

$$C^{\text{B-S}}(T, S) = (S - K)_+, \tag{1.5}$$

[3] In what follows, we will utilize the standard notation $(x)_+ := \max\{x, 0\}$.

where r denotes the instantaneous interest rate that is assumed to be constant, T is the maturity date of the option, and K is its exercise price.

Of course, the Black–Scholes–Merton rationale relies on a number of rather abstract assumptions that do not fully align with real-world market conditions. In particular, their reasoning required specific price dynamics, the absence of transaction costs, and the possibility to buy and sell any quantity of assets. However, no model captures reality in full detail, and some simplifications are inevitable—Black, Scholes and Merton themselves were well aware of the limitations of their approach. As quoted in [104], Fischer Black offered the following reflection on the matter:

> Yet that weakness is also its greatest strength. People like the model because they can easily understand its assumptions. The model is often good as a first approximation, and if you can see the holes in the assumptions you can use the model in more sophisticated ways.

What really mattered was the successful empirical performance of the vanilla Black–Scholes–Merton model; as it was noted by J. Wiggins, one of the pioneers of continuous-time stochastic volatility modeling, "*given the elegance and tractability of the Black-Scholes formula, profitable application of alternate models requires that economically significant valuation improvements can be obtained empirically*" [205].

However, the market crash of October 1987—known as *Black Monday*—exposed significant weaknesses in the log-normal paradigm that could no longer be dismissed as acceptable modeling inaccuracies. Jackwerth and Rubinstein [143] described the problem as follows:

> Following the standard paradigm, assume that stock market returns are lognormally distributed with an annualized volatility of 20% (near their historical realization). On October 19, 1987, the two month S&P 500 futures price fell 29%. Under the lognormal hypothesis, this is a -27 standard deviation event with probability 10^{-160}. Even if one were to have lived through the entire 20 billion year life of the universe and experienced this 20 billion times (20 billion big bangs), that such a decline could have happened even once in this period is a virtual impossibility.

Clearly, the occurrence of such a "*virtually impossible*" event—one that wiped out portfolios and left investors on the brink of ruin—was a compelling reason to reevaluate existing financial models. Yet beyond this single catastrophic episode—which, in principle, might have been dismissed as a statistical anomaly—a deeper and more persistent phenomenon emerged in the aftermath of Black Monday: **the volatility smile**.

To explain the problem, let us first discuss the parameters of the Black–Scholes model in more detail. Note that the sole parameter in (1.4)–(1.5) that is not directly observable is the volatility σ: indeed, the interest rate r is set, the maturity date T, and the exercise price K are written in the specifications of the given option contract, whereas the price $S(t)$ can be read from the market. The volatility σ is unknown and needs to be retrieved from data. In 1976, Latané and Rendelman [153] proposed an elegant method to do that. Note that the Eqs. (1.4)–(1.5) has an explicit solution of the form

$$C^{\text{B-S}}(t,T,S(t),K,\sigma) = S(t)\Phi\left(\frac{\log\frac{e^{r(T-t)}S(t)}{K} + \frac{\sigma^2}{2}(T-t)}{\sigma\sqrt{T-t}}\right) - Ke^{-r(T-t)}\Phi\left(\frac{\log\frac{e^{r(T-t)}S(t)}{K} - \frac{\sigma^2}{2}(T-t)}{\sigma\sqrt{T-t}}\right), \tag{1.6}$$

where $\Phi(x) := \frac{1}{\sqrt{2\pi}}\int_{-\infty}^{x} e^{-\frac{y^2}{2}}dy$. Now, take the *actual* market price $C_t(T, K)$ of the corresponding option with payoff K and maturity date T at some evaluation time $t \geq 0$ and notice that, since $C^{\text{B-S}}(t, T, S(t), K, \sigma)$ is supposed to coincide with $C_t(T, K)$, the volatility σ can be retrieved by solving the equation

$$C^{\text{B-S}}(t,T,S(t),K,\sigma) - C_t(T,K) = 0. \tag{1.7}$$

Note that Eq. (1.7) depends on t, T and K, so its solution $\widehat{\sigma} = \widehat{\sigma}_t(T, K)$, known in the literature as the *implied volatility*, is technically a function of t, T and K as well. If the stock price model (1.2)–(1.3) reflects the reality well enough, we would expect $\widehat{\sigma}(t, T, K)$ to closely match the constant model parameter σ, and thus to remain approximately constant across different strikes K and maturities T for options on the same underlying asset.

However, this is *not* the case! For example, $\widehat{\sigma}_t(T, K)$ turns out to change with T for fixed K (see Fig. 1.1).

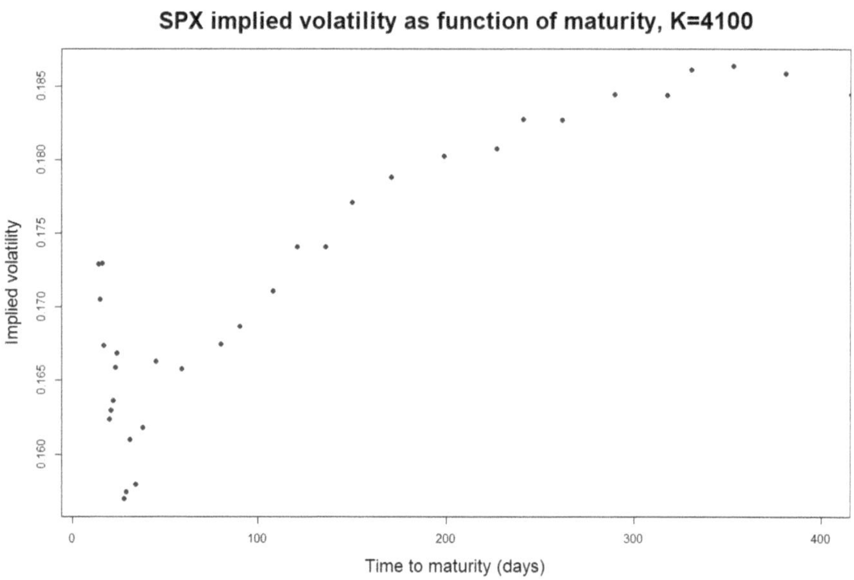

Fig. 1.1 Variation of implied volatility with time to maturity T for a fixed strike K. The values were calculated using S&P500 (SPX) option prices on May 3, 2023, retrieved from *Yahoo! Finance*

A natural extension of (1.2), first considered by Merton in his original paper [166], addresses this issue by allowing the volatility to be a deterministic function of time, $\sigma = \sigma(t)$. In that case, one can obtain a version of (1.6) of the form

$$\begin{aligned}
C^{\text{B-S}}(t, T, S(t), K, \sigma) &= S(t)\Phi\left(\frac{\log\frac{e^{r(T-t)}S(t)}{K} + \frac{1}{2}\int\limits_t^T \sigma^2(s)ds}{\sqrt{\int\limits_t^T \sigma^2(s)ds}}\right) \\
&\quad - Ke^{-r(T-t)}\Phi\left(\frac{\log\frac{e^{r(T-t)}S(t)}{K} - \frac{1}{2}\int_t^T \sigma^2(s)ds}{\sqrt{\int_t^T \sigma^2(s)ds}}\right) \\
&:= S(t)\Phi\left(\frac{\log\frac{e^{r(T-t)}S(t)}{K} + \frac{1}{2}\overline{\sigma}^2(t, T)(T-t)}{\overline{\sigma}(t, T)\sqrt{T-t}}\right) \\
&\quad - Ke^{-r(T-t)}\Phi\left(\frac{\log\frac{e^{r(T-t)}S(t)}{K} - \frac{1}{2}\overline{\sigma}^2(t, T)(T-t)}{\overline{\sigma}(t, T)\sqrt{T-t}}\right),
\end{aligned}$$

where $\overline{\sigma}^2(t, T) := \frac{1}{T-t}\int_t^T \sigma^2(s)ds$. Then, for a fixed evaluation time $t \geq 0$, the counterpart of the Eq. (1.7) gets the form

$$C^{\text{B-S}}(t, T, S(t), K, \overline{\sigma}(t, T)) - C_t(T, K) = 0,$$

so the solution $\widehat{\sigma}(t, T, K)$ approximates $\overline{\sigma}(t, T)$ and is therefore allowed to vary with T for a fixed K. One could even argue that this assumption of time-varying deterministic volatility is quite natural: as noted in [80, p. 144], "*there is nothing inconsistent about expecting high volatility this year and low volatility next year*".

As for the variation of implied volatility with respect to strike K for a fixed maturity T, it was long believed to be relatively insignificant. While *some* dependence on K was observed, it remained subtle enough to be largely ignored in practice—see, e.g., the discussion in [80, Chapter 1]. However, starting from the above-mentioned Black Monday crash in 1987, investors observed notable variability of the implied volatility with respect to K characterized by distinct convex patterns (see Fig. 1.2 as well as [74]) which were eventually called "*volatility smiles*" or "*volatility smirks*". This behavior was consistent, had a direct adverse impact on the empirical performance of the Black-Scholes formula, and could not be explained by the price dynamics (1.2)–(1.3).

Before we proceed further, let us provide a remark about the notation.

Notation 1.1 In the literature (see, e.g., [154, p. 244]), the implied volatility $\widehat{\sigma}$ is often written as a function $\widehat{\sigma}_{\text{log-m}}$ of time to maturity T and *log-moneyness* $\kappa := \log\frac{K}{e^{r(T-t)}S(t)}$, i.e.

$$\widehat{\sigma}_{\text{log-m}}(t, T, \kappa) := \widehat{\sigma}(t, T, S(t)e^{\kappa + r(T-t)}).$$

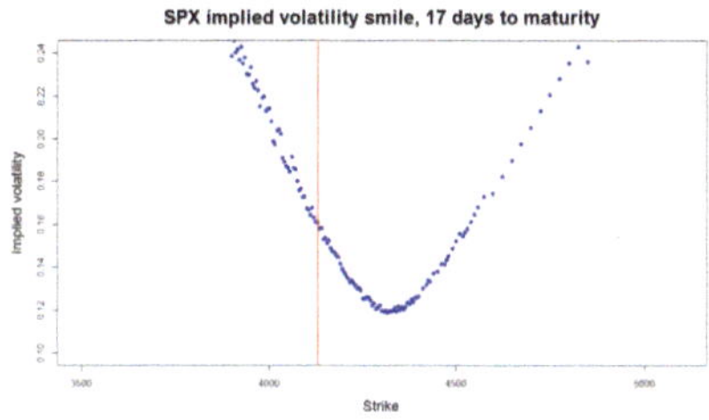

(a) 17 days to maturity

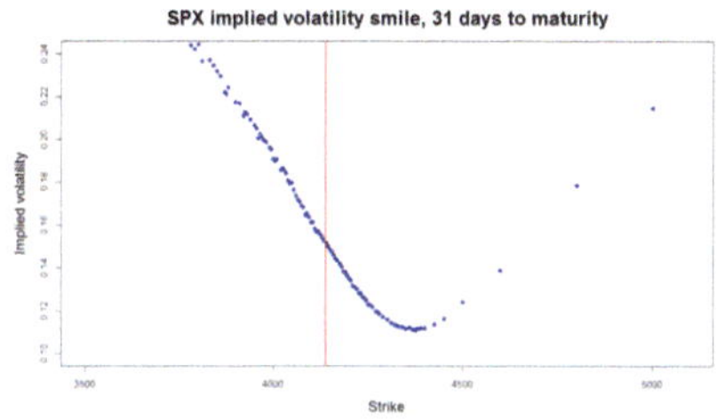

(b) 31 days to maturity

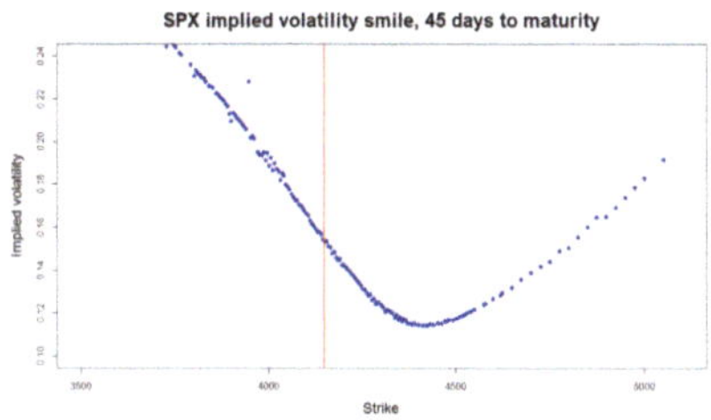

(c) 45 days to maturity

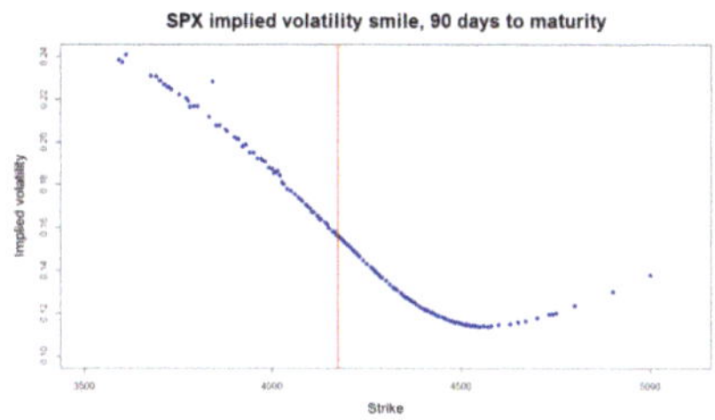

(d) 90 days to maturity

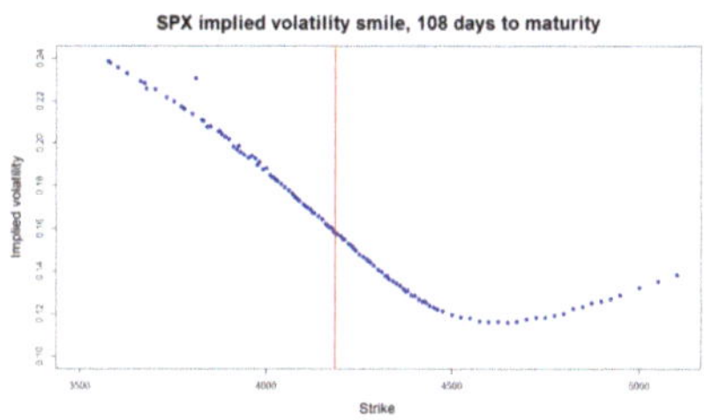

(e) 108 days to maturity

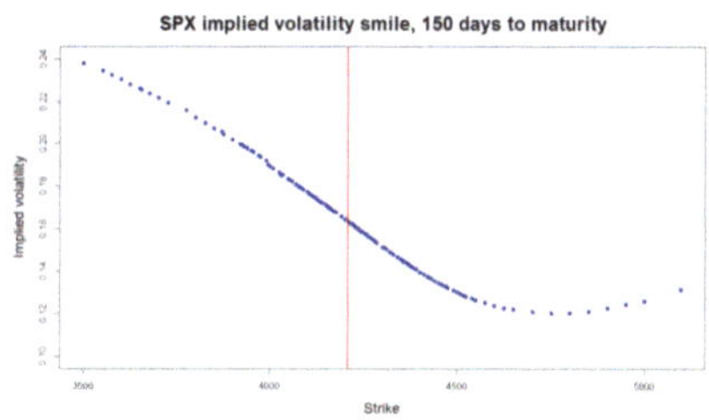

(f) 150 days to maturity

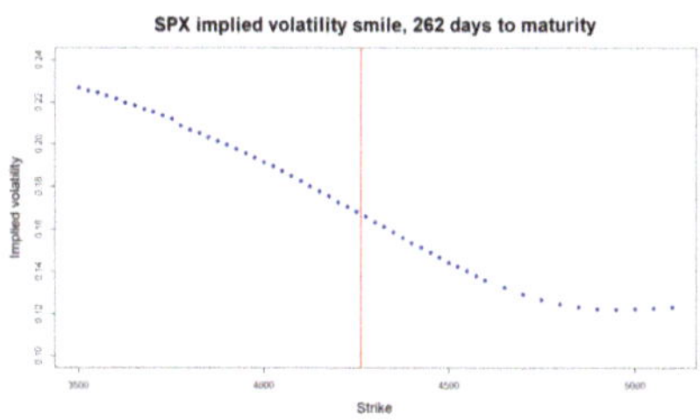

(g) 262 days to maturity

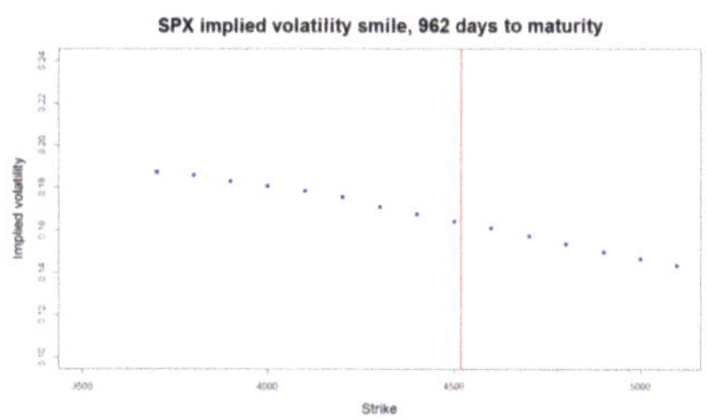

(h) 962 days to maturity

Fig. 1.2 Implied volatility smiles for different times to maturity. The values were calculated using S&P500 (SPX) option prices on May 3, 2023, retrieved from *Yahoo! Finance*. Vertical red lines correspond to the at-the-money level $K = S(0)e^{rT}$

In what follows, we will slightly abuse the notation and omit the subscript "log-m", using $\widehat{\sigma}(t, T, K)$ for the function of strike and $\widehat{\sigma}(t, T, \kappa)$ for the function of log-moneyness.

With these agreements in place, note that Fig. 1.2 also reveals two important typical characteristics of implied volatility smiles related to the change of the smile shape with T.

- First, the smile tends to flatten gradually as the time to maturity T increases. However, this flattening is fairly slow: for example, on Fig. 1.2, the curvature is noticeable at 262 days (Fig. 1.2g) and even at 962 days (Fig. 1.2h) before maturity. This long-range persistence of curvature in the smile requires a fairly nuanced treatment (see, e.g., [47, 57, 117]). We examine this issue in greater detail in Sect. 1.3.3 below.
- Second, observe the behavior of the smile *at-the-money*, i.e. when $K = S(0)e^{rT}$ (or, in terms of log-moneyness, when $\kappa := \log \frac{K}{e^{rT}S(0)} = 0$): as the expiry time T gets closer to the present moment $t = 0$, the smile at-the-money tends to become *steeper*. Figure 1.3 shows this phenomenon: to obtain it, we took the implied volatilities of 7 options with strikes closest to $S(0)e^{rT}$ for each T, performed the least squares linear fit and depicted the absolute values of the resulting slopes on Fig. 1.3a. It turns out that the variation of the absolute slopes with T for shorter maturities seems to be well-described by the *power law* $CT^{-\frac{1}{2}+H}$ with $H \approx 0$, *at least as the first approximation* (see the discussion at the end of Sect. 1.3.3 below as well as [132] and [78]). As an illustration, red lines on Fig. 1.3a and b depict power law fits for the SPX implied volatility slopes corresponding to May 3, 2023; for our dataset, $H = 0.06226572$. A similar type of behavior is reported by, e.g., Fouque et al. [107], Gatheral et al. [123] and, more recently, by Delemotte, De Marco and Segonne [78].

In principle, a "*perfect*" model should be able to reproduce[4] the shape of the entire *implied volatility surface* $(T, K) \mapsto \widehat{\sigma}_t(T, K)$ for any fixed $t \geq 0$ (or, equivalently, $(T, \kappa) \mapsto \widehat{\sigma}_t(T, \kappa)$, see Fig. 1.4). For example, if one wants to reflect the power law described above, one may demand that the model-generated *at-the-money volatility skew*

$$\Psi(t, T) := \left| \frac{\partial}{\partial \kappa} \widehat{\sigma}(t, T, \kappa) \right|_{\kappa=0} \tag{1.8}$$

[4] Here, by "*capturing the volatility surface*", we mean the established benchmark for assessing the performance of a model by its ability to reproduce the shape of $\widehat{\sigma}(T, K)$ or, equivalently, $\widehat{\sigma}(T, \kappa)$. As a rule, testing follows the following algorithm:

(1) the parameters of the model are calibrated with respect to the market data;
(2) the calibrated parameters are plugged into the model and the respective model no-arbitrage European option prices are computed;
(3) these "synthetic" prices are then used to construct the implied volatility surface by the procedure described here in Sect. 1.2.2;
(4) finally, the obtained model-generated surface is compared to the implied volatility surface extracted from real-world option prices.

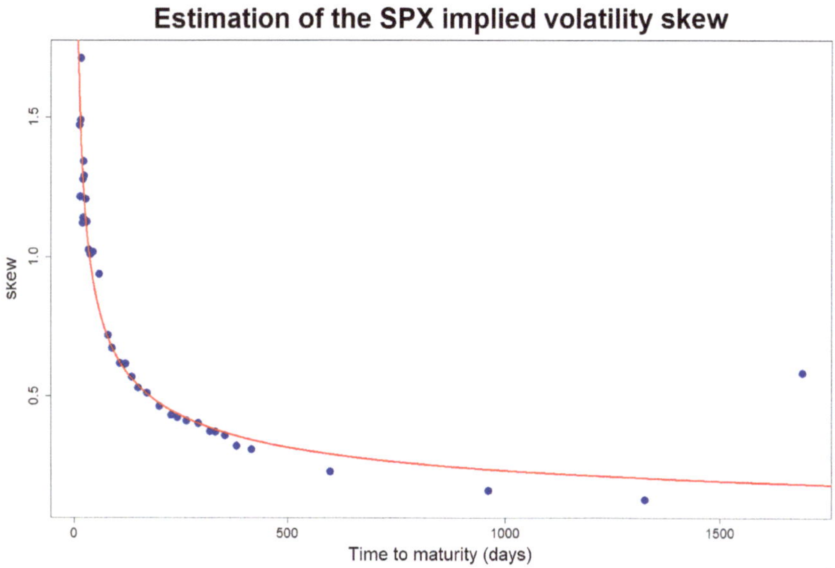

(a) Regular scale

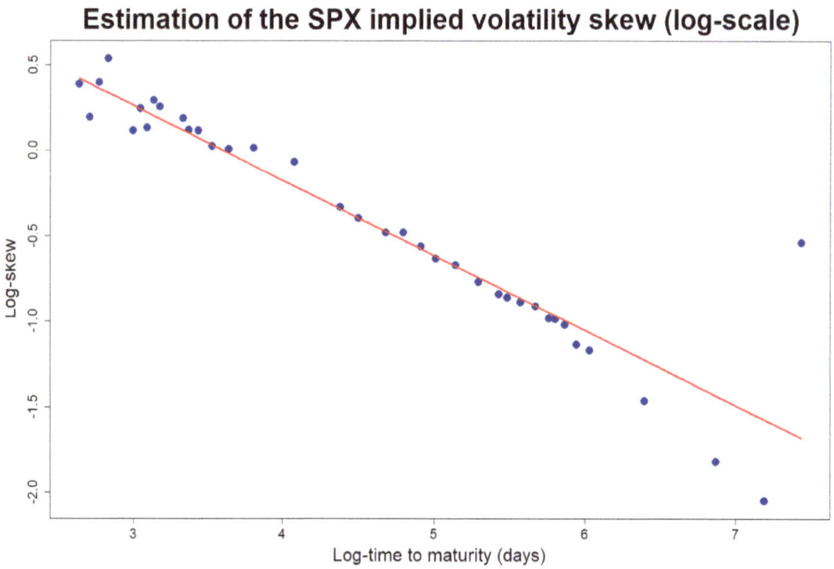

(b) Logarithmic scale

Fig. 1.3 Absolute values of at-the-money implied volatility slopes on regular (**a**) and log-log (**b**) scales. Blue dots denote the data and red lines denote the regression fits $e^{1.5753587}T^{-0.4377343}$ (**a**) and $1.5753587 - 0.4377343 \log T$ (**b**). The values were calculated using S&P500 (SPX) option prices on May 3, 2023, retrieved from *Yahoo! Finance*

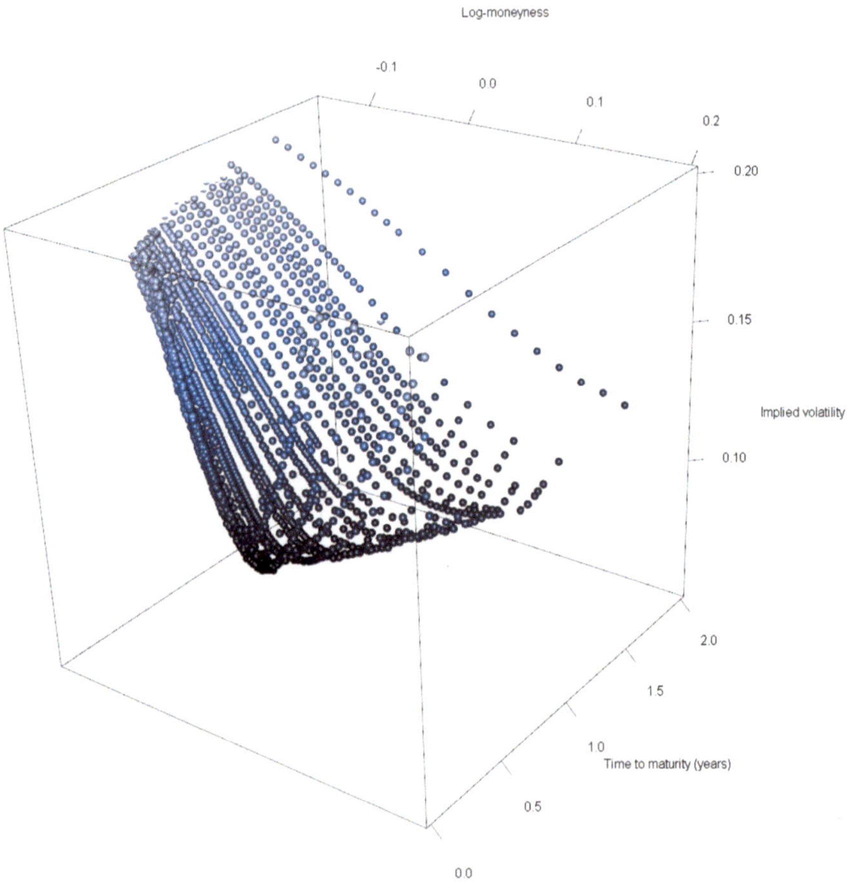

Fig. 1.4 Implied volatility surface $(T, \kappa) \mapsto \widehat{\sigma}(T, \kappa)$. The values were calculated using S&P500 (SPX) option prices on May 3, 2023, retrieved from *Yahoo! Finance*

has the power-law asymptotics $O((T - t)^{-\frac{1}{2}+H})$, $T \to t$. However, as we appreciate later, the task of constructing a model that reproduces all of the stylized facts simultaneously—both for long and short maturities—is not straightforward at all.

Notation 1.2 In the sequel, we often consider the case where the maturity time T approaches the fixed starting point t. In many applications, the primary focus is on the present moment $t = 0$, so we will frequently state our results for $t = 0$ fixed and $T \to 0$. To simplify our notation, we will omit the 0 from the list of arguments and simply write $C^{\text{B-S}}(T, S(0), K, \sigma)$, $C(T, K)$ and $\widehat{\sigma}(T, K)$ (or $\widehat{\sigma}(T, \kappa)$) for the Black–Scholes price, the observed option price and the implied volatility respectively. The same convention applies to the skew

$$\Psi(T) := \left| \frac{\partial}{\partial \kappa} \widehat{\sigma}(T, \kappa) \right|_{\kappa=0}, \tag{1.9}$$

for which we are interested in the asymptotic behavior as $T \to 0$.

The time parameter t will be written explicitly only when it is necessary to emphasize that the starting point may vary.

1.2.3 Fat Tails, Leverage, Clustering and Long Memory

The empirical evidence of volatility smiles makes a spectacular point against the geometric Brownian dynamics (1.2)–(1.3)—however, it is not the sole argument. In fact, objections to the log-normality of prices appeared as early as the log-normal model itself. In this section, we briefly enumerate several key stylized facts concerning log-returns and the volatility of financial time series. For a more detailed discussion, we also refer our readers to [125, Section 2.2], [112, Section 3], the well-known survey articles [60, 62] or the book [209].

Fat tails and non-Gaussian distribution of log-returns. One of the earliest and most well-documented empirical deviations from the geometric Brownian motion model concerns the distribution of asset returns. Statistical analyses have consistently shown that the distribution of log-returns exhibits fat tails–that is, extreme price movements occur far more frequently than predicted by the log-normal model. In this context, we mention the empirical studies of Mandelbrot [158] and Fama [103]; see also [64] for an early review. Overall, [62] gives the following summary regarding the properties of the "*true*" log-returns distribution: it tends to be non-Gaussian, sharp-peaked, and heavy-tailed. Numerous parametric models have been introduced to capture these characteristics and one can mention log-return models based on inverse Gaussian distributions [19], generalized hyperbolic distributions [178], truncated stable distributions [37, 59] and so on. For a more recent study on this topic, see also [102].

Leverage and Zumbach effects. Another interesting phenomenon not captured by the geometric Brownian motion is the so-called *leverage effect*: negative correlation between variance and returns of an asset. This empirical artefact, initially noticed by Black [30] and then pinpointed by Christie [54], Cheung and Ng [53] and Duffee [89], was explained by Black himself as follows: a decrease in a stock price results in a drop of firm's equity value, and hence increases its financial *leverage*—i.e., the ratio of debt to equity. This, according to Black, makes the stock riskier and hence more volatile. Although this explanation gave the leverage effect its name, subsequent research [5, 105, 140] pointed out that this correlation may not be connected to leverage at all. Zumbach in [209, Section 3.9.1] gives a different interpretation of this negative correlation: downward moves of stock prices are generally perceived as unfavorable, triggering many sales and thereby increasing volatility. Conversely, upward moves do not result in such drastic changes in investors' portfolios, given that the majority of market participants already hold long positions. Zumbach [208]

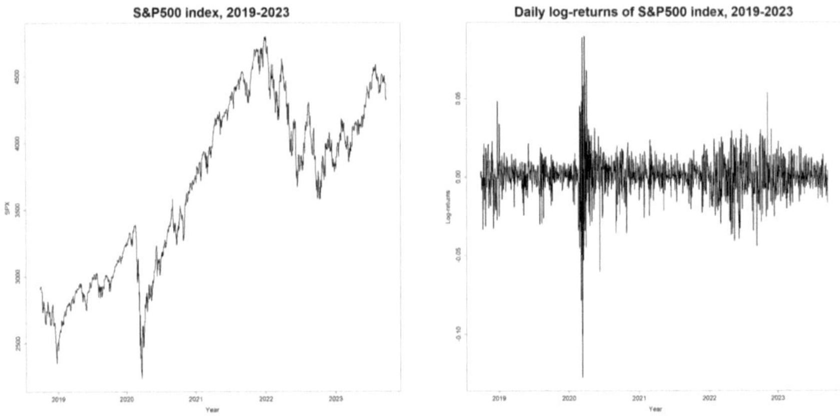

(a) Daily index data (b) Logarithmic returns

Fig. 1.5 Daily split-adjusted values of the S&P500 (SPX) index (**a**) and the corresponding logarithmic returns (**b**). Note that the amplitude of fluctuations in log-returns tends to form clusters over time; the period of the highest variation in March 2020 corresponds to the shock caused by the COVID-19 pandemic. The data is retrieved from *Yahoo! Finance*

also identified another behaviour of similar nature now recognized as the *Zumbach effect*: pronounced *trends* in stock price movement, irrespective of sign, increase the subsequent volatility. This effect stems from the fact that large price moves motivate investors to modify their portfolios, unlike scenarios where prices oscillate within a narrow range.

Volatility clustering. Another empirical challenge to the Black–Scholes–Merton framework arises from the observation of *volatility clustering* in financial time series. As noted by Mandelbrot [158], "*large changes tend to be followed by large changes, of either sign, and small changes tend to be followed by small changes*" (see Fig. 1.5b).

There are several ways to interpret this phenomenon. One common perspective (see, e.g., [109, Chapter 3]) is to link volatility clustering to the concept of *mean reversion*: in loose terms, this term refers to volatility's tendency to "return back" to the mean level of its long-run distribution (see, e.g., [109, Section 2.3.1 and Chap. 3]), with possible extended periods of staying above or below the latter. Another—and, perhaps, more nuanced—way to quantify clustering lies in analyzing the autocorrelation function of absolute log-returns (see, e.g., an excellent review [61] on the topic), i.e.

$$\operatorname{corr}(|R(t)|, |R(t+\tau)|), \tag{1.10}$$

where the log-return $R(t) := \log\left(\frac{S(t+\Delta)}{S(t)}\right)$ is defined for some given time scale Δ (which may vary between a fraction of a second for tick data to several days). Multiple empirical studies [34, 35, 39, 59–61, 87, 88, 128] report that the autocorrelation function (1.10) is consistently positive and, moreover, shows signs of slow decay of the type $O(\tau^{-\beta})$, $\tau \to \infty$, with an exponent $0 < \beta \le 0.5$.

Long range dependence. The $O(\tau^{-\beta})$ decay of (1.10) as $\tau \to \infty$ requires a separate discussion due to its far-reaching implications. For $\beta \in (0, 1)$, such behavior is often referred to as *long-range dependence* (see, e.g., [25, 190]). If statistically significant, it suggests the presence of *memory* in financial markets—i.e., that past volatility continues to influence the future over extended horizons.

However, confirming the long-range dependence in log returns is far from straightforward. By its nature, memory manifests itself when $\tau \to \infty$ which raises the concern that any statistical estimation procedure of (1.10) might be inconsistent due to e.g. the non-stationarity of financial time series over extended time periods (see, e.g., discussion in [167, Section 1.4]). Furthermore, as shown in [123, Section 4], model misspecification can produce *spurious* evidence of long memory, even if even if the "true" underlying process does not exhibit long memory at all. Nevertheless, several studies still report the presence of long range dependence on financial markets. For instance, Willinger et al. [206] apply the so-called *rescaled range* (R/S) *analysis* technique to the CRSP (Center for Research in Security Prices) daily stock time series and find some weak[5] evidence of memory in the data. Lobato and Velasco [157] analyze volatility in connection with trading volumes and find that both of these financial characteristics exhibit the same degree of long memory. Another interesting point comes from the analysis of the implied volatility surface: Comte and Renault [57] noticed that the decrease in the smile amplitude as the time to maturity increased was much slower than many advanced market models predicted. They argued that such an effect could be mimicked by having long memory in volatility (see also the simulation study [117] that directly confirms this claim).

1.3 Continuous Models of Volatility

To summarize, empirical evidence does not support the standard log-normal framework of (1.2)–(1.3). Fortunately, developments in option pricing theory subsequent to the seminal papers by Black, Scholes and Merton allowed for some decent flexibility in terms of the choice of price models. Namely, we refer to the gradual translation of the Black-Scholes-Merton approach into the language of martingale theory which evolved into the celebrated *Fundamental Theorem of Asset Pricing*—the result which connects no-arbitrage pricing and the existence of equivalent local martingale measures. In this regard, we mention the early research of Ross [187], Harrison and Kreps [138], Harrison and Pliska [139], Kreps [149] as well as subsequent seminal works of Delbaen and Schachermayer [75–77] (see also [194] for a detailed historical overview of the subject). This line of research has eventually evolved into a general

[5] As the authors write, "*...we find empirical evidence of long-range dependence in stock price returns, but because the corresponding degree of long-range dependence [...] is typically very low [...] the evidence is not absolutely conclusive*".

theory allowing for quite a broad variety of price models to choose from—hence giving researchers all the necessary tools to adjust the classical model (1.2) to account for volatility smiles and all other empirical inconsistencies.

As highlighted in Sect. 1.2.3, the particular problem of (1.3) lies in the fact that the volatility parameter σ cannot be constant. In reality, it varies unpredictably over time, exhibits correlation with the current asset price, forms persistent clusters of high and low values, and may even display signs of long-range dependence. One natural way to address this issue is to replace the constant volatility coefficient σ in (1.3) with a suitable *stochastic volatility* process $\{\sigma(t),\ t \geq 0\}$, i.e. by taking[6]

$$dS(t) = \mu(t)S(t)dt + \sigma(t)S(t)dW(t). \tag{1.11}$$

Then the problem seems to be reduced to the selection of a particular process $\{\sigma(t),\ t \geq 0\}$ that, once inserted in (1.11), can best reproduce the behavior of real-world prices. In this section, we list prominent continuous volatility modeling approaches and briefly characterize their performance.

1.3.1 Local Volatility

CEV model. The first cluster of models we cover are the so-called *local* or *deterministic* volatility models. The core idea behind this approach is the assumption that the process $\sigma = \{\sigma(t),\ t \geq 0\}$ in (1.11) is a non-random function of current price and time, i.e. $\sigma(t) = \sigma(t, S(t))$. Perhaps the first model of this kind was the *constant elasticity of variance (CEV)* model suggested by Cox in his 1975 note[7] [67]. CEV model assumes that $\sigma(t, S(t)) = \theta S^{\beta}(t)$, i.e. (1.11) takes the form

$$dS(t) = \mu S(t)dt + \theta S^{1+\beta}(t)dW(t), \tag{1.12}$$

where the parameter β is called the *elasticity parameter*. Initially, Cox considered $\beta \in [-1, 0)$: in such case, the increase in $S(t)$ decreases the value of $\sigma(t, S(t)) = \theta S^{\beta}(t)$ and vice versa hence reproducing the leverage effect. The case $\beta > 0$, more inherent to commodity prices, was discussed in [100]. For more details on the model (1.12), we also recommend a survey article [156].

As a side note regarding the CEV model, we observe that the SDE (1.12) exhibits substantially different behavior depending on the exact value of β. If $\beta \neq 0$, the diffusion coefficient in (1.12) is not Lipschitz and hence the existence and uniqueness of the solution cannot be guaranteed by the classical theory of stochastic differential

[6] In general, it is reasonable to treat the drift coefficient $\mu = \{\mu(t),\ t \geq 0\}$ as a stochastic process as well.

[7] The original 1975 note remained unpublished since J.C Cox viewed it as a simple extension of his earlier work with S. Ross [68]. The version [67] cited here appeared in the Journal of Portfolio Management in 1997 as a tribute to F. Black and his contribution to studying the leverage effect.

equations. For $\beta \in \left[-\frac{1}{2}, 0\right)$, the solution exists by the celebrated Yamada-Watanabe theorem (see, e.g., [207]), but the cases $\beta \in \left(-1, -\frac{1}{2}\right)$ and $\beta > 0$ require a separate and very careful treatment. For more details in that regard, we refer the reader to [52, Chapter 5] as well as to the discussion in [11].

Local volatility and Dupire formula. The general local volatility was initially considered by Derman and Kani [79] and Dupire [94]. Both assumed the risk-neutral dynamics of the form

$$dS(t) = rS(t)dt + \sigma(t, S(t))S(t)dW(t), \tag{1.13}$$

where r denotes the interest rate. Note that this model is far more flexible than it seems at first glance. Unlike the CEV approach, (1.13) does not specify any parametric form of the function σ: instead, the no-arbitrage principle and the properties of diffusions allow to fully recover σ from option prices using the celebrated Dupire formula (for its derivation, see the original paper [94] or [26, Section 2.2.1]): assuming no dividends,

$$\sigma(t, S) = \sqrt{2 \frac{\frac{\partial C}{\partial T}(T, K) + rS\frac{\partial C}{\partial K}(T, K)}{S^2 \frac{\partial^2 C}{\partial K^2}(T, K)}\Bigg|_{(T,K)=(t,S)}}, \tag{1.14}$$

where $C(T, K)$, as before, denotes the price at $t = 0$ of the European call-option with payoff K and maturity T. It is possible to prove (see, e.g., [26, Section 2.2.2]) that the no-arbitrage conditions imply that the expression in the right-hand side of (1.13) is well-defined. Moreover, recalling that

$$C(T, K) = C^{\text{B-S}}(T, S(0), K, \widehat{\sigma}(T, K)), \tag{1.15}$$

where $C^{\text{B-S}}$ is the Black-Scholes price, one can plug the right-hand side of (1.15) to (1.14) in place of $C(T, K)$ and obtain the theoretical relation between the implied volatility $\widehat{\sigma}$ and local volatility function σ that can then be used to accurately calibrate (1.13) to the current implied volatility smiles (see, e.g., [26, Section 2.3]). For more details on local volatility models, we refer the reader to [26, Chapter 2].

Shortcomings of local volatility. Local volatility models are perhaps the simplest and the most straightforward generalizations of the original log-normal model (1.2)–(1.3). In addition to their simplicity and tractability, they inherit another convenient property of the classical geometric Brownian motion: one can show that the market produced by (1.13) is complete, i.e. that any option with S as an underlying asset can be perfectly hedged by a self-financing portfolio composed exclusively of S and a riskless asset. However, despite the obvious mathematical attractiveness of this feature, some sources actually view completeness as a disadvantage from the modeling viewpoint. As noted in [202, Chapter 10],

> While absence of arbitrage is both a reasonable property to assume in a real market and a generic property of many stochastic models, market completeness is neither a financially

realistic nor a theoretically robust property. From a financial point of view, market completeness implies that options are redundant assets[8] and the very existence of the options market becomes a mystery, if not a paradox in such models.

This observation suggests that local volatility models may be too restrictive to reflect the complexity of the stock dynamics. And, importantly, such a claim is supported by multiple studies. For example, Buraschi and Jackwerth [43] conduct a formal statistical testing procedure to check whether options are indeed "*redundant*", and their findings strongly reject this hypothesis. Another empirical study by Dumas et al. [91] analyzes the predictive performance of (1.13) and concludes that its out-of-sample results are no better than ones of the standard Black-Scholes approach. Gatheral et al. [123] indicates that the dynamics (1.13) tends to generate future volatility surfaces "*completely unlike those we observe*". Aït-Sahalia and Jacod [4] utilize a statistical test based on local time to check whether stock prices follow[9] the SDE of the form

$$dS(t) = a(S(t))dt + b(S(t))dW(t),$$

where a and b are some deterministic functions, and report a "*clear rejection*" of such model, advocating for alternative approaches. For more details on the performance of local volatility models, see also an overview in [199, Section 4].

1.3.2 Stochastic Volatility Models

Another important modeling approach, broadly referred to as *stochastic volatility*[10], introduces a dedicated stochastic process that exhibits only partial dependence with S. For example, $\sigma = \{\sigma(t),\ t \geq 0\}$ may be another diffusion driven by a separate Brownian motion $B = \{B(t),\ t \geq 0\}$ correlated with W. Naturally, within this framework, there are countless candidates for σ, and choosing a particular one that characterizes the market well enough is a complex task. However, the stylized facts presented in Sect. 1.2.3 offer several points that are useful to keep in mind when selecting a stochastic volatility model.

- To begin with, given the nature of volatility, it seems reasonable to consider a non-negative process. In practice, this can be achieved by modeling log volatility and then taking an exponential. Alternatively, one may utilize non-negative diffusions such as Bessel-type processes (see, e.g., [182, Chapter XI]).

[8] Here "redundant" is understood in the sense that options are perfectly replicable in a complete market.

[9] In fact, they consider even more general framework with jumps and microstructure noise and still reject the local volatility hypothesis.

[10] A note on the terminology: the term "stochastic volatility" may seem too vague given that the volatility $\sigma(t, S(t))$ assumed in local models is also a stochastic process. Nevertheless, in the literature, the term "stochastic" generally refers to approaches that treat volatility separately from the price dynamics.

- Selecting an appropriate dependence structure between B and W from (1.11) provides all the necessary tools to account for the leverage effect: a typical assumption is $\mathbb{E}[W(t)B(t)] = \rho t$ with $-1 < \rho < 0$.
- Another property shared by multiple stochastic volatility models is *mean-reversion*. As mentioned above in Sect. 1.2.3, this behavior seems to be a common trait for real-life volatility. Moreover, a thoroughly chosen mean-reverting process can mimic the clustering effect to some extent (see, e.g., [109, Chapter 3]). A common approach to introduce mean reversion to the dynamics is to take a drift term of the form
$$\theta_1(\theta_2 - \sigma(t))$$
in the SDE for volatility. This drift "*pulls σ back*" to the level θ_2 whenever σ deviates from it. The parameter θ_1 calibrates the speed of mean-reversion.

Starting from the 1987 pioneering works of Hull and White [142], Wiggins [205] and Scott [196], multiple generations of approaches and numerous models have emerged in the literature. While not even attempting to provide an exhaustive list, we present below a selection of notable contributions.

- Hull and White [142] assumed that the squared volatility $\sigma^2 = \{\sigma^2(t),\ t \geq 0\}$ is itself a geometric Brownian motion, i.e. price and volatility satisfy stochastic differential equations of the form
$$\begin{aligned} dS(t) &= \mu S(t)dt + \sigma(t)S(t)dW(t),\\ d\sigma^2(t) &= \theta_1\sigma^2(t)dt + \theta_2\sigma^2(t)dB(t) \end{aligned}$$
respectively, where B and W are two Brownian motions that are allowed to be correlated to account for the leverage effect. Note that the volatility process is positive but not mean-reverting.
- Wiggins [205] suggested a slightly more general dynamics of the form
$$\begin{aligned} dS(t) &= \mu S(t)dt + \sigma(t)S(t)dW(t),\\ d\sigma(t) &= f(\sigma(t))dt + \theta\sigma(t)dB(t). \end{aligned}$$
- Scott [196] and Stein and Stein [201] considered the volatility to be an Ornstein-Uhlenbeck process, i.e.
$$\begin{aligned} dS(t) &= \mu S(t)dt + \sigma(t)S(t)dW(t),\\ d\sigma(t) &= \theta_1(\theta_2 - \sigma(t))dt + \theta_3 dB(t). \end{aligned} \tag{1.16}$$
In this case σ is not positive. Indeed, Ornstein-Uhlenbeck process is Gaussian and hence can take negative values with positive probability. In practice, this issue is treated by either taking the absolute value of σ or introducing a reflecting barrier to the volatility dynamics [195].
- Heston [141] introduced the SDE of the form

$$\begin{aligned} dS(t) &= \mu S(t)dt + \sqrt{\sigma(t)}S(t)dW(t), \\ d\sigma(t) &= \theta_1(\theta_2 - \sigma(t))dt + \theta_3\sqrt{\sigma(t)}dB(t). \end{aligned} \tag{1.17}$$

In this model, now commonly referred to as the *Heston model*, the volatility follows the so-called *Cox-Ingersoll-Ross* or *square root* process (see also [66]) which enjoys strict positivity provided that $2\theta_1\theta_2 \geq \theta_3^2$. Benhamou et al. [23] considered a modification of (1.17) of the form

$$\begin{aligned} dS(t) &= \sqrt{\sigma(t)}S(t)dW(t), \\ d\sigma(t) &= \theta_1(\theta_2(t) - \sigma(t))dt + \theta_3(t)\sqrt{\sigma(t)}dB(t), \\ d\langle W, B\rangle_t &= \rho(t)dt, \end{aligned}$$

with time-dependent θ_2, θ_3 and correlation ρ to account for structural changes on the market. Another modification of (1.17) was considered in [127]: there, the discounted price dynamics is assumed to follow

$$\begin{aligned} dS(t) &= \sqrt{\sigma(t)}S(t)dW(t), \\ d\sigma(t) &= \theta_1(Z_t)(\theta_2(Z_t) - \sigma(t))dt + \theta_3(Z_t)\sqrt{\sigma(t)}dB(t), \\ d\langle W, B\rangle_t &= \rho(Z_t)dt, \end{aligned}$$

where Z is a homogeneous continuous-time Markov chain that represents market switching between different regimes.

- Melino and Turnbull [163, 164] considered the model of the form

$$\begin{aligned} dS(t) &= (a + bS(t))dt + \sigma(t)S^\beta dW(t), \\ d\log(\sigma(t)) &= (\theta_1 + \theta_2\log(\sigma(t)))\,dt + \theta_3 dB(t), \end{aligned} \tag{1.18}$$

where $\beta \in \left[\frac{1}{2}, 1\right]$, i.e. the log volatility is a mean-reverting Ornstein-Uhlenbeck process and S can be regarded as a combination of the CEV Eq. (1.12) with stochastic volatility.

- Hagan et al. [136] proposed the *stochastic alpha-beta-rho (SABR) model* of the form

$$\begin{aligned} dS(t) &= \sigma(t)S^\beta(t)dW(t), \\ d\sigma(t) &= \alpha\sigma(t)dB(t). \end{aligned} \tag{1.19}$$

where $\alpha \geq 0$, $0 \leq \beta \leq 1$ and $\mathbb{E}[W(t)B(t)] = \rho t$, $-1 < \rho < 1$. Note that the case $\beta \in \left(0, \frac{1}{2}\right)$ is quite intricate from the mathematical perspective and requires a special treatment; see [52] for more details.

- Lewis [155] and, later, Carr and Sun [46] (see also Baldeaux and Badran [17, Section 2]) considered the so-called *3/2-model*

$$
\begin{aligned}
dS(t) &= S(t)\sqrt{\sigma(t)}dW(t), \\
d\sigma(t) &= \kappa\sigma(t)(\theta - \sigma(t))dt + \varepsilon\sigma^{3/2}(t)dB(t).
\end{aligned} \tag{1.20}
$$

The motivation behind this model partially comes from the statistical analysis of volatility models performed by Jawaheri [145] and Bakshi et al. [16] as well as from its ability to replicate the VIX skew (see Sect. 1.4 for further details). For more details on the existence and properties of the solution to (1.20), we refer the reader to [52].

- Gatheral [121] presented the double CEV dynamics of the form

$$
\begin{aligned}
dS(t) &= S(t)\sqrt{\sigma(t)}dW(t), \\
d\sigma(t) &= -\kappa(\sigma(t) - \sigma'(t))dt + \eta\sigma^{\alpha}(t)dB(t), \\
d\sigma'(t) &= -\kappa'(\sigma'(t) - \theta)dt + \eta'\sigma'^{\beta}(t)dB'(t),
\end{aligned} \tag{1.21}
$$

where $\alpha, \beta \in \left[\frac{1}{2}, 1\right]$;

- Fouque et al. [110] propose a *multiscale* stochastic volatility. Their approach lies in modeling volatility with two diffusions, one fluctuating on a fast "time scale", and the other fluctuating on a "slow" time scale. Namely, their model takes the form

$$
\begin{aligned}
dS(t) &= \mu S(t)dt + \sigma(t)S(t)dW(t), \\
\sigma(t) &= f(Y(t), Z(t)), \\
dY(t) &= \frac{1}{\varepsilon}(\theta_1 - Y(t))dt + \frac{\theta_2\sqrt{2}}{\sqrt{\varepsilon}}dB_1(t) \\
dZ(t) &= \delta c(Z(t)) + \sqrt{\delta}g(Z(t))dB_2(t),
\end{aligned}
$$

where f is a bounded positive function, $\delta, \varepsilon > 0$ are assumed to be small, Y is the fast scale volatility factor, i.e. a fast mean-reverting process, and Z is the slow scale volatility factor. Later in [108], Fouque and Saporito employ the same multiscale paradigm to model the Heston-type stochastic volatility-of-volatility parameter:

$$
\begin{aligned}
dS(t) &= S(t)\sqrt{\sigma(t)}dW(t), \\
d\sigma(t) &= \kappa(\theta - \sigma(t))dt + \eta(t)\sqrt{\sigma(t)}dW'(t), \\
\eta(t) &= f(Y(t), Z(t)), \\
dY(t) &= \frac{\sigma(t)}{\varepsilon}\alpha(Y(t))dt + \sqrt{\frac{\sigma(t)}{\varepsilon}}\beta(Y(t))dB(t), \\
dZ(t) &= \sigma(t)\delta c(Z(t))dt + \sqrt{\delta\sigma(t)}g(Z(t))dB'(t).
\end{aligned} \tag{1.22}
$$

In addition to reproducing the leverage effect and clustering via mean reversion, Brownian stochastic volatility models turn out to have an additional important advantage. To some extent, they have the ability to capture "smiley" patterns of the implied volatility (see, e.g., [181], [109, Section 2.8.2] or [122]). However, by design, parametric stochastic volatility models impose structural constraints on the relationship between the dynamics of the spot and implied volatilities. Hence they may not be able to capture the exact shape of the implied volatility surface (see, e.g., [122, Figure 3.6]).

A detailed analysis of this matter can be found in [154] and [8]. In particular, as stated in [8, Section 7.1] or [154, Remark 11.3.21], Brownian diffusion models of volatility tend to produce the at-the-money skew $\Psi(T) = O(1)$, $T \to 0$, where Ψ is defined by (1.9). This directly contradicts the empirical behavior of $O(T^{-\beta})$, $\beta \approx \frac{1}{2}$, mentioned in Sect. 1.2.3. In addition, as highlighted by Comte and Renault [57] (see also [47]), the decrease of the real-life volatility smile amplitude as $T \to \infty$ seems to be much slower than predicted by the classical Brownian diffusions. Comte and Renault connect this phenomenon to the *long-range dependence* in the volatility dynamics, which is well-aligned with several empirical studies listed above in Sect. 1.2.3 that also report long memory in the market.

In other words, the inherent properties of Brownian motion limit the modeling capabilities of diffusion models and it is no surprise that a lot of effort was made to advance the stochastic volatility framework further to account for the mentioned inconsistencies.

1.3.3 Fractional and Rough Models

One of the ways to extend stochastic volatility models described in Sect. 1.3.2 involves substituting the standard Brownian driver B with an alternative process possessing the attributes to capture the intended stylized facts. Perhaps the most common option utilized in the literature is *fractional Brownian motion* $B^H = \{B^H(t), t \geq 0\}$ defined as a Gaussian process with $B^H(0) = 0$ a.s., $\mathbb{E}[B^H(t)] = 0$ for all $t \geq 0$ and

$$\mathbb{E}\left[B^H(t)B^H(s)\right] = \frac{1}{2}\left(t^{2H} + s^{2H} - |t-s|^{2H}\right), \quad s, t \geq 0, \tag{1.23}$$

where H can take values in $(0, 1)$ and is called the *Hurst index*[11]. Fractional Brownian motion was initially considered by Kolmogorov [148] and later reintroduced by Mandelbrot and van Ness [159] who also obtained its Volterra-type representation

[11] An interesting historical remark that highlights the spectrum of different applications of mathematics: this parameter is named after Harold Edwin Hurst (1880–1978) who studied long-range dependence in fluctuations of the water level in the Nile River.

$$B^H(t) := \frac{1}{\Gamma\left(H+\frac{1}{2}\right)} \int_{-\infty}^{0} \left((t-s)^{H-\frac{1}{2}} - (-s)^{H-\frac{1}{2}}\right) dB(s) + \frac{1}{\Gamma\left(H+\frac{1}{2}\right)} \int_{0}^{t} (t-s)^{H-\frac{1}{2}} dB(s), \tag{1.24}$$

where $B = \{B(t),\ t \geq 0\}$ is a standard Brownian motion. In the literature, a truncated version of (1.24) called the *Riemann-Liouville fractional Brownian motion* is often used:

$$B^H_{RL}(t) := \frac{1}{\Gamma\left(H+\frac{1}{2}\right)} \int_{0}^{t} (t-s)^{H-\frac{1}{2}} dB(s), \tag{1.25}$$

and the Volterra kernel

$$\mathcal{K}(t,s) := \frac{1}{\Gamma\left(H+\frac{1}{2}\right)} (t-s)^{H-\frac{1}{2}} \mathbb{1}_{s<t} \tag{1.26}$$

in (1.25) is called the *fractional* kernel.

Fractional Brownian motion is extremely convenient from the mathematical perspective: it is the only stochastic process that simultaneously [171]

- is Gaussian,
- has stationary increments and
- is *self-similar*, i.e.

$$B^H(at) \overset{\text{Law}}{=} a^H B^H(t), \quad \forall a \geq 0. \tag{1.27}$$

In addition, if $H = 1/2$, B^H coincides with a standard Brownian motion and hence can be viewed as a broad generalization of the latter. Interestingly, the value $H = 1/2$ turns out to be the boundary between two distinct volatility modeling paradigms: one favoring $H > 1/2$ and the other advocating for $H < 1/2$, and the aim of this subsection is to characterize both of them.

Fractional models with long memory: $H > 1/2$. First of all, observe that

$$\begin{aligned} \mathbb{E}\left[B^H(1)\left(B^H(n) - B^H(n-1)\right)\right] &= \frac{1}{2}\left(n^{2H} - 2(n-1)^{2H} + (n-2)^{2H}\right) \\ &\sim H(2H-1)n^{2H-2}, \quad n \to \infty. \end{aligned} \tag{1.28}$$

Therefore, if $H \in (1/2, 1)$, the autocorrelation function behaves as $O(n^{-\beta})$ with $\beta \in (0, 1)$, revealing the long memory property. In particular, if $H \in (3/4, 1)$, the behavior of (1.28) matches the empirical estimates $0 < \beta \leq \frac{1}{2}$ for absolute log-returns highlighted in Sect. 1.2.3.

Historically, the long memory of fractional Brownian motion for $H > 1/2$ was the original reason to employ the latter in volatility modeling. Namely, in 1998, Comte and Renault [57] suggested the first continuous time fractional volatility model of the form

$$\sigma(t) = \theta_1 \exp\left\{\theta_2 \int_0^t e^{-\theta_3(t-s)} dB^H(s)\right\}, \tag{1.29}$$

which can mimic volatility persistence and explain slow decays in the smile amplitude of implied volatility surfaces when $T \to \infty$ (in this regard, we also recommend the simulation study [117] that illustrates this phenomenon numerically). Other contributions studying stochastic volatility driven by fractional Brownian motion with $H > 1/2$ include:

- Reference [185], which considers the model of the form

$$dS(t) = \sigma(t) dW(t),$$
$$\sigma(t) = F\left(\int_0^t a(t, u) dB^H(u) + f(t)\xi_0\right)$$

  where W and B^H are independent, F, a and f are nuisance parameters and ξ_0 is a random initial condition;
- References [55] and [27], which discuss a fractional counterpart of the model (1.16):

$$\begin{aligned} dS(t) &= \mu S(t)dt + \sigma(Y(t))S(t)dW(t),\\ Y(t) &= -\theta_1 Y(t)dt + \theta_2 dB^H(t), \end{aligned} \tag{1.30}$$

  where $\mu, \theta_1, \theta_2 > 0$ and σ is a deterministic function that is additionally assumed to have sublinear growth in [27];
- the series of papers [168, 169, 172] that proposes

$$\begin{aligned} dS(t) &= \mu S(t)dt + \sigma(Y(t))S(t)dW(t),\\ dY(t) &= \frac{1}{2}\left(\frac{\theta_1}{Y(t)} - \theta_2 Y(t)\right)dt + \frac{\theta_3}{2} dB^H(t), \end{aligned} \tag{1.31}$$

  where $\mu, \theta_1, \theta_2, \theta_3 > 0$ and σ is a given function with sublinear growth. Note that (1.31) can be regarded as a fractional extension of the Heston model (1.17) since the process $X(t) = Y^2(t)$ satisfies the pathwise SDE

$$dX(t) = (\theta_1 - \theta_2 X(t))dt + \theta_3\sqrt{X(t)} dB^H(t).$$

Note that such models also allow to capture the leverage effect by imposing a suitable dependence structure between the Brownian motion W driving the price and the fractional Brownian motion B^H driving the volatility. A typical strategy (used, for example, in [27] or [172]) is to start from a Volterra-type representation of B^H with respect to a standard Brownian motion B and to assume that B and W are correlated.

In some models, long memory is incorporated not directly through fractional Brownian motion, but rather by combining standard Brownian diffusions with the fractional kernel (1.26). Examples of this approach can be found in e.g. [21], where the prices are assumed to follow

$$dS(t) = \mu(t)S(t)dt + \sigma(t)S(t)dW(t),$$

$$\sigma^2(t) = \theta + \frac{1}{\Gamma\left(H+\frac{1}{2}\right)}\int_{-\infty}^{t}(t-s)^{H-\frac{1}{2}}X(s)ds,$$

$$dX(t) = \theta_1(\theta_2 - X(t))dt + \theta_3\sqrt{X(t)}dB(t),$$

where W and B are correlated standard Brownian motions.

As a final remark, we mention the special interplay between long memory and self-similarity (1.27) properties of fractional Brownian motion. As previously discussed in Sect. 1.2.3, any measures of long-range dependence are hard to estimate statistically: the analysis of time series during overly extended time periods naturally raises concerns about data non-stationarity. However, if data are additionally assumed to display self-similarity, their long-term behavior can be inferred from high-frequency observations over shorter periods. For a detailed discussion of self-similarity in financial time series as well as its connection with long memory, we refer the reader to [60, Sects. 2.3 and 2.4].

Rough revolution: $H < 1/2$. As highlighted in the simulation study [117], fractional Brownian motion with $H > 1/2$ indeed allows to capture the behavior of implied volatility surfaces for longer maturities. However, a high Hurst index does not seem to have any positive impact on the short-term fit. For example, paper [8] analyzes the behavior of the short-term implied volatility skew (1.8) generated by a variation of the model (1.30) with $H > 1/2$. They analytically prove that this skew evaluated at a fixed time $t \geq 0$ does not behave as $O((T-t)^{-\beta})$, $\beta \approx \frac{1}{2}$, for $T \downarrow t$. Interestingly, in Sect. 7.2.2 of the same paper, they notice that the stochastic volatility model driven by a Riemann-Liouville fractional Brownian motion with $H \in (0, 1/2)$ produces the required skew asymptotics $O((T-t)^{-\frac{1}{2}+H})$ as $T \downarrow t$.

The arguments in [8] were based on *Malliavin calculus* (see, e.g., [7, 175]): under certain regularity assumptions, the short-term explosion of the implied volatility skew translates to the explosion of the Malliavin derivative of volatility, the property which holds for e.g. fractional Brownian motion with $H < 1/2$. In 2014[12], Gatheral, Jaisson and Rosenbaum [123] advocated for $H < 1/2$ from a different perspective: using estimation techniques based on power variations, they came to a conclusion that volatility must have Hölder regularity of order ≈ 0.1. Using (1.23), it is easy to check that

[12] Although paper [123] appeared in *Quantitative finance* in 2018, the preprint had been available on SSRN and ArXiv since 2014.

$$\mathbb{E}[|B^H(t) - B^H(s)|^2] = |t - s|^{2H}, \quad s, t \geq 0,$$

so, by the Kolmogorov-Chentsov theorem (see, e.g., [126, p. 192]), the paths of B^H are Hölder continuous up to the order H. In other words, Ref. [123] concluded that fractional Brownian motions with very small Hurst index H are preferable choices for stochastic volatility modeling. Later in 2021, Fukasawa [114] re-visited the interplay between roughness and power-law of at-the-money volatility skew considering the maturity date $T > 0$ fixed and the evaluation time $t > 0$ approaching T. The main theoretical result presented in [114] is as follows: *if* the price S is a positive *continuous* semimartingale and *if* the at-the-money implied volatility skew (1.8) evaluated at $t > 0$ exhibits the power-law behavior of the form $(T - t)^{-\frac{1}{2}+H}$ as $t \uparrow T$ with $H \approx 0$, then H_0-Hölder continuity of the quadratic variation derivative

$$\frac{d}{dt}\langle \log S \rangle_t \tag{1.32}$$

leads to an arbitrage opportunity if $H_0 > H$. Note that (1.32) exactly coincides with $\sigma^2 = \{\sigma^2(t),\ t \geq 0\}$ in the general stochastic volatility model (1.11) and hence [114] implies that, in a continuous setting with power law of the implied volatility skew, the volatility process "*has to*" be rough to avoid arbitrage.

These collective observations rapidly developed into a vast research field known as "*rough volatility*", with hundreds of papers published over the years. For an extensive literature list on the topic that is regularly updated by specialists in the field, we refer the reader to [189]. Some notable models include (in all cases, $H \in \left(0, \frac{1}{2}\right)$ and B, W are correlated Brownian motions):

- the rough fractional stochastic volatility (RFSV) model [123]

$$\sigma(t) = \exp\left\{\theta_1 + \theta_2 \int_{-\infty}^{t} e^{-\theta_3(t-s)} dB^H(s)\right\},$$

  which can be regarded as a rough counterpart of the model (1.29);
- the rough Bergomi model [22, 144]

$$\begin{aligned} dS(t) &= \sqrt{\sigma(t)} S(t) dW(t), \\ \sigma(t) &= \theta_0(t) \exp\left\{2\theta_1 \int_0^t (t-s)^{H-\frac{1}{2}} dB(s) - \theta_1^2 t^{2H}\right\}, \end{aligned} \tag{1.33}$$

  where $\theta_1 > 0$ and θ_0 is a deterministic function;
- the mixed rough Bergomi model [129, 152]

$$
\begin{aligned}
dS(t) &= S(t)\sqrt{\sigma(t)}dB(t),\\
\sigma(t) &= \theta_0(t)\Bigg(\delta \exp\left\{2\theta_1\int_0^t (t-s)^{H-\frac{1}{2}}dW(s) - \theta_1^2 t^{2H}\right\}\\
&\qquad + (1-\delta)\exp\left\{2\theta_2\int_0^t (t-s)^{H-\frac{1}{2}}dW(s) - \theta_2^2 t^{2H}\right\}\Bigg),
\end{aligned}
\tag{1.34}
$$

where $\delta \in [0, 1]$, $\theta_1, \theta_2 > 0$ and θ_0 is a deterministic function;

- the rough SABR model [115] where the price is given by

$$
\begin{aligned}
dS(t) &= \sqrt{\sigma(t)}\beta(S(t))dW(t),\\
\sigma(t) &= \theta_0(t)\exp\left\{\theta\sqrt{2H}\int_0^t (t-u)^{H-\frac{1}{2}}dB(s) - \frac{1}{2}\theta^2 t^{2H}\right\}dB(t),
\end{aligned}
$$

with $0 \le t \le s$, θ_0 being a positive random process and β being a positive continuous function;

- the rough Stein-Stein model [137]

$$
\begin{aligned}
dS(t) &= \theta\sigma(t)S(t)dW(t),\\
\sigma(t) &= \frac{1}{\Gamma\left(H+\frac{1}{2}\right)}\int_0^t (t-s)^{H-\frac{1}{2}}dB(s);
\end{aligned}
$$

- the fast-varying rough volatility [119]

$$
\begin{aligned}
dS(t) &= \mu(t)dt + \sigma^{\varepsilon}(t)dW(t),\\
\sigma^{\varepsilon}(t) &= F(Z^{\varepsilon}(t)),\\
Z^{\varepsilon}(t) &= \frac{\theta}{\sqrt{\varepsilon}}\int_{-\infty}^t \mathcal{K}_H\left(\frac{t-s}{\varepsilon}\right)dB(s),
\end{aligned}
\tag{1.35}
$$

where $\mathcal{K}_H(t) := C_H\left(t^{H-\frac{1}{2}} - \int_0^t (t-s)^{H-\frac{1}{2}}e^{-s}ds\right)$, $C_H > 0$, F is positive and bounded one-to-one function, $\theta > 0$ and $\varepsilon > 0$ is assumed to be small;

- the rough Heston model [97]

$$
\begin{aligned}
dS(t) &= \sqrt{\sigma(t)}S(t)dW(t),\\
\sigma(t) &= \sigma(0) + \int_0^t \frac{(t-s)^{H-\frac{1}{2}}}{\Gamma\left(H+\frac{1}{2}\right)}\left(\theta_1(\theta_2 - \sigma(s))ds + \theta_3\sqrt{\sigma(s)}dB(s)\right)
\end{aligned}
\tag{1.36}
$$

and its modification, the quadratic rough Heston model [124, 186],

$$
\begin{aligned}
dS(t) &= S(t)\sqrt{\sigma(t)}dW(t),\\
\sigma(t) &= a(Z(t)-b)^2 + c,\\
Z(t) &= \int_0^t \theta_1 \frac{(t-s)^{H-\frac{1}{2}}}{\Gamma\left(H+\frac{1}{2}\right)}(\theta_0(s) - Z(s))ds\\
&\quad + \int_0^t \theta_2 \frac{(t-s)^{H-\frac{1}{2}}}{\Gamma\left(H+\frac{1}{2}\right)}\sqrt{a(Z(s)-b)^2 + c}dW(s),
\end{aligned}
\tag{1.37}
$$

where $a, b, c, \theta_1, \theta_2, \theta_3 > 0$ and θ_0 is a deterministic function.

The rough Heston approach (1.36) is especially interesting from the modeling perspective: in addition to the power law behavior of (1.9), it can reproduce a special form of Zumbach effect [73, 99] and can be interpreted as the limit of a reasonable tick-by-tick price model based on two-dimensional Hawkes processes [98], i.e. deduces roughness of the volatility directly from the market microstructure.

Puzzles of fractionality and roughness. In principle, one can divide the majority of arguments in favor of rough volatility into two distinct categories:

- econometric studies like [123] which involve regularity estimations of volatility from historical high-frequency samples of an asset (e.g. S&P500 index);
- analyses from the options pricing perspective such as [8, 114] that justify roughness by its ability to reproduce e.g. the power law behavior of implied volatility skews (1.9).

However, neither argument is sufficiently persuasive to be exempt from criticism when modeling extremely complex systems such as financial markets. For example, the methodology used in [123] was examined by Rogers [183], Cont and Das [58] as well as Fukasawa, Takabatake and Westphal [113, 116]: they apply the same regularity estimation procedure as in [123] to *synthetic* datasets and report that the estimator tends to produce low values of H *regardless of the true parameter* used in the simulation. This issue is consensually explained as follows. Since the volatility σ is not observable directly, one must first extract some volatility proxy from the stock data and then use *it* as the sample for regularity estimation. However, this procedure results in additional approximation errors that bias the estimator of H towards zero. The authors of [123] acknowledge this issue themselves writing that "*estimation errors when estimating volatility can be quite significant for some models, leading to downward biases in the measurement of the smoothness*" [123, p. 946].

Interestingly, there were several subsequent attempts to account for the problem described above. For example, Fukasawa et al. [113, 116] develop a more robust estimation technique and still report that $H < 0.1$ is indeed the best fit for the volatility model

$$
d\log\sigma^2(t) = \theta_1(t)dt + \theta_2 dB^H(t) \tag{1.38}
$$

with θ_1 being an unknown adapted càdlàg process. Bolko et al. [32] perform the generalized method of moment (GMM) estimation and also confirm the roughness of volatility for thirty-one leading stock indexes under the assumption that $\log \sigma^2$ is a fractional Ornstein-Uhlenbeck process.

Next, the claim that the implied volatility skew follows the power law can also be challenged. For example, Guyon and El Amrani [132] investigate the volatility skew behavior in more detail and conclude that

> ...power law fits the term-structure of ATM skew of equity indexes quite well over a large range of maturities, from 1 month to a few years[13]. But this should not lead one to conclude that the ATM skew blows up like this power law at zero maturity. Zooming in on the short maturities shows that the power-law extrapolation fits the data poorly.

In summary, Guyon and El Amrani [132, p. 14] argue that "*far from being infinite, the zero-maturity extrapolation of the skew*[14] *is distributed around a typical value of 1.5*". On the contrary, Delemotte et al. [78] investigate the behavior of average volatility skews between 2007 and 2015 (a period that is different from 2020–2022 of Guyon and El Amrani [132]) and find that a power-law-type explosive behavior of the implied volatility skew when $T \to 0$ is entirely appropriate! In addition, they argue [78, Section 3] that the average skew behaves differently for shorter and longer maturities: their analysis suggests that

$$\mathbb{E}[\Psi(T)] \propto \begin{cases} T^{-\frac{1}{2}+H_1}, & T < T_1, \\ T^{-\frac{1}{2}+H_2}, & T > T_1, \end{cases} \tag{1.39}$$

where $H_1 > H_2$ and T_1 is roughly 2 months.

What is the reason for such a discrepancy in conclusions regarding very similar (and sometimes *literally the same*) datasets? Features like the behavior of the volatility skew are very intricate and their assessment based on discrete data leaves much space for interpretation and extrapolation. For example, Delemotte et al. [78, pp. 2–3] compare their result to [132] and summarize the difference as follows:

> As a result, both models could very reasonably be used to extrapolate the ATM skew for very short maturities, and in any case below the first maturity quoted on the market, while leading to different skew asymptotics: finite limiting skew in the 2-factors Bergomi model[15], *and exploding skew in the 2PL model*[16]*–indicating that* ***the question of the explosive or non-explosive nature of the short-end of the skew curve might be (at least in the case of the SP500 index) hard to disambiguate***.

In some sense, this ambiguity is very natural: the volatility itself is a purely theoretical (albeit meaningful) concept, and hence there is no single "*ultimately true volatility model*". In this context, the question "*is volatility* ***truly*** *rough*?" lacks a definitive "*correct*" answer. Rough volatility is a valid modeling framework with its advantages

---

[13] In the dataset we used to produce Fig. 1.3a, the shortest maturity was 10 days.

[14] I.e. the value $\Psi(0)$ in (1.9).

[15] *The one of* [132].

[16] *The one of* [78].

and limitations which indeed mimics some important features of the data: for example, the empirically observed steepness of smiles at-the-money does increase close to maturity (as clearly visible on Fig. 1.2), and rough models seem to capture this effect better than classical Brownian diffusions, irrespective of whether this increase "truly" follows the power law. In turn, econometric analyses like [32, 113, 116] can be regarded as a form of reality check revealing possible contradictions with data and testing specific models against each other. For example, the results of [32, 113, 116] could mean that (1.38) with small H seems to be more consistent with some features of the data than e.g. (1.18) with $\beta = 1$ or (1.29) with $H > \frac{1}{2}$. When dealing with complex phenomena, one must bear in mind that more than one model may be compatible with observations. After all, according to a famous aphorism commonly attributed to the statistician George Box, *all models are wrong, but some are useful.*

With this disclaimer in mind, let us finish this Subsection with a discussion on another issue coming from the interplay between roughness and long memory *in the specific context of fractional Brownian motion* B^H. More precisely, *long memory* requires the Hurst index $H > 1/2$ whereas *roughness* demands $H < 1/2$. Despite some studies that demonstrate spurious long memory appearing due to model misspecifications (see, e.g., [123, Section 4]), roughness alone cannot explain the behavior of the entire volatility surface. For example, Funahashi and Kijima [118] demonstrate that volatility models based on fractional Brownian motion with $H < 1/2$ do not give the required rate of decrease in the smile amplitude as $T \to \infty$ whereas $H > 1/2$ gives this effect (see also [117]).

It should be noted that this "contradiction", referred to as the "*fractional modeling puzzle*" in [118], is somewhat artificial. In general, long memory and roughness do not depend on each other and can co-exist within a single stochastic process. In other words, the reason for this "*fractional puzzle*" comes from the choice to use fractional Brownian motion and a different modeling framework may utilize both of these features without any structural contradictions (see, e.g., the model based on Brownian semistationary processes by Bennedsen, Lunde and Pakkanen [24]). Nevertheless, in a continuous setting, B^H seems to be an extremely convenient and valuable asset from the mathematical perspective:

- it is Gaussian, enabling numerous methods from Gaussian process theory including efficient numerical methods, Malliavin calculus etc. (see, e.g., [175, Chapter 5]);
- it has stationary increments, which facilitates various statistical estimation techniques (see, e.g., [150]);
- despite being non-semimartingale, fractional Brownian motion has a developed stochastic integration theory [171].

In the literature, there are several possible approaches to incorporate long memory and roughness within the fractional framework. The most straightforward methodology is to use two fractional Brownian motions with different Hurst indices. Such model was utilized in e.g. [118] (see also [7, Section 7.7]) in the form

$$
\begin{aligned}
dS(t) &= \mu S(t)dt + \sigma(X^1(t), X^2(t))dW(t), \\
dX^i(t) &= (\theta_1^i - \theta_2^i X^i(t))dt + \theta_3^i dB^{H_i}(t), \quad i = 1, 2,
\end{aligned}
\tag{1.40}
$$

with B^{H_1}, B^{H_2} being two fractional Brownian motions with $H_1 > 1/2$ and $H_2 < 1/2$. The authors report that this model indeed manages to grasp the implied volatility surface with the power law for short maturities and slower decay of the smile amplitude for long maturities. Such a suggestion somehow resonates with the observations made in [78]: the authors also discuss the possibility of introducing several factors with different regularity into the model in order to reproduce (1.39) (although they consider $0 < H_2 < H_2 < \frac{1}{2}$).

Another interesting possibility—*a multifractional Brownian motion*—is advocated in Corlay et al. [65] (see also [13]). There, the authors estimate the local roughness of the volatility and conclude that it is heavily variable and has periods of low (≈ 0.1) and high (≈ 0.8) regularity (see [65, Figure 2]). It is also important to note that [123, Section 2.6] also reports some dependence of the volatility roughness on time.

As a generalization of both approaches mentioned above, one can also consider the usage of *Gaussian Volterra processes* $Z(t) := \int_0^t \mathcal{K}(t,s)dB(s)$. Such volatility drivers were considered in e.g. [49, 165] or the series of papers [82, 85].

1.4 VIX and the Joint Calibration Puzzle

As previously mentioned in the Introduction, volatility is an important tool for quantifying risk, in addition to being a critical variable in option pricing. Therefore, it is no surprise that market actors have been seeking dedicated financial instruments to hedge against drastic volatility changes and capitalize on overall volatility trends. Nowadays, derivatives written on various volatility proxies are extremely popular among investors: for instance, according to the Cboe Annual Report 2022 [51], the *CBOE Volatility Index* (*VIX*, see Fig. 1.6) is one of the most traded underlying assets of the Chicago Board Options Exchange, alongside the S&P 500 (SPX) index.

Since volatility is not a directly observable value and, in some sense, derives from the theoretical domain, its quantification is a necessary step before considering its use as an underlying asset. Naturally, various methodologies have been applied for such quantification; a detailed historical overview of this subject can be found in e.g. [45, Section 3]. Here we mention the early approaches of Gastineau [120] and Cox and Rubinstein [69, Appendix 8A], who proposed indices based on implied volatility, as well as Brenner and Galai [40], who suggested a metric emerging from the realized volatility and considered derivatives employing it as the underlying assets. Another index utilizing an average of S&P100 option implied volatilities was introduced by Fleming et al. [106], and this methodology was used by the Chicago Board Options Exchange for the computation of their early version of VIX between 1993 and 2003. The general idea behind the modern computation of VIX can be traced back to

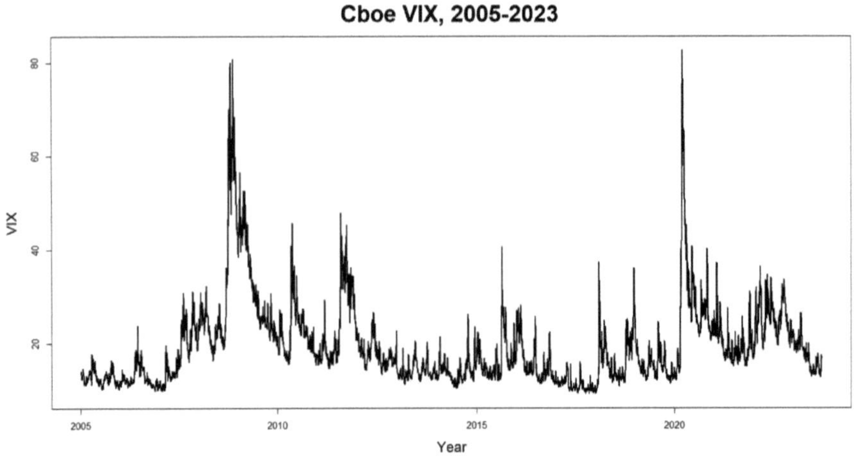

Fig. 1.6 Daily values of the CBOE VIX index, 2005–2023. Note the two spikes in 2008 and 2020 which correspond to the respective economic crises. The data is retrieved from CBOE.com [204]

Breeden and Litzenberger [38], but the exact formulas were crystalized in the works of Dupire [93] and Carr and Madan [44].

Nowadays, VIX is the preeminent volatility proxy on a global scale, closely connected to a volatility modeling challenge occasionally referred to as "*The Holy Grail of volatility modeling*" [124, 131]. In order to outline this problem in more detail and prevent any ambiguity, let us first dedicate some time and delve into the specifics of VIX computation and interpretation.

1.4.1 Intermezzo: VIX and Its Interpretation

Assume that the SPX forward price S is a continuous martingale satisfying

$$dS(t) = \sigma(t)S(t)dW(t), \tag{1.41}$$

on a filtered probability space $(\Omega, \mathcal{F}, \{\mathcal{F}_t\}, \mathbb{Q})$ under the risk-neutral probability measure $\mathbb{Q}$, where W is a $\mathbb{Q}$-Brownian motion and $\sigma = \{\sigma(t),\ t \geq 0\}$ is an adapted square integrable stochastic process. It is easy to verify that, under some mild assumptions on σ, (1.41) has a unique solution of the form

$$S(t) = S(0) \exp\left\{ -\frac{1}{2} \int_0^t \sigma^2(s)ds + \int_0^t \sigma(s)dW(s) \right\}.$$

Next, fix a period of time T (in the case of CBOE VIX, $T = 30$ days) and consider the value

$$-\frac{2}{T}\mathbb{E}_{\mathbb{Q}}\left[\log\frac{S(t+T)}{S(t)}\,\Big|\,\mathcal{F}_t\right]. \tag{1.42}$$

One the one hand, since, by the martingale property,

$$\mathbb{E}_{\mathbb{Q}}\left[\int_t^{t+T}\sigma(s)dW(s)\,\Big|\,\mathcal{F}_t\right] = 0,$$

one can observe that

$$\begin{aligned}
-\frac{2}{T}\mathbb{E}_{\mathbb{Q}}&\left[\log\frac{S(t+T)}{S(t)}\,\Big|\,\mathcal{F}_t\right]\\
&= -\frac{2}{T}\mathbb{E}_{\mathbb{Q}}\left[-\frac{1}{2}\int_t^{t+T}\sigma^2(s)ds + \int_t^{t+T}\sigma(s)dW(s)\,\Big|\,\mathcal{F}_t\right]\\
&= \frac{1}{T}\int_t^{t+T}\mathbb{E}_{\mathbb{Q}}\left[\sigma^2(s)\,|\,\mathcal{F}_t\right]ds.
\end{aligned} \tag{1.43}$$

On the other hand, by Taylor's formula with integral remainder,

$$\begin{aligned}
\log\frac{S(t+T)}{S(t)} &= \frac{S(t+T)-S(t)}{S(t)} - \int_{S(t)}^{S(t+T)}\frac{(S(t+T)-K)}{K^2}dK\\
&= \frac{S(t+T)-S(t)}{S(t)} - \int_0^{S(t)}\frac{(K-S(t+T))_+}{K^2}dK\\
&\quad - \int_{S(t)}^{\infty}\frac{(S(t+T)-K)_+}{K^2}dK,
\end{aligned}$$

and hence, since

$$\mathbb{E}_{\mathbb{Q}}\left[\frac{S(t+T)-S(t)}{S(t)}\,\Big|\,\mathcal{F}_t\right] = 0$$

by the martingale property, (1.42) can be re-written as

$$
\begin{aligned}
-\frac{2}{T}\mathbb{E}_{\mathbb{Q}}&\left[\log\frac{S(t+T)}{S(t)}\,\Big|\,\mathcal{F}_t\right]\\
&=\mathbb{E}_{\mathbb{Q}}\left[\frac{2}{T}\int_0^{S(t)}\frac{(K-S(t+T))_+}{K^2}dK\,\Big|\,\mathcal{F}_t\right]\\
&\quad+\mathbb{E}_{\mathbb{Q}}\left[\int_{S(t)}^{\infty}\frac{(S(t+T)-K)_+}{K^2}dK\,\Big|\,\mathcal{F}_t\right]\\
&=\frac{2}{T}\int_0^{S(t)}\frac{\mathbb{E}_{\mathbb{Q}}\left[(K-S(t+T))_+\mid\mathcal{F}_t\right]}{K^2}dK\\
&\quad+\int_{S(t)}^{\infty}\frac{\mathbb{E}_{\mathbb{Q}}\left[(S(t+T)-K)_+\mid\mathcal{F}_t\right]}{K^2}dK\\
&=\frac{2e^{rT}}{T}\left(\int_0^{S(t)}\frac{P_t(K,T)}{K^2}dK+\int_{S(t)}^{\infty}\frac{C_t(K,T)}{K^2}dK\right),
\end{aligned}
\tag{1.44}
$$

where r is the instantaneous interest rate and $P_t(K,T)$, $C_t(K,T)$ are, respectively, the prices (at moment t) of put and call options with strike K and expiry date $t+T$. To summarize, under the setting specified above, (1.43) and (1.44) yield

$$
\begin{aligned}
\frac{1}{T}\int_t^{t+T}&\mathbb{E}_{\mathbb{Q}}\left[\sigma^2(s)\mid\mathcal{F}_t\right]ds\\
&=\frac{2e^{rT}}{T}\left(\int_0^{S(t)}\frac{P_t(K,T)}{K^2}dK+\int_{S(t)}^{\infty}\frac{C_t(K,T)}{K^2}dK\right).
\end{aligned}
\tag{1.45}
$$

The VIX is computed (see, e.g., [50] or [151, Section 1.5.2]) by discretizing the right-hand side of (1.45) (note that the latter relies exclusively on the empirically observable values, SPX forward $S(t)$ and put/call option prices):

$$
\begin{aligned}
\mathrm{VIX}_t^2(T):=\frac{2e^{rT}}{T}&\left(\sum_{j=-M}^{-1}\frac{P_t(K_j,T)}{K_j^2}(K_{j+1}-K_j)+\sum_{i=1}^{N}\frac{C_t(K_i,T)}{K_i^2}(K_i-K_{i-1})\right)\\
&-\frac{1}{T}\left(\frac{S(t)}{K_0}-1\right)^2,
\end{aligned}
$$

where $K_{-M} < K_{-(M-1)} < \cdots < K_0$ are payoffs of all listed out-of-the-money put options with maturity at $t+T$, $K_0 \le S(t)$, and $K_1 < K_2 < \cdots < K_N$ are payoffs of all listed out-of-the-money call options with maturity at $t+T$, $K_1 > S(t)$. In the meanwhile, the left-hand side of (1.45) is a standard notion of the forward variance, precisely the value one intends to capture.

1.4.2 VIX Smile and the Joint Calibration Puzzle

Despite any potential inaccuracies, VIX stands as the most popular proxy for market volatility in the world and, therefore, it is unsurprising that derivatives with VIX as an underlying are actively traded on the market. In particular, in March 2004, CBOE launched VIX futures and, in February 2006, VIX options were introduced—and, naturally, a problem of pricing such derivatives emerged.

First of all, having VIX option prices, one can use the procedure described in Sect. 1.2.2 and construct the corresponding implied volatility surface $\widehat{\sigma}(T, K)$, with T denoting the time to maturity and K being the strike. Contrary to the convex "smiles" for derivatives written on classical stocks and stock indices, the VIX implied volatility $\widehat{\sigma}(T, K)$ consistently exhibits concave behavior in K for every fixed T with positive slope at the money (see [9, Figure 1] as well as [17, 108] or [147]).

It turns out that continuous stochastic volatility models have mixed success in reproducing this phenomenon, especially for shorter maturities—in this regard, we recommend an excellent review [184] on the topic. Alòs et al. [9] (see also [7, Chapter 10]) use Malliavin calculus tools and analyze selected models to check whether they produce the positive slope at the money for short maturities. Their results show that

- SABR model (1.19) gives the flat skew, but its modification called the *mixed SABR model*

$$\sigma(t) = \theta_0 \left(\delta \exp\left\{2\theta_1 W(t) - \theta_1^2 t\right\} + (1-\delta) \exp\left\{2\theta_2 W(t) - \theta_2^2 t\right\}\right),$$

  where $\delta \in [0, 1], \theta_0, \theta_1, \theta_2 > 0$, generates the required positive slope provided that $\delta \ne 0, 1$ and $\theta_1 \ne \theta_2$;
- Heston model (1.17) normally generates negative VIX skew (see also [127, Figure 3]); however, under some conditions on coefficients (e.g. when Feller condition is violated, see [147]), the slope can be positive;
- the rough Bergomi model (1.33) as well as its mixed modification (1.34) both give the required slope (see also [144]).

Other notable continuous stochastic volatility that are able to capture the upward smirk include:

- the 3/2-model (1.20) [17, Section 2];
- double CEV-model (1.21) [122];
- Heston with stochastic vol-of-vol model (1.22) [108];

- quintic Ornstein-Uhlenbeck model [2]

$$\sigma(t) = \sqrt{\theta_0(t)} \frac{p(X(t))}{\sqrt{\mathbb{E}\left[p^2(X(t))\right]}}, \\ X(t) = \varepsilon^{-\theta} \int_0^t e^{-\frac{\theta}{\varepsilon}(t-s)} dB(s), \tag{1.46}$$

and, more generally, a class of Gaussian polynomial models [3]

$$\sigma(t) = \sqrt{\theta_0(t)} \frac{p(X(t))}{\sqrt{\mathbb{E}\left[p^2(X(t))\right]}}, \\ X(t) = \int_0^t \mathcal{K}(t-s) dB(s), \tag{1.47}$$

where $\theta > 0$, p is a polynomial and $\mathcal{K}$ is a square integrable Volterra kernel;
- quadratic rough Heston model (1.37).

As one can see, multiple continuous models, both standard Brownian diffusions and rough volatility models, are able to reproduce the shape of the VIX smile. However, there is an additional dimension to that problem: since there is a direct connection between SPX and VIX, models calibrated with respect to the SPX data should be consistent with VIX and vice versa. In other words, one seeks a market model that would *jointly calibrate to both VIX and SPX simultaneously*. As noted by Guyon in [131], "*Without such models, financial institutions could possibly arbitrage each other, and even market making desks within the same institution could do so...*".

It turns out that this problem, known as the "*joint calibration puzzle*", is extremely difficult to solve; so much so that some researchers label it "*the Holy Grail of volatility modeling*" [124, 131]. For a long time, continuous-time stochastic volatility models without jumps failed to produce the perfect SPX-VIX joint calibration. Some notable attempts include the 3/2 model [17], double CEV-model [122], Heston with stochastic vol-of-vol model [108] and rough Berghomi model [144]—in all cases, the joint fit was only partially successful, predominantly failing for short maturities (see also an excellent presentation by J. Guyon [129] on the topic). The most fruitful approaches in solving the joint calibration puzzle were outside of the continuous-time-continuous-price paradigm:

- Cont and Kokholm [63] use jump-diffusions to directly model forward variances and SPX; jumps indeed allow to de-couple the ATM S&P500 skew and the ATM VIX implied volatility giving a good joint fit;
- Guyon [130, 131] utilizes nonparametric discrete time model building joint probability measure on SPX and VIX on a discrete set of dates; this allows to obtain an almost perfect joint calibration.

However, in recent years there have been several successful attempts inside the fully continuous framework, all with a higher number of parameters:

- Gatheral et al. [124] and Rosenbaum and Zhang [186] show that quadratic rough Heston model (1.37) calibrates very well to SPX and VIX smiles;
- Abi Jaber et al. [2, 3] provide a good fit with Gaussian polynomial models (1.47);
- Bourgey and Guyon [133, Section 4] extend the discrete time model of [130, 131] to a continuous-time setting;
- Guyon and Mustapha [135] achieve an impressive calibration with the model

$$\begin{aligned} dS_t &= -\frac{1}{2}\sigma_S^2(t, S_t, Y_t)dt + \sigma_S(t, S_t, Y_t)dW_t, \\ dY_t &= \mu_Y(t, S_t, Y_t)dt + \sigma_Y(t, S_t, Y_t)\left(\rho(t, S_t, Y_t)dW_t + \sqrt{1 - \rho^2(t, S_t, Y_t)}dB_t\right), \end{aligned}$$

where the drift, volatility and correlation functions σ_S, μ_Y, σ_Y and ρ are modeled with neural networks.

1.4.3 VIX and Continuous Models: Interpretation Issues

We finalize this section by highlighting a significant caveat related to the interpretation and tractability of VIX within the context of continuous stochastic volatility models.

Despite the elegance of VIX computation, the precision of VIX in capturing the volatility of SPX heavily depends on the underlying assumptions. Namely, it is crucial that SPX is assumed to follow the continuous dynamics (1.41) and any deviations from this model lead to some deterioration of the VIX tractability. In particular, if there are jumps in the SPX forward dynamics, (1.43) does not hold with an additional term appearing on the right-hand side of the expression (see, e.g., [6]). Therefore, if (1.41) is violated, (1.45) does not hold either, and VIX loses its original interpretation.

It is important to note that the viability of the model (1.41) can be tested statistically by comparing VIX to *variance swap contracts*. By definition, a *variance swap* is a futures contract with a payoff at the moment T of the form

$$\frac{1}{T}\sum_{k=1}^{n}\left(\log\frac{S(t_k)}{S(t_{k-1})}\right)^2 - VS,$$

where $0 = t_0 < t_1 < \cdots < t_n = T$ is a partition of $[0, T]$ and VS denotes the corresponding swap rate that is determined by the market. No-arbitrage arguments imply that, if the dynamics (1.41) indeed holds and some mild assumptions on σ are satisfied,

$$VS = \mathbb{E}_{\mathbb{Q}}\left[\frac{1}{T}\sum_{k=1}^{n}\left(\log\frac{S(t_k)}{S(t_{k-1})}\right)^2\right] \to \frac{1}{T}\int_0^T E_{\mathbb{Q}}\left[\sigma^2(s)\right]ds$$

as $\max_k |t_k - t_{k-1}| \to 0$. In other words, under the model (1.41), market swap rates $\sqrt{VS}$ should approximately coincide with VIX. However, as reported by Aït-Sahalia et al. in their analysis [6] using the over-the-counter swap rate data, there is a statistically significant gap between VS and VIX^2: the values of $VS - \text{VIX}^2$ are mostly positive, larger during market turmoils but sizable even in quiet times. For more details in that regard as well as for possible alternatives to the current VIX computation method, we also refer the reader to the discussions in [12, 48, 162].

In other words, there is some evidence showing that the tractability of VIX as a volatility index is not that straightforward and the real-world observations of VIX are not sampled from

$$\sqrt{\frac{1}{T}\int_t^{t+T} \mathbb{E}_{\mathbb{Q}}\left[\sigma^2(s) \mid \mathcal{F}_t\right]ds} \tag{1.48}$$

as models with continuous price dynamics assume[17].

1.5 Challenges and New Directions

Armed with empirical insights, with the stylized facts detected, and with the modeling frameworks discussed above, we conclude this chapter by outlining some challenges in volatility modeling—challenges that we aim to address in the subsequent chapters of this book.

- **Positivity of volatility**. As mentioned in Sect. 1.3.2, one of the natural expectations from a "*reasonable*" volatility model is the positivity of its paths. This requirement actually goes beyond the simple consideration that $\sigma = \{\sigma(t),\ t \geq 0\}$ should resemble its original proxy (1.1) and is connected to the procedure of transition between the physical and the pricing measures. In the Black-Scholes-type models (1.11), martingale densities usually involve terms of the form $\int_0^T \frac{1}{\sigma(s)}dW(s)$ and $\int_0^T \frac{1}{\sigma^2(s)}ds$ (see, e.g., [28, Proposition 1.11]) which can be poorly defined, if the volatility hits zero with positive probability. Of course, one may model the market under the pricing measure in the first place, but, in this situation, one sacrifices the ambition to justify their approach with econometric analysis based on historical time series a la [123] which, of course, should be performed under the physical measure.

[17] As suggested in [162], in the presence of jumps, VIX measures *the risk-neutral entropy* of the simple return on the S&P 500 index rather than (1.48).

- **Possibility of moment explosions**. A common issue for the stochastic volatility framework is the possibility of moment explosions in price [11]. That means that the moment of the price $\mathbb{E}[S^r(t)]$ may be infinite for all time points t after some t_*. Moment explosions can be a notable drawback from at least two perspectives. First, numerical schemes involving L^r-convergence become inaccessible. Another perspective is asset pricing: as it is noted in [11, Section 8], "*several actively traded fixed-income derivatives require at least L^2 solutions to avoid infinite model prices*". In principle, both positivity and absence of moment explosions can be achieved by assuming bounds on the volatility; see, e.g., [10, 27, 82, 85, 110, 119, 176] and the footnote on p. 2 in [186]. It seems that this assumption is not too aggravating: in industry, there seems to be a consensus [95] that real-world proxies of volatility are typically range-bound.
- **Numerical methods**. As noted in [122, p. 24], the standard Heston model is still widely used despite all the empirical inconsistencies outlined in Sect. 1.3.2. The reason for that is mainly the availability of algorithms for practically all possible applications. Stochastic volatility models normally do not have closed-form expressions for option pricing, portfolio optimization, hedging etc., and therefore the development of efficient numerics for them is of acute importance. It is especially relevant for fractional/rough models which are predominantly non-Markovian and hence cannot utilize various stochastic optimization techniques developed for Markov processes. In recent years, some work in this direction was done in e.g. [1, 3, 186].

Naturally, the development of volatility models remains a dynamic and ongoing process. New methodologies and modeling paradigms continue to appear, pushing the boundaries of what is possible. Notably, recent work by Guyon and Lekeufack [134] presents strong arguments in favor of fully path-dependent volatility models. In parallel, machine learning and signature-based approaches–such as those introduced in [41, 42] and [70, 71]–offer promising new directions. Interestingly, the latter class of models encompasses the Gaussian polynomial model (1.47) as a special case.

The overarching goal of this book is to contribute to this evolving landscape by constructing a stochastic volatility model that addresses the challenges mentioned above: it ensures a consistent transition between the physical and risk-neutral measures, generates implied volatility surfaces with realistic skews and smile amplitudes across different maturities, and supports numerically efficient methods for simulation, pricing, and hedging.

The results presented in this book stem from our long-term research project, and the chapters are based on systematically organized and revised material available in articles developed during the research period.

- **Chapter** 2, based on [170, 173], begins with a closer examination of the well-known Cox–Ingersoll–Ross (CIR) process used as the volatility component in the Heston model. We analyze the square root of the CIR process under various parameter configurations and observe that its stochastic differential equation includes a drift component that diverges to $+\infty$ as the process approaches zero. This creates a "repelling" effect, ensuring the almost sure positivity of the solution. As a side note, we also explore connections between the CIR process and Skorokhod problems.

- **Chapter** 3, based on [81], introduces the central object of our study: *sandwiched processes*. More specifically, we define a class of stochastic integro-differential equations characterized by an explosive drift component, replicating the repelling behavior observed in Chapter 2. Solutions to these equations turn out to be bounded between two pre-specified functions, φ and ψ—hence the term "sandwiched". A notable advantage of this setup is its flexibility with respect to the driving noise: the model accommodates *any* Hölder continuous process, which can be used to replicate roughness, long memory *or both*.
- **Chapter** 4, based on [168, 169], applies the theory from Chapter 3 to construct generalized versions of the CIR and Chan–Karolyi–Longstaff–Sanders (CKLS) processes driven by noises with memory.
- **Chapter** 5, based on [84], presents a numerical scheme for simulating sandwiched processes. We use the so-called *drift-implicit method* that preserves the "sandwiched" property of the solution.
- **Chapter** 6, drawing on the series of papers [82, 85, 86], considers sandwiched equations driven by Hölder continuous Gaussian Volterra processes. We study the Malliavin differentiability of the solutions and develop their Markovian approximations that are further used in Chapter 9 for numerical methods.
- **Chapter** 7, based on [85, 86], finally introduces a market model built upon sandwiched processes, which we call the Sandwiched Volterra Volatility (SVV) model. This model offers several advantages: it avoids moment explosions, supports efficient numerical schemes and provides a complete characterization of the set of martingale measures. We also analyze the implied volatility surfaces generated by the SVV model and show that an appropriate choice of the Gaussian Volterra driver allows to "solve" the fractional modeling puzzle introduced earlier.
- **Chapter** 8, based on [85, 172], further explores numerical techniques for option pricing within the SVV framework. In particular, we propose a Malliavin-duality-based algorithm for pricing options with discontinuous payoffs.
- **Chapter** 9, based on [82], concludes by examining quadratic hedging within the SVV model.
- The **Appendix** collects several useful results from stochastic analysis that are referenced in the main text.

We invite our readers to join us on this journey and hope that the theory and techniques presented in the chapters ahead will prove both interesting and useful.

References

1. CBOE: Cboe Global Markets Annual Report 2022 (2023). https://ir.cboe.com/sites/cboe-ir-v1/files/2022-annual-report.pdf
2. Abi Jaber, E., El Euch, O.: Multifactor approximation of rough volatility models. SIAM J. Fin. Math. **10**(2), 309–349 (2019). https://doi.org/10.1137/18m1170236

3. Abi Jaber, E., Illand, C., Li, S.: The quintic Ornstein-Uhlenbeck volatility model that jointly calibrates SPX & VIX smiles. ArXiv:2212.10917 (2022). https://doi.org/10.48550/ARXIV.2212.10917
4. Abi Jaber, E., Illand, C., Li, S.: Joint SPX-VIX calibration with Gaussian polynomial volatility models: deep pricing with quantization hints. ArXiv:2212.08297 (2022). https://doi.org/10.48550/ARXIV.2212.08297
5. Aït-Sahalia, Y., Jacod, J.: Testing whether volatility can be written as a function of the asset price. In: Presented at Princeton-QUT-SJTU-SMU Conference on Econometrics (2019)
6. Aït-Sahalia, Y., Fan, J., Li, Y.: The leverage effect puzzle: disentangling sources of bias at high frequency. J. Financ. Econ. **109**(1), 224–249 (2013). https://doi.org/10.1016/j.jfineco.2013.02.018
7. Aït-Sahalia, Y., Karaman, M., Mancini, L.: The term structure of equity and variance risk premia. J. Econ. **219**(2), 204–230 (2020). https://doi.org/10.1016/j.jeconom.2020.03.002
8. Alòs, E., Garcia Lorite, D.: Malliavin calculus in finance: theory and practice. CRC Press, London (2021)
9. Alòs, E., León, J.A., Vives, J.: On the short-time behavior of the implied volatility for jump-diffusion models with stochastic volatility. Finance Stoch. **11**(4), 571–589 (2007). https://doi.org/10.1007/s00780-007-0049-1
10. Alòs, E., García-Lorite, D., Gonzalez, A.M.: On smile properties of volatility derivatives: understanding the VIX skew. SIAM J. Fin. Math. **13**(1), 32–69 (2022). https://doi.org/10.1137/19m1269981
11. Alòs, E., Rolloos, F., Shiraya, K.: Forward start volatility swaps in rough volatility models. Asia-Pac. Financ. Mark. (2025). https://doi.org/10.1007/s10690-025-09545-3
12. Andersen, L.B.G., Piterbarg, V.V.: Moment explosions in stochastic volatility models. Finance Stoch. **11**(1), 29–50 (2006). https://doi.org/10.1007/s00780-006-0011-7
13. Andersen, T.G., Bondarenko, O., Gonzalez-Perez, M.T.: Exploring return dynamics via corridor implied volatility. Rev. Financ. Stud. **28**(10), 2902–2945 (2015). https://doi.org/10.1093/rfs/hhv033
14. Ayache, A., Peng, Q.: Stochastic volatility and multifractional Brownian motion. In: Stochastic Differential Equations and Processes, pp. 211–237. Springer, Berlin Heidelberg (2012)
15. Bachelier, L.: Théorie de la spéculation. Ann. Sci. l'Ecole Norm. Sup. Quatrieme Serie **17**, 21–86 (1900). https://doi.org/10.24033/asens.476
16. Bachelier, L.: Louis Bachelier's theory of speculation: the origins of modern finance. Princeton University Press, Princeton, NJ (2006)
17. Bakshi, G., Ju, N., Ou-Yang, H.: Estimation of continuous-time models with an application to equity volatility dynamics. J. Financ. Econ. **82**(1), 227–249 (2006). https://doi.org/10.1016/j.jfineco.2005.09.005
18. Baldeaux, J., Badran, A.: Consistent modelling of VIX and equity derivatives using a 3/2 plus jumps model. Appl. Math. Finance **21**(4), 299–312 (2014). https://doi.org/10.1080/1350486X.2013.868631
19. Barndorff-Nielsen, O.E., Shephard, N.: Non-Gaussian Ornstein-Uhlenbeck-based models and some of their uses in financial economics. J. R. Stat. Soc. Ser. B Stat. Methodol. **63**(2), 167–241 (2001). http://www.jstor.org/stable/2680596
20. Barndorff-Nielsen, O.E.: Normal inverse Gaussian distributions and stochastic volatility modelling. Scand. J. Stat. Theory Appl. **24**(1), 1–13 (1997)
21. Barndorff-Nielsen, O.E., Shephard, N.: Lévy driven volatility models (2011). https://pure.au.dk/ws/files/193874249/levybook.pdf
22. Bäuerle, N., Desmettre, S.: Portfolio optimization in fractional and rough Heston models. SIAM J. Financ. Math. **11**(1), 240–273 (2020). https://doi.org/10.1137/18m1217243
23. Bayer, C., Friz, P., Gatheral, J.: Pricing under rough volatility. Quant. Finance **16**(6), 887–904 (2016). https://doi.org/10.1080/14697688.2015.1099717
24. Benhamou, E., Gobet, E., Miri, M.: Time dependent Heston model. SIAM J. Financ. Math. **1**(1), 289–325 (2010). https://doi.org/10.1137/090753814

25. Bennedsen, M., Lunde, A., Pakkanen, M.S.: Decoupling the short- and long-term behavior of stochastic volatility. J. Financ. Economet. **20**(5), 961–1006 (2022). https://doi.org/10.1093/jjfinec/nbaa049
26. Beran, J.: Statistics for Long-Memory Processes. Chapman and Hall/CRC (1994). https://www.routledge.com/Statistics-for-Long-Memory-Processes/Beran/p/book/9780412049019
27. Bergomi, L.: Stochastic Volatility Modeling. Chapman & Hall/CRC, Oakville, MO (2016)
28. Bezborodov, V., Persio, L.D., Mishura, Y.: Option pricing with fractional stochastic volatility and discontinuous payoff function of polynomial growth. Methodol. Comput. Appl. Probab. **21**(1), 331–366 (2018). https://doi.org/10.1007/s11009-018-9650-3
29. Biagini, F., Guasoni, P., Pratelli, M.: Mean-variance hedging for stochastic volatility models. Math. Financ. **10**(2), 109–123 (2000). https://doi.org/10.1111/1467-9965.00084
30. Black, F., Scholes, M.: The pricing of options and corporate liabilities. J. Polit. Econ. **81**(3), 637–654 (1973). http://www.jstor.org/stable/1831029
31. Black, F.: Studies of stock price volatility changes. In: Proceedings of the Business and Economics Section of the American Statistical Association, pp. 177–181 (1976)
32. Black, F.: Citation classic: the pricing of options and corporate liabilities. Citation Classic **33**, 16 (1987)
33. Bolko, A.E., Christensen, K., Pakkanen, M.S., Veliyev, B.: A GMM approach to estimate the roughness of stochastic volatility. J. Econom. **235**(2), 745–778 (2023). https://doi.org/10.1016/j.jeconom.2022.06.009
34. Bollerslev, T.: Generalized autoregressive conditional heteroskedasticity. J. Econom. **31**(3), 307–327 (1986). https://doi.org/10.1016/0304-4076(86)90063-1
35. Bollerslev, T., Ole Mikkelsen, H.: Modeling and pricing long memory in stock market volatility. J. Econom. **73**(1), 151–184 (1996). https://doi.org/10.1016/0304-4076(95)01736-4
36. Bollerslev, T., Chou, R.Y., Kroner, K.F.: ARCH modeling in finance. J. Econom. **52**(1–2), 5–59 (1992). https://doi.org/10.1016/0304-4076(92)90064-x
37. Boness, A.J.: Elements of a theory of stock-option value. J. Polit. Econ. **72**(2), 163–175 (1964). http://www.jstor.org/stable/1828962
38. Bouchaud, J.P., Potters, M.: Theory of Financial Risks: From Statistical Physics to Risk Management. Cambridge University Press, Cambridge (2000)
39. Breeden, D.T., Litzenberger, R.H.: Prices of state-contingent claims implicit in option prices. J. Bus. **51**(4), 621–651 (1978). http://www.jstor.org/stable/2352653
40. Breidt, F.J., Crato, N., de Lima, P.: The detection and estimation of long memory in stochastic volatility. J. Econom. **83**(1), 325–348 (1998)
41. Brenner, M., Galai, D.: New financial instruments for hedging changes in volatility. Financ. Anal. J. **45**(4), 61–65 (1989). http://www.jstor.org/stable/4479241
42. Buehler, H., Gonon, L., Teichmann, J., Wood, B.: Deep hedging. Quant. Finance **19**(8), 1271–1291 (2019). https://doi.org/10.1080/14697688.2019.1571683
43. Bühler, H., Horvath, B., Lyons, T., Arribas, I.P., Wood, B.: A data-driven market simulator for small data environments. ArXiv:2006.14498 (2020). https://doi.org/10.48550/ARXIV.2006.14498
44. Buraschi, A., Jackwerth, J.: The price of a smile: hedging and spanning in option markets. Rev. Financ. Stud. **14**(2), 495–527 (2001). http://www.jstor.org/stable/2696749
45. Carr, P., Madan, D.: Towards a theory of volatility trading. In: Jarrow, R. (ed.) Volatility, pp. 417–427. Risk Publications (1998)
46. Carr, P., Lee, R.: Volatility derivatives. Annu. Rev. Financ. Econ. **1**(1), 319–339 (2009). https://doi.org/10.1146/annurev.financial.050808.114304
47. Carr, P., Sun, J.: A new approach for option pricing under stochastic volatility. Rev. Deriv. Res. **10**(2), 87–150 (2007). https://doi.org/10.1007/s11147-007-9014-6
48. Carr, P., Wu, L.: The finite moment log stable process and option pricing. J. Financ. **58**(2), 753–777 (2003). https://doi.org/10.1111/1540-6261.00544

49. Carr, P., Lee, R., Wu, L.: Variance swaps on time-changed lévy processes. Finance Stoch. **16**(2), 335–355 (2012). https://doi.org/10.1007/s00780-011-0157-9
50. Catalini, G., Pacchiarotti, B.: Asymptotics for multifactor Volterra type stochastic volatility models. Stoch. Anal. Appl. 1–31 (2022). https://doi.org/10.1080/07362994.2022.2120012
51. CBOE: Volatility index methodology: Cboe volatility index. https://cdn.cboe.com/api/global/us_indices/governance/Volatility_Index_Methodology_Cboe_Volatility_Index.pdf (2023). Accessed 30 Aug 2023
52. Cherny, A.S., Engelbert, H.J.: Singular Stochastic Differential Equations. Springer, Berlin, Heidelberg (2005)
53. Cheung, Y.W., Ng, L.K.: Stock price dynamics and firm size: an empirical investigation. J. Financ. **47**(5), 1985–1997 (1992). https://doi.org/10.2307/2329006
54. Christie, A.: The stochastic behavior of common stock variances: value, leverage and interest rate effects. J. Financ. Econ. **10**(4), 407–432 (1982). https://doi.org/10.1016/0304-405x(82)90018-6
55. Chronopoulou, A., Viens, F.G.: Estimation and pricing under long-memory stochastic volatility. Ann. Finance **8**(2–3), 379–403 (2012)
56. Clark, P.K.: A subordinated stochastic process model with finite variance for speculative prices. Econom. J. Econom. Soc. **41**(1), 135 (1973). https://doi.org/10.2307/1913889
57. Comte, F., Renault, E.: Long memory in continuous-time stochastic volatility models. Math. Finance Int. J. Math. Statist. Financ. Econ. **8**(4), 291–323 (1998). https://doi.org/10.1111/1467-9965.00057
58. Cont, R., Das, P.: Rough volatility: Fact or artefact? Sankhya Series B (2024). https://doi.org/10.1007/s13571-024-00322-2
59. Cont, R., Potters, M., Bouchaud, J.P.: Scaling in stock market data: stable laws and beyond. In: Scale Invariance and Beyond, pp. 75–85. Springer, Berlin, Heidelberg (1997)
60. Cont, R.: Long range dependence in financial markets. In: Fractals in Engineering, pp. 159–179. Springer, London (2005)
61. Cont, R.: Volatility clustering in financial markets: empirical facts and agent-based models. In: Long Memory in Economics, pp. 289–309. Springer, Berlin Heidelberg (2006)
62. Cont, R.: Empirical properties of asset returns: stylized facts and statistical issues. Quant. Finance **1**(2), 223–236 (2001). https://doi.org/10.1080/713665670
63. Cont, R., Kokholm, T.: A consistent pricing model for index options and volatility derivatives. Math. Financ. **23**(2), 248–274 (2013). https://doi.org/10.1111/j.1467-9965.2011.00492.x
64. Cootner, P.: Random Character of Stock Market Prices. MIT Press, London (1967)
65. Corlay, S., Lebovits, J., Lévy Véhel, J.: Multifractional stochastic volatility models. Math. Finance Int. J. Math. Stat. Financ. Econom. **24**(2), 364–402 (2014). https://doi.org/10.1111/mafi.12024
66. Cox, J.C., Ingersoll, J.E., Ross, S.A.: A theory of the term structure of interest rates. Econom. J. Econom. Soc. **53**(2), 385–407 (1985). https://doi.org/10.2307/1911242
67. Cox, J.C.: Notes on option pricing I: the constant elasticity of variance option pricing model. J. Portfolio Manage. **23**(5), 15–17 (1997). https://doi.org/10.3905/jpm.1996.015
68. Cox, J.C., Ross, S.A.: The valuation of options for alternative stochastic processes. J. Financ. Econ. **3**(1–2), 145–166 (1976). https://doi.org/10.1016/0304-405x(76)90023-4
69. Cox, J.C., Rubinstein, M.: Options Markets. Pearson, Upper Saddle River, NJ (1985)
70. Cuchiero, C., Gazzani, G., Svaluto-Ferro, S.: Signature-based models: theory and calibration. SIAM J. Financ. Math. **14**(3), 910–957 (2023). https://doi.org/10.1137/22m1512338
71. Cuchiero, C., Gazzani, G., Möller, J., Svaluto-Ferro, S.: Joint calibration to SPX and VIX options with signature-based models. Math. Financ. **35**(1), 161–213 (2025). https://doi.org/10.1111/mafi.12442
72. Cutler, D., Poterba, J., Summers, L.: What moves stock prices? J. Portfolio Manage. 4–12 (1989)
73. Dandapani, A., Jusselin, P., Rosenbaum, M.: From quadratic hawkes processes to super-Heston rough volatility models with Zumbach effect. Quant. Finance **21**(8), 1235–1247 (2021). https://doi.org/10.1080/14697688.2020.1841906

74. Das, S.R., Sundaram, R.K.: Of smiles and smirks: a term structure perspective. J. Financ. Quant. Anal. **34**(2), 211 (1999). https://doi.org/10.2307/2676279
75. Delbaen, F., Schachermayer, W.: A general version of the fundamental theorem of asset pricing. Math. Ann. **300**(1), 463–520 (1994). https://doi.org/10.1007/bf01450498
76. Delbaen, F., Schachermayer, W.: The fundamental theorem of asset pricing for unbounded stochastic processes. Math. Ann. **312**(2), 215–250 (1998). https://doi.org/10.1007/s002080050220
77. Delbaen, F., Schachermayer, W.: Mathematics of Arbitrage. Springer, New York, NY (2006)
78. Delemotte, J., De Marco, S., Segonne, F.: Yet another analysis of the SP500 at-the-money skew: crossover of different power-law behaviours. SSRN Electron. J. (2023). https://doi.org/10.2139/ssrn.4428407
79. Derman, E., Kani, I.: Riding on the smile. Risk **7**(2), 32–39 (1994)
80. Derman, E., Miller, M.B., Park, D.: The volatility smile. Wiley, Nashville, TN (2016)
81. Di Nunno, G., Mishura, Y., Yurchenko-Tytarenko, A.: Sandwiched SDEs with unbounded drift driven by Hölder noises. Adv. Appl. Probab. **55**(3) (2023)
82. Di Nunno, G., Yurchenko-Tytarenko, A.: Sandwiched Volterra volatility model: Markovian approximations and hedging. Finance Stoch. **30**(1), 277–325 (2026). https://doi.org/10.1007/s00780-025-00584-2
83. Di Nunno, G., Kubilius, K., Mishura, Y., Yurchenko-Tytarenko, A.: From constant to rough: a survey of continuous volatility modeling. Mathematics **11**(19), 4201 (2023). https://doi.org/10.3390/math11194201
84. Di Nunno, G., Mishura, Y., Yurchenko-Tytarenko, A.: Drift-implicit Euler scheme for sandwiched processes driven by Hölder noises. Numer. Algorithms **93**(2), 459–491 (2023). https://doi.org/10.1007/s11075-022-01424-6
85. Di Nunno, G., Mishura, Y., Yurchenko-Tytarenko, A.: Option pricing in sandwiched Volterra volatility model. SIAM J. Financ. Math. **15**(3), 824–882 (2024). https://doi.org/10.1137/22m1521328
86. Di Nunno, G., Yurchenko-Tytarenko, A.: Power law in Sandwiched Volterra Volatility model. Mod. Stoch. Theory Appl. 169–194 (2024). https://doi.org/10.15559/24-vmsta246
87. Ding, Z., Granger, C.W.J.: Modeling volatility persistence of speculative returns: a new approach. J. Econom. **73**(1), 185–215 (1996). https://doi.org/10.1016/0304-4076(95)01737-2
88. Ding, Z., Granger, C.W.J., Engle, R.F.: A long memory property of stock market returns and a new model. J. Empir. Financ. **1**(1), 83–106 (1993). https://doi.org/10.1016/0927-5398(93)90006-d
89. Duffee, G.: Stock returns and volatility a firm-level analysis. J. Financ. Econ. **37**(3), 399–420 (1995). https://doi.org/10.1016/0304-405x(94)00801-7
90. Duffie, D., Pan, J., Singleton, K.: Transform analysis and asset pricing for affine jump-diffusions. Econom. J. Econom. Soc. **68**(6), 1343–1376 (2000). http://www.jstor.org/stable/3003992
91. Dumas, B., Fleming, J., Whaley, R.E.: Implied volatility functions: empirical tests. J. Financ. **53**(6), 2059–2106 (1998). https://doi.org/10.1111/0022-1082.00083
92. Duong, D., Swanson, N.R.: Volatility in discrete and continuous-time models: a survey with new evidence on large and small jumps. In: Missing Data Methods: Time-Series Methods and Applications, pp. 179–233. Emerald Group Publishing Limited (2011)
93. Dupire, B.: Arbitrage pricing with stochastic volatility. https://cims.nyu.edu/~essid/ctf/stochvol.pdf (1993). Accessed 30 Aug 2023
94. Dupire, B.: Pricing with a smile. Risk **7**, 18–20 (1994)
95. Edwards, T., Lazzara, C.: Realized Volatility Indices: Measuring Market Risk (2016). Research for S&P Dow Jones Indices, McGraww Hill Financial
96. Einstein, A.: Über die von der molekularkinetischen Theorie der Wärme geforderte Bewegung von in ruhenden Flüssigkeiten suspendierten Teilchen. Ann. Phys. **322**(8), 549–560 (1905). https://doi.org/10.1002/andp.19053220806

97. El Euch, O., Rosenbaum, M.: The characteristic function of rough Heston models. Math. Finance Int. J. Math. Stat. Financ. Econom. **29**(1), 3–38 (2019). https://doi.org/10.1111/mafi.12173
98. El Euch, O., Fukasawa, M., Rosenbaum, M.: The microstructural foundations of leverage effect and rough volatility. Finance Stoch. **22**(2), 241–280 (2018). https://doi.org/10.1007/s00780-018-0360-z
99. El Euch, O., Gatheral, J., Radoičić, R., Rosenbaum, M.: The Zumbach effect under rough Heston. Quant. Finance **20**(2), 235–241 (2020). https://doi.org/10.1080/14697688.2019.1658889
100. Emanuel, D.C., MacBeth, J.D.: Further results on the constant elasticity of variance call option pricing model. J. Financ. Quant. Anal. **17**(4), 533 (1982). https://doi.org/10.2307/2330906
101. Engle, R.F.: Autoregressive conditional heteroscedasticity with estimates of the variance of United Kingdom inflation. Econom. J. Econom. Soc. **50**(4), 987 (1982). https://doi.org/10.2307/1912773. http://www.jstor.org/stable/1912773
102. Eom, C., Kaizoji, T., Scalas, E.: Fat tails in financial return distributions revisited: evidence from the Korean stock market. Phys. A **526**(121055), 121055 (2019). https://doi.org/10.1016/j.physa.2019.121055
103. Fama, E.F.: The behavior of stock-market prices. J. Bus. **38**(1), 34–105 (1965). http://www.jstor.org/stable/2350752
104. Fengler, M.: Semiparametric modeling of implied volatility, 2005th edn. Springer, Berlin (2005)
105. Figlewski, S., Wang, X.: Is the “leverage effect” a leverage effect? SSRN Electron. J. (2001). https://doi.org/10.2139/ssrn.256109
106. Fleming, J., Ostdiek, B., Whaley, R.E.: Predicting stock market volatility: a new measure. J. Fut. Mark. **15**(3), 265–302 (1995). https://doi.org/10.1002/fut.3990150303
107. Fouque, J.P., Papanicolaou, G., Sircar, R., Sølna, K.: Maturity cycles in implied volatility. Finance Stoch. **8**(4) (2004). https://doi.org/10.1007/s00780-004-0126-7
108. Fouque, J.P., Saporito, Y.F.: Heston stochastic vol-of-vol model for joint calibration of VIX and S&P 500 options. Quant. Finance **18**(6), 1003–1016 (2018). https://doi.org/10.1080/14697688.2017.1412493
109. Fouque, J.P., Papanicolaou, G., Sircar, K.R.: Derivatives in financial markets with stochastic volatility. Cambridge University Press, Cambridge (2000)
110. Fouque, J.P., Papanicolaou, G., Sircar, R., Sølna, K.: Singular perturbations in option pricing. SIAM J. Appl. Math. **63**(5), 1648–1665 (2003). https://doi.org/10.1137/s0036139902401550
111. Francq, C.: GARCH models: structure, statistical inference and financial applications, 2nd edn. Wiley-Blackwell, Hoboken, NJ (2019)
112. Frey, R.: Derivative asset analysis in models with level-dependent and stochastic volatility. Res. Pap. Econ. (1997)
113. Fukasawa, M., Takabatake, T., Westphal, R.: Consistent estimation for fractional stochastic volatility model under high-frequency asymptotics. Math. Finance Int. J. Math. Stat. Financ. Econ. **32**(4), 1086–1132 (2022). https://doi.org/10.1111/mafi.12354
114. Fukasawa, M.: Volatility has to be rough. Quant. Finance **21**(1), 1–8 (2021). https://doi.org/10.1080/14697688.2020.1825781
115. Fukasawa, M., Gatheral, J.: A rough SABR formula. Front. Math. Finance **1**(1), 81 (2022). https://doi.org/10.3934/fmf.2021003
116. Fukasawa, M., Takabatake, T., Westphal, R.: Is volatility rough? ArXiv:1905.04852 (2019). http://arxiv.org/abs/1905.04852
117. Funahashi, H., Kijima, M.: Does the Hurst index matter for option prices under fractional volatility? Ann. Finance **13**(1), 55–74 (2017). https://doi.org/10.1007/s10436-016-0289-1
118. Funahashi, H., Kijima, M.: A solution to the time-scale fractional puzzle in the implied volatility. Fractal Fract. **1**(1), 14 (2017). https://doi.org/10.3390/fractalfract1010014
119. Garnier, J., Sølna, K.: Optimal hedging under fast-varying stochastic volatility. SIAM J. Financ. Math. **11**(1), 274–325 (2020). https://doi.org/10.1137/18m1221655
120. Gastineau, G.L.: An index of listed option premiums. Financ. Anal. J. **33**(3), 70–75 (1977)

121. Gatheral, J.: Consistent modeling of SPX and VIX options (2008). http://faculty.baruch.cuny.edu/jgatheral/Bachelier2008.pdf. Presented at The 5th World Congress of the Bachelier Finance Society
122. Gatheral, J.: The Volatility Surface: A Practitioner's Guide. Wiley, Nashville, TN (2006)
123. Gatheral, J., Jaisson, T., Rosenbaum, M.: Volatility is rough. SSRN Electron. J. (2014). https://doi.org/10.2139/ssrn.2509457
124. Gatheral, J., Jusselin, P., Rosenbaum, M.: The quadratic rough Heston model and the joint S&P 500/VIX smile calibration problem. SSRN Electron. J. (2020). https://doi.org/10.2139/ssrn.3514894
125. Ghysels, E., Harvey, A.C., Renault, E.: Stochastic volatility. In: Handbook of Statistics, pp. 119–191. Elsevier (1996)
126. Gihman, I., Skorokhod, A.: The Theory of Stochastic Processes I. Springer, Berlin, Heidelberg (2004)
127. Goutte, S., Ismail, A., Pham, H.: Regime-switching stochastic volatility model: estimation and calibration to VIX options. Appl. Math. Finance **24**(1), 38–75 (2017). https://doi.org/10.1080/1350486x.2017.1333015
128. Guillaume, D.M., Dacorogna, M.M., Davé, R.R., Müller, U.A., Olsen, R.B., Pictet, O.V.: From the bird's eye to the microscope: a survey of new stylized facts of the intra-daily foreign exchange markets. Finance Stoch. **1**(2), 95–129 (1997). https://doi.org/10.1007/s007800050018
129. Guyon, J.: On the joint calibration of SPX and VIX options: a dispersion-constrained martingale transport approach. In: Presented at Research in Options 2019. IMPA, Rio de Janeiro (2019)
130. Guyon, J.: The joint S&P 500/VIX smile calibration puzzle solved. SSRN Electron. J. (2019). https://doi.org/10.2139/ssrn.3397382
131. Guyon, J.: Dispersion-constrained martingale Schrödinger problems and the exact joint S&P 500/VIX smile calibration puzzle. SSRN Electron. J. (2021). https://doi.org/10.2139/ssrn.3853237
132. Guyon, J., El Amrani, M.: Does the term-structure of equity at-the-money skew really follow a power law? SSRN Electron. J. (2022). https://doi.org/10.2139/ssrn.4174538
133. Guyon, J., Bourgey, F.: Fast exact joint S&P 500/VIX smile calibration in discrete and continuous time. SSRN Electron. J. (2022). https://doi.org/10.2139/ssrn.4315084
134. Guyon, J., Lekeufack, J.: Volatility is (mostly) path-dependent. Quant. Finance **23**(9), 1221–1258 (2023). https://doi.org/10.1080/14697688.2023.2221281
135. Guyon, J., Mustapha, S.: Neural joint S&P 500/VIX smile calibration. SSRN Electron. J. (2022). https://doi.org/10.2139/ssrn.4309576
136. Hagan, P., Kumar, D., Lesniewski, A., Woodward, D.: Managing smile risk. Wilmott Mag. **1**, 84–108 (2002)
137. Harms, P., Stefanovits, D.: Affine representations of fractional processes with applications in mathematical finance. Stoch. Process. Appl. **129**(4), 1185–1228 (2019). https://doi.org/10.1016/j.spa.2018.04.010
138. Harrison, J.M., Kreps, D.M.: Martingales and arbitrage in multiperiod securities markets. J. Econ. Theory **20**(3), 381–408 (1979). https://doi.org/10.1016/0022-0531(79)90043-7
139. Harrison, J.M., Pliska, S.R.: Martingales and stochastic integrals in the theory of continuous trading. Stoch. Process. Appl. **11**(3), 215–260 (1981). https://doi.org/10.1016/0304-4149(81)90026-0
140. Hasanhodzic, J., Lo, A.W.: On Black's leverage effect in firms with no leverage. J. Portfolio Manage. **46**(1), 106–122 (2019). https://doi.org/10.3905/jpm.2019.46.1.106
141. Heston, S.L.: A closed-form solution for options with stochastic volatility with applications to bond and currency options. Rev. Financ. Stud. **6**(2), 327–343 (1993)
142. Hull, J., White, A.: The pricing of options on assets with stochastic volatilities. J. Financ. **42**(2), 281–300 (1987). https://doi.org/10.1111/j.1540-6261.1987.tb02568.x
143. Jackwerth, J.C., Rubinstein, M.: Recovering probability distributions from option prices. J. Financ. **51**(5), 1611 (1996). https://doi.org/10.2307/2329531

144. Jacquier, A., Martini, C., Muguruza, A.: On VIX futures in the rough Bergomi model. Quant. Finance **18**(1), 45–61 (2018). https://doi.org/10.1080/14697688.2017.1353127
145. Javaheri, A.: The volatility process: a study of stock market dynamics via para- metric stochastic volatility models and a comparison to the information embedded in option prices. Ph.D. thesis, Ecole de Mines de Paris, Paris (2004)
146. Jovanovic, F., Le Gall, P.: Does god practice a random walk? The “financial physics” of a nineteenth-century forerunner, Jules Regnault. Eur. J. History Econ. Thought **8**(3), 332–362 (2001). https://doi.org/10.1080/09672560110062960
147. Kokholm, T., Stisen, M.: Joint pricing of VIX and SPX options with stochastic volatility and jump models. J. Risk Finance **16**(1), 27–48 (2015)
148. Kolmogorov, A.: Wiener spirals and some other interesting curves in Hilbert space. Doklady AN SSSR 26(2) (1940) (in Russian)
149. Kreps, D.M.: Arbitrage and equilibrium in economies with infinitely many commodities. J. Math. Econ. **8**(1), 15–35 (1981). https://doi.org/10.1016/0304-4068(81)90010-0
150. Kubilius, K., Mishura, Y., Ralchenko, K.: Parameter Estimation in Fractional Diffusion Models. Springer, Cham (2017)
151. Kwok, Y.K., Zheng, W.: Pricing Models of Volatility Products and Exotic Variance Derivatives. Chapman & Hall/CRC, Philadelphia, PA (2022)
152. Lacombe, C., Muguruza, A., Stone, H.: Asymptotics for volatility derivatives in multi-factor rough volatility models. Math. Financ. Econ. **15**(3), 545–577 (2021). https://doi.org/10.1007/s11579-020-00288-5
153. Latané, H.A., Richard, R., J., J.: Standard deviations of stock price ratios implied in option prices. J. Financ. **31**(2), 369–381 (1976). https://doi.org/10.1111/j.1540-6261.1976.tb01892.x
154. Lee, R.W.: Implied volatility: statics, dynamics, and probabilistic interpretation. In: Recent Advances in Applied Probability, pp. 241–268. Kluwer Academic Publishers, Boston (2006)
155. Lewis, A.L.: Option Valuation Under Stochastic Volatility. Finance Press (2000)
156. Linetsky, V., Mendoza, R.: Constant elasticity of variance (CEV) diffusion model. In: Encyclopedia of Quantitative Finance (2010). https://doi.org/10.1002/9780470061602.eqf08015
157. Lobato, I.N., Velasco, C.: Long memory in stock-market trading volume. J. Bus. Econom. Stat. **18**(4), 410–427 (2000). https://doi.org/10.2307/1392223
158. Mandelbrot, B.: The variation of certain speculative prices. J. Bus. **36**(4), 394–419 (1963). http://www.jstor.org/stable/2350970
159. Mandelbrot, B.B., Van Ness, J.W.: Fractional Brownian motions, fractional noises and applications. SIAM Rev. **10**(4), 422–437 (1968). https://doi.org/10.1137/1010093
160. Mariani, M.C., Florescu, I.: Quantitative Finance. Wiley, Nashville, TN (2020)
161. Markowitz, H.: Portfolio selection. J. Finance **7**(1), 77 (1952). https://doi.org/10.2307/2975974
162. Martin, I.: Simple variance swaps. In: NBER Working Paper Series (2011)
163. Melino, A., Turnbull, S.M.: Pricing foreign currency options with stochastic volatility. J. Econom. **45**(1–2), 239–265 (1990). https://doi.org/10.1016/0304-4076(90)90100-8
164. Melino, A., Turnbull, S.M.: Misspecification and the pricing and hedging of long-term foreign currency options. J. Int. Money Financ. **14**(3), 373–393 (1995). https://doi.org/10.1016/0261-5606(95)00003-w
165. Merino, R., Pospíšil, J., Sobotka, T., Sottinen, T., Vives, J.: Decomposition formula for rough Volterra stochastic volatility models. Int. J. Theor. Appl. Finance **24**(02), 2150008 (2021)
166. Merton, R.C.: Theory of rational option pricing. Bell J. Econom. Manage. Sci. **4**(1), 141 (1973). https://doi.org/10.2307/3003143
167. Mikosch, T., Starica, C.: Is it really long memory we see in financial returns? In: Extremes and Integrated Risk Management, pp. 149–168. Risk Books (2000)
168. Mishura, Y., Yurchenko-Tytarenko, A.: Fractional Cox–Ingersoll–Ross process with non-zero “mean”. Mod. Stoch. Theory Appl. **5**(1), 99–111 (2018). https://doi.org/10.15559/18-VMSTA97

169. Mishura, Y., Yurchenko-Tytarenko, A.: Fractional Cox–Ingersoll–Ross process with small Hurst indices. Mod. Stoch. Theory Appl. **6**(1), 13–39 (2018). https://doi.org/10.15559/18-VMSTA126
170. Mishura, Y., Yurchenko-Tytarenko, A.: Standard and fractional reflected Ornstein–Uhlenbeck processes as the limits of square roots of Cox–Ingersoll–Ross processes. Stochastics 1–19 (2022). https://doi.org/10.1080/17442508.2022.2047188
171. Mishura, Y.: Stochastic calculus for fractional Brownian motion and related processes, 2008th edn. Springer, Berlin (2007)
172. Mishura, Y., Yurchenko-Tytarenko, A.: Approximating expected value of an option with non-Lipschitz payoff in fractional Heston-type model. Int. J. Theor. Appl. Finance **23**(05), 2050031 (2020). https://doi.org/10.1142/s0219024920500314
173. Mishura, Y., Pilipenko, A., Yurchenko-Tytarenko, A.: Low-dimensional Cox–Ingersoll–Ross process. Stochastics **96**(5), 1530–1550 (2024). https://doi.org/10.1080/17442508.2023.2300291
174. Nelson, D.B.: Conditional heteroskedasticity in asset returns: a new approach. Econom. J. Econom. Soc. **59**(2), 347–370 (1991). https://doi.org/10.2307/2938260. http://www.jstor.org/stable/2938260
175. Nualart, D.: The Malliavin Calculus and Related Topics. Springer, Berlin/Heidelberg (2006)
176. Ocone, D.L., Karatzas, I.: A generalized Clark representation formula, with application to optimal portfolios. Stoch. Rep. **34**(3–4), 187–220 (1991). https://doi.org/10.1080/17442509108833682
177. Osborne, M.F.M.: Brownian motion in the stock market. Oper. Res. **7**(2), 145–173 (1959). http://www.jstor.org/stable/167153
178. Prause, K.: The generalized hyperbolic model: estimation, financial derivatives, and risk measures
179. Rachev, S.T., Kim, Y.S., Bianchi, M.L., Fabozzi, F.J.: Financial Models with Lévy Processes and Volatility Clustering. Wiley, Chichester (2011)
180. Regnault, J.: Calcul des Chances et Philosophie de la Bourse. Mallet Bachelier and Castel (1863)
181. Renault, E., Touzi, N.: Option hedging and implied volatilities in a stochastic volatility model. Math. Finance Int. J. Math. Stat. Financ. Econ. **6**(3), 279–302 (1996). https://doi.org/10.1111/j.1467-9965.1996.tb00117.x
182. Revuz, D., Yor, M.: Continuous Martingales and Brownian Motion. Springer, Berlin, Heidelberg (1999)
183. Rogers, L.C.G.: Things We Think We Know (2019). https://www.skokholm.co.uk/wp-content/uploads/2019/11/TWTWKpaper.pdf
184. Rømer, S.E.: Empirical analysis of rough and classical stochastic volatility models to the SPX and VIX markets. Quant. Finance **22**(10), 1805–1838 (2022). https://doi.org/10.1080/14697688.2022.2081592
185. Rosenbaum, M.: Estimation of the volatility persistence in a discretely observed diffusion model. Stoch. Process. Appl. **118**(8), 1434–1462 (2008). https://doi.org/10.1016/j.spa.2007.09.004
186. Rosenbaum, M., Zhang, J.: Deep calibration of the quadratic rough Heston model. ArXiv:2107.01611 (2021). https://doi.org/10.48550/ARXIV.2107.01611
187. Ross, S.A.: A simple approach to the valuation of risky streams. J. Bus. **51**(3), 453–475 (1978). http://www.jstor.org/stable/2352277
188. Ross, S.A.: Finance. In: Eatwell, J., Milgate, M., Newman, P. (eds.) The New Palgrave Dictionary of Economics, vol. 2, pp. 322–336. Macmillan, London (1987)
189. Rough volatility literature. https://sites.google.com/site/roughvol/home/risks-1. Accessed 15 Sept 2025
190. Samorodnitsky, G.: Stochastic processes and long range dependence. Springer, Cham (2016)
191. Samuelson, P.A.: Mathematics of speculative price. SIAM Rev. **15**(1), 1–42 (1973)
192. Samuelson, P.A.: Rational theory of warrant pricing. Indus. Manage. Rev. **6**(2), 13–39 (1959)

193. Samuelson, P.A., Merton, R.C.: A complete model of warrant pricing that maximizes utility. Indus. Manage. Rev. **10**, 17–46 (1972)
194. Schachermayer, W.: The fundamental theorem of asset pricing. In: Handbook of the Fundamentals of Financial Decision Making, pp. 31–48. World Scientific (2013)
195. Schöbel, R., Zhu, J.: Stochastic volatility with an Ornstein-Uhlenbeck process: an extension. Rev. Finance **3**(1), 23–46 (1999)
196. Scott, L.O.: Option pricing when the variance changes randomly: theory, estimation, and an application. J. Financ. Quant. Anal. **22**(4), 419 (1987). https://doi.org/10.2307/2330793
197. Shephard, N., Andersen, T.G.: Stochastic volatility: origins and overview. In: Handbook of Financial Time Series, pp. 233–254. Springer, Berlin, Heidelberg (2009)
198. Shiryaev, A.N.: Essentials of Stochastic Finance: Facts, Models, Theory. World Scientific Publishing, Singapore (1999)
199. Skiadopoulos, G.: Volatility smile consistent option models: a survey. Int. J. Theor. Appl. Finance **04**(03), 403–437 (2001). https://doi.org/10.1142/s021902490100105x
200. Sprenkle, C.M.: Warrant prices as indicators of expectations and preferences. Yale Econom. Essays **1**(2), 179–231 (1959)
201. Stein, E.M., Stein, J.C.: Stock price distributions with stochastic volatility: an analytic approach. Rev. Financ. Stud. **4**, 727–752 (1991). http://rfs.oxfordjournals.org/
202. Tankov, P., Cont, R.: Financial Modelling with Jump Processes. Chapman & Hall/CRC, Philadelphia, PA (2003)
203. Thorp, E.O., Kassouf, S.T.: Beat the Market: A Scientific Stock Market System. Random House, New York (1967)
204. VIX index historical data. https://www.cboe.com/tradable_products/vix/vix_historical_data/. Accessed 26 Sept 2023
205. Wiggins, J.B.: Option values under stochastic volatility: theory and empirical estimates. J. Financ. Econ. **19**(2), 351–372 (1987). https://doi.org/10.1016/0304-405x(87)90009-2
206. Willinger, W., Taqqu, M.S., Teverovsky, V.: Stock market prices and long-range dependence. Finance Stoch. **3**(1), 1–13 (1999). https://doi.org/10.1007/s007800050049
207. Yamada, T., Watanabe, S.: On the uniqueness of solutions of stochastic differential equations. Kyoto J. Math. **11**(1), 155–167 (1971)
208. Zumbach, G.: Volatility conditional on price trends. Quant. Finance **10**(4), 431–442 (2010). https://doi.org/10.1080/14697680903266730
209. Zumbach, G.: Discrete time series, processes, and applications in finance, 2013th edn. Springer, Berlin (2012)

Chapter 2
Cox–Ingersoll–Ross Process and Skorokhod Problems

Abstract In this chapter, we begin our exploration of volatility models by revisiting the classical Cox–Ingersoll–Ross (CIR) process, a key component of the classical Heston model. Although the CIR process is Markovian and thus lacks memory effects, it exhibits very interesting structural features. In particular, under the Feller condition, the square root of the CIR process satisfies a stochastic differential equation with a drift that diverges near zero—this creates a "repelling" effect that ensures a.s. positivity of the solution. We argue that this mechanism can be extended to construct broader classes of positive volatility models with memory. To complete the picture, we also examine the behavior of CIR processes that do not satisfy the Feller condition and establish connections between this regime and the reflected Ornstein–Uhlenbeck (ROU) model.

2.1 Introduction

We begin the main part of this book by examining a model that is fairly well-known in the financial community: the Cox-Ingersoll-Ross (CIR) process

$$X(t) = x_0 + \int_0^t (a - bX(s))\, ds + \sigma \int_0^t \sqrt{X(s)} dW(s) \tag{2.1}$$

driven by standard Brownian motion W. An attentive reader may argue that this topic has no direct relation to *Volterra volatility* as announced in the book title or modeling *memory* in general: indeed, X defined by (2.1) is Markovian and hence, by its nature, it is fundamentally unable to reflect any non-trivial dependence on the past. Nevertheless, the CIR process has a number of properties that are extremely valuable for a volatility model: most notably, if $2a \geq \sigma^2$, the trajectories of the CIR process are strictly positive with probability 1. It turns out that structural features of the CIR dynamics can be exploited in a far more general setting with non-Markovian drivers and hence (2.1) serves as an excellent starting point for the introduction of a more complex setting. In this chapter, we will analyze (2.1) to identify such structural

G. Di Nunno et al., *Volterra Volatility Models*, Springer Finance,
https://doi.org/10.1007/978-3-032-26576-0_2

features and get some important insights into what the dynamics of a positive model with memory may look like. The present chapter is based on a considerably revised and extended material from [8, 9].

2.2 CIR Process: Path Properties and Dynamics of Its Square Root

Let $X = \{X(t),\ t \geq 0\}$ be a continuous modification of a CIR process defined by (2.1) with $x_0 > 0$, $a, \sigma > 0$, $b \geq 0$, and $W = \{W(t),\ t \geq 0\}$ being a Brownian motion.

Remark 2.1 The existence and uniqueness of the solution to (2.1) follows from the celebrated Yamada-Watanabe theorem [17, Theorem 1].

Remark 2.2 As explained in Appendix A.1, Brownian motion has a modification that is Hölder-continuous of any order strictly less than $1/2$. Throughout this chapter (and indeed throughout the entire book) we *always* work with this continuous modification when referring to Brownian motion.

As we will see later, the behavior of the process X is substantially different depending on the values of a and σ. Let us provide a useful definition to grasp the interplay between these parameters.

Definition 2.1 The value $d := \frac{4a}{\sigma^2}$ is called a *dimension* of the CIR process.

Remark 2.3 The choice of the term "*dimension*" in Definition 2.1 is meaningful. Indeed, take an arbitrary $b \geq 0$ and let a and σ be such that $d = \frac{4a}{\sigma^2} \in \mathbb{N}$. Let $B = (B_1, \dots, B_d)$ be a d-dimensional Brownian motion and $U = (U_1, \dots, U_d)$ be a standard d-dimensional Ornstein-Uhlenbeck process given by

$$U_i(t) = U_i(0) - \frac{b}{2}\int_0^t U_i(s)ds + \frac{\sigma}{2}B_i(t), \quad t \geq 0, \quad i = 1, \dots, d.$$

Using Itô's formula, it is then easy to see that the process $\sum_{i=1}^d U_i^2(t)$, $t \geq 0$, is the CIR process satisfying (2.1) with

$$W(t) := \sum_{i=1}^d \int_0^t \frac{U_i(s)}{\sqrt{\sum_{j=1}^d U_j^2(s)}} dB_i(s),$$

which is a standard Brownian motion by Levy's characterization.

It is well-known (see e.g. [6, Chapter IV, Example 8.2]) that the positivity of a, and consequently of the dimension $d = \frac{4a}{\sigma^2}$, implies that the CIR process (2.1) has non-negative trajectories. Moreover, as discussed rigorously in [7, Section 6.3.1], d fully determines whether X hits zero or not:

- if $d \geq 2$, X remains *strictly positive* with probability 1 and never hits zero[1];
- if $0 < d < 2$, X eventually hits zero with probability 1.

Since $X \geq 0$ when $d > 0$, it is possible to consider the process $Y = \{Y(t),\ t \geq 0\} := \{\sqrt{X(t)},\ t \geq 0\}$. As mentioned above, X can hit zero when $0 < a < \frac{\sigma^2}{2}$, so we cannot directly use Itô's formula to $Y = \sqrt{X}$ to analyze its dynamics. Instead, picking an arbitrary $\varepsilon > 0$, we can apply Itô's formula to $\sqrt{X(t)+\varepsilon}$ and write

$$\begin{aligned}\sqrt{X(t)+\varepsilon} &= \sqrt{x_0+\varepsilon} + \frac{1}{2}\int_0^t \left(\frac{a}{\sqrt{X(s)+\varepsilon}} - \frac{\sigma^2}{4}\frac{X(s)}{(X(s)+\varepsilon)^{\frac{3}{2}}}\right) ds \\ &\quad - \frac{1}{2}\int_0^t \frac{bX(s)}{\sqrt{X(s)+\varepsilon}} ds + \frac{\sigma}{2}\int_0^t \frac{\sqrt{X(s)}}{\sqrt{X(s)+\varepsilon}} dW(s).\end{aligned} \tag{2.2}$$

Fix an arbitrary $T > 0$ and note that the left-hand side of (2.2) converges to $Y(t)$ uniformly on $[0, T]$ with probability 1 as $\varepsilon \downarrow 0$. It is also evident that

$$\sup_{t\in[0,T]}\left|\int_0^t \frac{X(s)}{\sqrt{X(s)+\varepsilon}} ds - \int_0^t Y(s)ds\right| \to 0 \quad a.s., \qquad \varepsilon \downarrow 0. \tag{2.3}$$

Next, by [12, Chapter XI], $\mathbb{E}\int_0^\infty \mathbb{1}_{\{X(s)=0\}}ds = 0$ and hence, by the Burkholder-Davis-Gundy inequality (see [4, Theorem 11.5.5]), for any $T > 0$,

$$\begin{aligned}&\mathbb{E}\left(\sup_{t\in[0,T]}\left|\int_0^t \frac{\sqrt{X(s)}}{\sqrt{X(s)+\varepsilon}} dW(s) - W(t)\right|\right)^2 \\ &\leq 4\mathbb{E}\int_0^T \left(\frac{\sqrt{X(s)}}{\sqrt{X(s)+\varepsilon}} - 1\right)^2 ds \\ &= 4\mathbb{E}\int_0^T \left(\frac{\sqrt{X(s)}}{\sqrt{X(s)+\varepsilon}} - 1\right)^2 \mathbb{1}_{\{X(s)>0\}}ds + 4\mathbb{E}\int_0^T \mathbb{1}_{\{X(s)=0\}}ds \\ &= 4\mathbb{E}\int_0^T \left(\frac{\sqrt{X(s)}}{\sqrt{X(s)+\varepsilon}} - 1\right)^2 \mathbb{1}_{\{X(s)>0\}}ds \to 0, \quad \varepsilon \downarrow 0.\end{aligned}$$

This yields that for each $T > 0$

---

[1] The condition $d \geq 2$, equivalently written as $2a \geq \sigma^2$, is commonly referred to as the *Feller condition*.

$$\sup_{t\in[0,T]}\left|\int_0^t \frac{\sqrt{X(s)}}{\sqrt{X(s)+\varepsilon}}dW(s)-W(t)\right| \xrightarrow{\mathbb{P}} 0, \quad \varepsilon\downarrow 0. \tag{2.4}$$

In particular, Eq. (2.2) and the convergences (2.3) and (2.4) imply that the left-hand side of

$$\sqrt{X(t)+\varepsilon}-\sqrt{x_0+\varepsilon}+\frac{1}{2}\int_0^t \frac{bX(s)}{\sqrt{X(s)+\varepsilon}}ds-\frac{\sigma}{2}\int_0^t \frac{\sqrt{X(s)}}{\sqrt{X(s)+\varepsilon}}dW(s)$$
$$=\frac{1}{2}\int_0^t\left(\frac{a}{\sqrt{X(s)+\varepsilon}}-\frac{\sigma^2}{4}\frac{X(s)}{(X(s)+\varepsilon)^{\frac{3}{2}}}\right)ds$$

converges uniformly on compacts in probability as $\varepsilon\downarrow 0$. Therefore, there exists the ucp-limit

$$L(t):=\lim_{\varepsilon\downarrow 0}\frac{1}{2}\int_0^t\left(\frac{a}{\sqrt{X(s)+\varepsilon}}-\frac{\sigma^2}{4}\frac{X(s)}{(X(s)+\varepsilon)^{\frac{3}{2}}}\right)ds, \tag{2.5}$$

and the process $Y=\sqrt{X}$ satisfies the SDE of the form

$$Y(t)=Y(0)-\frac{b}{2}\int_0^t Y(s)ds+\frac{\sigma}{2}W(t)+L(t),$$

where $Y(0)=\sqrt{x_0}$.

Remark 2.4 Ucp convergence in (2.5) implies that for an arbitrary $T>0$ there exists a sequence $\{\varepsilon_n,\ n\ge 1\}$ (depending on T) such that, for any $t\in[0,T]$,

$$\begin{aligned}&\left|\int_0^t \frac{\sqrt{X(s)}}{\sqrt{X(s)+\varepsilon_n}}dW(s)-W(t)\right|\to 0,\\ &\left|L(t)-\frac{1}{2}\int_0^t\left(\frac{a}{\sqrt{X(s)+\varepsilon_n}}-\frac{\sigma^2}{4}\frac{X(s)}{(X(s)+\varepsilon_n)^{\frac{3}{2}}}\right)ds\right|\to 0\end{aligned} \tag{2.6}$$

with probability 1 as $n\to\infty$. Later on, we will see that the a.s. convergence (2.6) holds for an *arbitrary* sequence $\{\varepsilon_n,\ n\ge 1\}$ such that $\varepsilon_n\downarrow 0$ as $n\to\infty$.

Remark 2.5 Note that the process L defined by (2.5) is a.s. continuous in t due to the equality

$$L(t)=Y(t)-Y(0)+\frac{b}{2}\int_0^t Y(s)ds-\frac{\sigma}{2}W(t) \tag{2.7}$$

and the fact that the right-hand side of (2.7) is a.s. continuous in t.

To summarize, in order to identify the dynamics of $Y = \sqrt{X}$, it is sufficient to analyze the limit (2.5). Clearly, if it is true that

$$\int_0^t \frac{1}{\sqrt{X(s)}} ds = \int_0^t \frac{1}{Y(s)} ds < \infty \quad a.s., \tag{2.8}$$

then the situation simplifies substantially: the limit in (2.5) is

$$L(t) = \frac{1}{2}\left(a - \frac{\sigma^2}{4}\right) \int_0^t \frac{1}{Y(s)} ds \quad a.s.$$

by monotone convergence. This is exactly the case when the CIR dimension $d \geq 2$: indeed, in this case both X and Y have paths that are bounded away from zero and therefore (2.8) holds for all $t \geq 0$ with probability 1. It turns out that (2.8) also holds if $1 < d < 2$, i.e. when $\frac{\sigma^2}{4} < a < \frac{\sigma^2}{2}$.

Theorem 2.1 *Let $a > \frac{\sigma^2}{4}$. Then, for any $t \geq 0$,*

$$\int_0^t \frac{1}{Y(s)} ds < \infty \quad a.s., \tag{2.9}$$

and Y a.s. satisfies the SDE of the form

$$Y(t) = \sqrt{x_0} + \frac{1}{2}\left(a - \frac{\sigma^2}{4}\right) \int_0^t \frac{1}{Y(s)} ds - \frac{b}{2} \int_0^t Y(s) ds + \frac{\sigma}{2} W(t). \tag{2.10}$$

Proof Denote $p := a - \frac{\sigma^2}{4} > 0$, define

$$A(t) := \left\{ \omega \in \Omega : \int_0^t \frac{1}{\sqrt{X(s)}} ds = +\infty \right\},$$

and assume that $\mathbb{P}(A(t)) > 0$ for some $t > 0$. Fix $T > t$ and let $\{\varepsilon_n,\ n \geq 1\}$ be such that the convergences in (2.6) hold for any $t \in [0, T]$ on some $\Omega' \subset \Omega$, $\mathbb{P}(\Omega') = 1$. Fix an arbitrary $\omega \in A(t) \cap \Omega'$ (in what follows, ω in brackets will be omitted) and write

$$\int_0^t \left(\frac{a}{\sqrt{X(s) + \varepsilon_n}} - \frac{\sigma^2}{4} \frac{X(s)}{(X(s) + \varepsilon_n)^{\frac{3}{2}}} \right) ds$$

$$= \frac{\sigma^2}{4}\int_0^t \left(\frac{1}{\sqrt{X(s)+\varepsilon_n}} - \frac{X(s)}{(X(s)+\varepsilon_n)^{\frac{3}{2}}}\right) ds + p\int_0^t \frac{1}{\sqrt{X(s)+\varepsilon_n}} ds.$$

Obviously, for all $s \in [0, t]$

$$\frac{1}{\sqrt{X(s)+\varepsilon_n}} \geq \frac{X(s)}{(X(s)+\varepsilon_n)^{\frac{3}{2}}} \quad a.s.,$$

so, for $\omega \in A(t) \cap \Omega'$,

$$\int_0^t \left(\frac{a}{\sqrt{X(s)+\varepsilon_n}} - \frac{\sigma^2}{4}\frac{X(s)}{(X(s)+\varepsilon_n)^{\frac{3}{2}}}\right) ds \to \infty \quad a.s., \quad n \to \infty.$$

Hence, taking into account (2.2)–(2.3), we obtain that, for any $\omega \in A(t) \cap \Omega'$,

$$\sqrt{X(t)} - \sqrt{X(0)} + \frac{b}{2}\int_0^t \sqrt{X(s)}ds - \frac{\sigma}{2}W(t) = \infty.$$

However, obviously, with probability 1,

$$\sqrt{X(t)} - \sqrt{X(0)} + \frac{b}{2}\int_0^t \sqrt{X(s)}ds - \frac{\sigma}{2}W(t) < \infty,$$

which immediately implies that $\mathbb{P}(A(t)) = 0$, for all $t \geq 0$, and $\int_0^t \frac{1}{\sqrt{X(s)}}ds = \int_0^t \frac{1}{Y(s)}ds < \infty$ a.s. By going to the limit in (2.2), we immediately get (2.10).

Concerning the uniqueness of the solution to (2.10), let $\widetilde{Y}(t)$ be any of its non-negative solutions. Then, by Itô's formula,

$$\widetilde{Y}^2(t) = X(0) + \int_0^t \left(a - b\widetilde{Y}^2(s)\right) ds + \sigma\int_0^t \widetilde{Y}(s)dW(s),$$

so $\widetilde{Y}$ satisfies equation (2.1) and thus coincides with X. Therefore,

$$\widetilde{Y}(t) = \sqrt{X(t)} = Y(t) \quad a.s., \quad t \geq 0.$$

□

Remark 2.6 Using the same arguments as in [2, Theorem 3.2], it is possible to prove that for $a > \frac{\sigma^2}{4}$ the process $Y = \sqrt{X}$ is the unique *non-negative* strong solution to the SDE (2.10). However, if $\frac{\sigma^2}{4} < a < \frac{\sigma^2}{2}$, (2.10) has other strong solutions; moreover, the uniqueness in law does not hold for (2.10). For a more detailed discussion of this phenomenon, we refer the reader to [1] whereas a comprehensive overview of SDEs of the type (2.10) can be found in [3].

The shape of Eq. (2.10) provides a valuable intuition on why the process Y is positive for bigger values of a. Namely, whenever the solution Y approaches 0, the integrand

$$\frac{1}{2}\left(a - \frac{\sigma^2}{4}\right)\frac{1}{Y}$$

explodes, "*repelling*" Y from 0. In the meanwhile, the condition $2a \geq \sigma^2$ turns out to be just enough for this effect to become dominant over any possible downward movements caused by the noise σW. This structural observation is crucial in our further analysis and will be used in our models in the chapters to come.

2.3 CIR Process of Dimension 1 and Reflected Ornstein-Uhlenbeck Process

The case $d = 1$ (corresponding to $a = \frac{\sigma^2}{4}$) turns out to be significantly different from the one described above in Sect. 2.2. In this regime, X reaches zero and, as we will see later, (2.9) *does not* hold after hitting zero for the first time. However, the limit L from (2.5) has a simple interpretation in terms of reflected processes and the main goal of this section is to establish this connection.

We shall start from the definition of a *reflection function* following the classical works [14, 15] (for more details on reflected processes see also [10]).

Definition 2.2 Let $\xi = \{\xi(t),\ t \geq 0\}$ be some stochastic process. The process $\zeta = \{\zeta(t),\ t \geq 0\}$ is called *a Skorokhod reflection function* for ξ, if, with probability 1, ζ is a continuous non-decreasing process such that $\zeta(0) = 0$ and the points of growth of ζ can occur only at the zeros of ξ.

Definition 2.3 Stochastic process $\widetilde{Y} = \{\widetilde{Y}(t),\ t \geq 0\}$ is called a *reflected Ornstein-Uhlenbeck (ROU) process* if it satisfies a stochastic differential equation of the form

$$\widetilde{Y}(t) = Y(0) - \widetilde{b}\int_0^t \widetilde{Y}(s)ds + \widetilde{\sigma}W(t) + \widetilde{L}(t), \quad t \geq 0, \tag{2.11}$$

where $Y(0)$, $\widetilde{b}$ and $\widetilde{\sigma}$ are positive constants, $W = \{W(t),\ t \geq 0\}$ is a standard Brownian motion, $\{\widetilde{L}(t),\ t \geq 0\}$ is a reflection function for $\widetilde{Y}$ and $\widetilde{Y} \geq 0$ a.s.

Remark 2.7 The ROU process is well studied in the literature (see e.g. [16]). Moreover, it fits into the broader topic of the so-called *Skorokhod reflections*—in this regard, we refer our reader to the recent book [10] containing a general overview of the subject. Here, we note that the solution to (2.11) is unique despite having two unknown functions $\widetilde{Y}$ and $\widetilde{L}$ in the equation. Indeed, let $\widetilde{Y}$ and $\widehat{Y}$ be two stochastic processes satisfying

$$\widetilde{Y}(t) = Y(0) - \widetilde{b} \int_0^t \widetilde{Y}(s)ds + \widetilde{\sigma} W(t) + \widetilde{L}(t)$$

and

$$\widehat{Y}(t) = Y(0) - \widetilde{b} \int_0^t \widehat{Y}(s)ds + \widetilde{\sigma} W(t) + \widehat{L}(t),$$

where $\widetilde{L}$ and $\widehat{L}$ are the corresponding reflection functions. Assume that on some $\omega \in \Omega$ such that both $\widetilde{Y}$ and $\widehat{Y}$ are continuous, it holds that

$$\widetilde{Y}(t) - \widehat{Y}(t) > 0 \tag{2.12}$$

for some $t > 0$. Consider $\tau(t) := \sup\{s \in [0, t) : \widetilde{Y}(s) - \widehat{Y}(s) = 0\}$. Then $\widetilde{Y}(u) - \widehat{Y}(u) > 0$ for all $u \in (\tau(t), t]$; moreover, $\widetilde{Y}(u) > 0$ for $u \in (\tau(t), t]$, therefore $\widetilde{L}$ is non-increasing on $(\tau(t), t]$. It means that the difference $\widetilde{Y}(u) - \widehat{Y}(u)$ is also non-increasing on $(\tau(t), t]$ since

$$\widetilde{Y}(u) - \widehat{Y}(u) = -\widetilde{b} \int_{\tau(t)}^u (\widetilde{Y}(s) - \widehat{Y}(s))ds + \left(\widetilde{L}(u) - \widehat{L}(u)\right) - \left(\widetilde{L}(\tau(t)) - \widehat{L}(\tau(t))\right)$$

and the right-hand side is non-increasing w.r.t. u. Hence, taking into account that $\widetilde{Y}(\tau(t)) - \widehat{Y}(\tau(t)) = 0$ due to the definition of $\tau(t)$ and to the continuity of both $\widetilde{Y}$ and $\widehat{Y}$, the difference $\widetilde{Y}(u) - \widehat{Y}(u)$ cannot be positive for any $u \in (\tau(t), t]$. This contradicts (2.12). Interchanging the roles of $\widetilde{Y}$ and $\widehat{Y}$, one can easily verify that $\widetilde{Y}(t) - \widehat{Y}(t)$ cannot be negative either and whence $\widehat{Y} = \widetilde{Y}$, $\widehat{L} = \widetilde{L}$.

Theorem 2.2 *Let* $a = \frac{\sigma^2}{4}$ *and denote* $\tau := \inf\{s \geq 0 \mid X(s) = 0\}$.

(1) For all $\gamma > 0$,

$$\int_0^{\tau+\gamma} \frac{1}{Y(s)} ds = \infty \quad a.s.$$

(2) The pair (Y, L)*, where* $Y := \sqrt{X}$ *and* L *is defined by* (2.5)*, is the (unique) solution to*

$$Y(t) = \sqrt{x_0} - \frac{b}{2}\int_0^t Y(s)ds + \frac{\sigma}{2}W(t) + L(t), \tag{2.13}$$

with L being the corresponding Skorokhod reflection function for Y.

Proof Fix $T > 0$ and take a sequence $\{\varepsilon_n,\ n \geq 1\}$ such that (2.6) holds on some Ω', $\mathbb{P}(\Omega') = 1$. Since $a = \frac{\sigma^2}{4}$, one can rewrite the a.s. limit (2.5) along the sequence $\{\varepsilon_n,\ n \geq 1\}$ as

$$L(t) := \lim_{n\to\infty} \frac{\sigma^2}{8} \int_0^t \frac{\varepsilon_n}{(X(s) + \varepsilon_n)^{\frac{3}{2}}} ds, \quad t \in [0, T].$$

It is evident that $L(0) = 0$ a.s. thanks to the continuity of X and the fact that $X(0) > 0$. Moreover, L is continuous in t by Remark 2.5. Next,

$$\int_0^{t_1} \frac{\varepsilon_n}{(X(s) + \varepsilon_n)^{\frac{3}{2}}} ds \leq \int_0^{t_2} \frac{\varepsilon_n}{(X(s) + \varepsilon_n)^{\frac{3}{2}}} ds$$

for all $t_1 < t_2$, and hence L is non-decreasing in t a.s. Finally, if $X(t) = x > 0$, there exists an interval $[t_1, t_2]$ containing t such that $X(s) > \frac{x}{2}$ for all $s \in [t_1, t_2]$ and thus

$$L(t_1) - L(t_2) = \lim_{n\to\infty} \frac{\sigma^2}{8} \int_{t_1}^{t_2} \frac{\varepsilon_n}{(X(s) + \varepsilon_n)^{\frac{3}{2}}} ds \to 0, \quad n \to \infty.$$

Therefore, L can increase only at the points where X hits zero, and they coincide with the ones of Y. Taking into the account all of the above as well as an arbitrary choice of T, L is the reflection function for Y and the latter is indeed a ROU process.

Now, let us prove that $\int_0^\tau \frac{1}{Y_s} ds < \infty$ a.s. Consider a standard Ornstein-Uhlenbeck process $U = \{U(t),\ t \geq 0\}$ of the form

$$U(t) = \sqrt{X(0)} - \frac{b}{2}\int_0^t U(s)ds + \frac{\sigma}{2}W(t), \tag{2.14}$$

with W being the same Brownian motion that drives X. We can see that Y coincides with U until τ a.s. Thus it is sufficient to prove that $\int_0^\tau \frac{1}{U(s)} ds < \infty$ a.s. For any $\varepsilon > 0$ consider

$$\int_0^\tau \frac{1}{U(s)} \mathbb{1}_{\{\varepsilon < U(s) < 1\}} ds = \int_\varepsilon^1 \frac{L^U(\tau, x)}{x} dx,$$

where L^U denotes the local time of U (see Example A.4 in Appendix A.2), and observe that

$$\int_0^\tau \frac{1}{U(s)}ds \le \lim_{\varepsilon\downarrow 0}\int_0^\tau \frac{1}{U(s)}\mathbb{1}_{\{\varepsilon<U(s)<1\}}ds = \int_0^1 \frac{L^U(\tau,x)}{x}dx.$$

As noted in Example A.4 from Appendix A.2, the local time $L^U(t,x)$ of U is Hölder continuous in x up to order $\frac{1}{2}$ over bounded time intervals and thus $\int_0^\tau \frac{1}{U(s)}ds = \int_0^\tau \frac{1}{Y(s)}ds < \infty$ a.s.

Finally, assume that for some $\gamma > 0$

$$\int_0^{\tau+\gamma} \frac{1}{Y(s)}ds < \infty$$

with positive probability. On $\omega \in \Omega$ where this property holds, we have that

$$\int_0^{\tau+\gamma} \frac{ds}{\sqrt{X(s)+\varepsilon}} - \int_0^{\tau+\gamma} \frac{X(s)}{(X(s)+\varepsilon)^{\frac{3}{2}}}ds$$
$$\to \int_0^{\tau+\gamma} \frac{1}{Y(s)}ds - \int_0^{\tau+\gamma} \frac{1}{Y(s)}ds = 0, \quad \varepsilon \to 0.$$

Therefore, for such ω, Y satisfies the equation of the form

$$Y(t) = Y(0) - \frac{b}{2}\int_0^t Y(s)ds + \frac{\sigma}{2}W(t)$$

on the interval $[0, \tau+\gamma]$, i.e. such paths of Y coincide with the corresponding paths of the Ornstein-Uhlenbeck process U defined by (2.14) up until $\tau+\gamma$. This implies that $U(\tau) = 0$ and U is nonnegative on the interval $[\tau, \tau+\gamma]$ for such ω, which is impossible due to the non-tangent property of Gaussian processes stated by [18] (see also [11]). □

As a corollary of Theorem 2.2, we have a representation of the reflection function of the ROU process as the limit of integral functionals of the CIR processes. It is interesting that the reflection function is singular w.r.t. the Lebesgue measure (see Remark 2.8) while the processes that converge to it are absolutely continuous a.s.

Theorem 2.3 *Let $Y(0), b, \sigma > 0$ be given constants and $\{\varepsilon_n,\ n \geq 1\}$ be an arbitrary sequence such that $\varepsilon_n \downarrow 0$, $n \to \infty$. For any ε_n from this sequence, consider the CIR process $X_{\varepsilon_n} = \{X_{\varepsilon_n}(t),\ t \geq 0\}$ given by*

$$X_{\varepsilon_n}(t) = X(0) + \int_0^t \left(\frac{\sigma^2}{4} + \varepsilon_n - bX_{\varepsilon_n}(s)\right) ds + \sigma \int_0^t \sqrt{X_{\varepsilon_n}(s)} dW(s),$$

and denote its square root by $Y_{\varepsilon_n}(t) := \sqrt{X_{\varepsilon_n}(t)}$. Then, with probability 1,

(1) the limit $\lim_{n\to\infty} Y_{\varepsilon_n}(t) =: Y(t)$ is well-defined, finite and non-negative for any $t \geq 0$;

(2) the limit process $Y = \{Y(t),\ t \geq 0\}$ is a ROU process satisfying the equation of the form

$$Y(t) = Y(0) - \frac{b}{2} \int_0^t Y(s) ds + \frac{\sigma}{2} W(t) + L(t), \quad t \geq 0,$$

with $Y(0) = \sqrt{X(0)} > 0$ and L being the reflection function for Y;

3) for any $T > 0$, we have

$$\sup_{t\in[0,T]} |Y(t) - Y_{\varepsilon_n}(t)| \to 0, \quad n \to \infty, \tag{2.15}$$

and

$$\sup_{t\in[0,T]} \left| L(t) - \frac{1}{2} \int_0^t \frac{\varepsilon_n}{Y_{\varepsilon_n}(s)} ds \right| \to 0, \quad n \to \infty. \tag{2.16}$$

Proof Let X denote the CIR process of the form

$$X(t) = X(0) + \int_0^t \left(\frac{\sigma^2}{4} - bX(s)\right) ds + \sigma \int_0^t \sqrt{X(s)} dW(s).$$

By Theorem 2.2, there exists $\Omega' \subset \Omega$ with $\mathbb{P}(\Omega') = 1$ such that, for all $\omega \in \Omega'$, the processes X and each Y_{ε_n}, $n \geq 1$, are continuous and the latter satisfy the equations of the form

$$Y_{\varepsilon_n}(t) = Y(0) + \frac{1}{2} \int_0^t \frac{\varepsilon_n}{Y_{\varepsilon_n}(s)} ds - \frac{b}{2} \int_0^t Y(s) ds + \frac{\sigma}{2} W(t), \quad t \geq 0,$$

with the integral $\int_0^t \frac{1}{Y_{\varepsilon_n}(s)} ds < \infty$. Furthermore, since each $X_{\varepsilon_n} = Y_{\varepsilon_n}^2$ is a CIR process that satisfies conditions of the comparison theorem from [5], this Ω' can be chosen such that for all $\omega \in \Omega'$

$$Y_{\varepsilon_n}(\omega, t) \ge Y_{\varepsilon_{n+1}}(\omega, t) \ge \sqrt{X(t)} \ge 0, \quad t \ge 0, \quad n \ge 1. \tag{2.17}$$

Fix $\omega \in \Omega'$ (note that, in what follows, we will omit ω in brackets for notational simplicity). Since the sequence $\{Y_{\varepsilon_n}(t),\ n \ge 1\}$ is non-increasing for each $t \ge 0$, there exists a pointwise limit $Y(t) := \lim_{n\to\infty} Y_{\varepsilon_n}(t) \in [0, \infty)$. Moreover, it is evident that $\lim_{n\to\infty} \int_0^t Y_{\varepsilon_n}(s)ds = \int_0^t Y(s)ds$. In addition,

$$\begin{aligned} Y(t) &= \lim_{n\to\infty} Y_{\varepsilon_n}(t) \\ &= Y(0) - \lim_{n\to\infty} \frac{b}{2} \int_0^t Y_{\varepsilon_n}(s)ds + \frac{\sigma}{2} W(t) + \lim_{n\to\infty} \frac{1}{2} \int_0^t \frac{\varepsilon_n}{Y_{\varepsilon_n}(s)} ds \\ &= Y(0) - \frac{b}{2} \int_0^t Y(s)ds + \frac{\sigma}{2} W(t) + \lim_{n\to\infty} \frac{1}{2} \int_0^t \frac{\varepsilon_n}{Y_{\varepsilon_n}(s)} ds, \end{aligned}$$

therefore the limit $L(t) := \lim_{n\to\infty} \frac{1}{2} \int_0^t \frac{\varepsilon_n}{Y_{\varepsilon_n}(s)} ds$ is well-defined, nonnegative and finite.

In order to complete the proof of the theorem, it is sufficient to check that the function L defined above is indeed a reflection function for Y, i.e. is a continuous and nondecreasing process that starts at zero with the points of growth occurring only at zeros of Y. Note that the continuity of L would also automatically imply the uniform convergences (2.15) and (2.16) on each compact $[0, T]$. Indeed, recall that $Y_{\varepsilon_n}(t) \ge Y_{\varepsilon_{n+1}}(t)$ for all $t \ge 0$, $n \ge 1$, and, moreover, if L is continuous, Y is continuous as well. Therefore, if we prove the continuity of L, we can use Dini's theorem to guarantee the uniform convergence (2.15). The same argument applies to (2.16). In fact, the right-hand side of

$$\frac{1}{2} \int_0^t \frac{\varepsilon_n}{Y_{\varepsilon_n}(s)} ds = Y_{\varepsilon_n}(t) - Y(0) + \frac{b}{2} \int_0^t Y_{\varepsilon_n}(s)ds - \frac{\sigma}{2} W(t)$$

is non-increasing w.r.t. t, therefore, for each $t \ge 0$ and $n \ge 1$, the inequality

$$\frac{1}{2} \int_0^t \frac{\varepsilon_n}{Y_{\varepsilon_n}(s)} ds \ge \frac{1}{2} \int_0^t \frac{\varepsilon_{n+1}}{Y_{\varepsilon_{n+1}}(s)} ds$$

holds and Dini's theorem would imply (2.16).

By (2.17), the continuity of X and the fact that $X(0) > 0$, there exists an interval $[0, t_0)$ such that $Y_{\varepsilon_n}(t) \geq \frac{Y(0)}{2}$ for all $t \in [0, t_0)$ and $n \geq 1$. Thus, for any $t \in [0, t_0)$, we have

$$L(t) = \lim_{n\to\infty} \frac{1}{2} \int_0^t \frac{\varepsilon_n}{Y_{\varepsilon_n}(s)} ds \leq \lim_{n\to\infty} \frac{t_0 \varepsilon_n}{Y(0)} = 0,$$

i.e. $L(t) = 0$ for all $t \in [0, t_0]$.

For the reader's convenience, we organize the remaining part of the proof in four steps.

Step 1. *L in non-decreasing.* Monotonicity of L is obvious since, for any fixed $n \geq 1$ and $t_1 < t_2$,

$$\int_0^{t_1} \frac{\varepsilon_n}{Y_{\varepsilon_n}(s)} ds \leq \int_0^{t_2} \frac{\varepsilon_n}{Y_{\varepsilon_n}(s)} ds.$$

Step 2. L **is right-continuous**. For any fixed $t \geq 0$, denote $L(t+) := \lim_{\delta\downarrow 0} L(t+\delta)$ (the right limit exists since L is non-decreasing) and assume that $L(t+) - L(t) = \alpha > 0$. Due to the monotonicity of L, this implies that, for all $\delta > 0$,

$$L(t+\delta) - L(t) \geq \alpha > 0. \tag{2.18}$$

Now, take n_0 such that, for all $n \geq n_0$, we have

$$\frac{1}{2} \int_0^t \frac{\varepsilon_n}{Y_{\varepsilon_n}(s)} ds \in \left[L(t), L(t) + \frac{\alpha}{4}\right)$$

and $\delta_0 > 0$ such that

$$\frac{1}{2} \int_0^{t+\delta_0} \frac{\varepsilon_{n_0}}{Y_{\varepsilon_{n_0}}(s)} ds \in \left[L(t), L(t) + \frac{\alpha}{2}\right).$$

As it was previously remarked, for each $s \geq 0$, the values of $\frac{1}{2}\int_0^s \frac{\varepsilon_n}{Y_{\varepsilon_n}(u)} du$ are non-increasing when $n \to \infty$. Thus, for any $n \geq n_0$,

$$\frac{1}{2} \int_0^{t+\delta_0} \frac{\varepsilon_n}{Y_{\varepsilon_n}(s)} ds \leq \frac{1}{2} \int_0^{t+\delta_0} \frac{\varepsilon_{n_0}}{Y_{\varepsilon_{n_0}}(s)} ds \leq L(t) + \frac{\alpha}{2},$$

i.e.

$$L(t+) \leq \lim_{n\to\infty} \frac{1}{2} \int_0^{t+\delta_0} \frac{\varepsilon_n}{Y_{\varepsilon_n}(s)} ds < L(t) + \frac{\alpha}{2},$$

which contradicts (2.18). Therefore, $L(t+) - L(t) = 0$, i.e. L is right-continuous.

Step 3. *L is left-continuous*. Assume that the left-continuity is not true and that there exists $t > 0$ such that $L(t) - L(t-) > 0$ (note that $L(t-) = \lim_{\delta \downarrow 0} L(t-\delta)$ is well-defined due to the monotonicity of L). Since L may have only positive jumps, so does Y. Moreover, the points of jumps of L and Y coincide. This all implies that $Y(t) - Y(t-) > 0$ and we now consider two cases.

Case 1. $Y(t-) = y > 0$. Then $Y(t) = Y(t+) > y$ (note that Y is right-continuous by Step 2) and there exists an interval $[t-\delta, t+\delta]$ such that $Y_{\varepsilon_n}(s) \geq Y(s) > \frac{y}{2}$, for all $s \in [t-\delta, t+\delta]$ and all ε_n, $n \geq 1$. This implies that

$$L(t+\delta) - L(t-\delta) = \lim_{n\to\infty} \frac{1}{2} \int_{t-\delta}^{t+\delta} \frac{\varepsilon_n}{Y_{\varepsilon_n}(s)} ds \leq \lim_{n\to\infty} \frac{2\delta\varepsilon_n}{y} = 0,$$

i.e. L cannot have a jump at t. This means that Y cannot have a jump at point t either and we get a contradiction.

Case 2. $Y(t-) = 0$ *and* $Y(t+) = Y(t) = y > 0$. Fix $T > t$, $\lambda \in \left(0, \frac{1}{2}\right)$ and let Λ be a random variable such that, for all $t_1, t_2 \in [0, T]$,

$$|W(t_1) - W(t_2)| \leq \Lambda |t_1 - t_2|^\lambda.$$

Take $n_1 \geq 1$ and $\delta_1 > 0$ such that $\varepsilon_{n_1} + \delta_1 + \frac{\sigma\Lambda}{2}\delta_1^\lambda < y$ and note that there exists $\delta_2 < \delta_1$ such that $Y(t-\delta_2) < \varepsilon_{n_1}$. Since $Y_{\varepsilon_n}(t-\delta_2) \downarrow Y(t-\delta_2)$ as $n \to \infty$, there exists $n_2 > n_1$ such that $Y_{\varepsilon_{n_2}}(t-\delta_2) < \varepsilon_{n_1} < y$. Moreover, $Y_{\varepsilon_{n_2}}(t) \geq Y(t) = y$. Therefore, one can define

$$\tau := \sup\{s \in (t-\delta_2, t),\ Y_{\varepsilon_{n_2}}(s) = \varepsilon_{n_1}\}.$$

Observe that $Y_{\varepsilon_{n_2}}(\tau) = \varepsilon_{n_1}$ and $Y_{\varepsilon_{n_2}}(s) \geq \varepsilon_{n_1}$ for all $s \in [\tau, t]$, hence

$$\begin{aligned} Y_{\varepsilon_{n_2}}(t) &= Y_{\varepsilon_{n_2}}(\tau) + \frac{1}{2}\int_\tau^t \frac{\varepsilon_{n_2}}{Y_{\varepsilon_{n_2}}(s)} ds - \frac{b}{2}\int_\tau^t Y_{\varepsilon_{n_2}}(s) ds + \frac{\sigma}{2}(W(t) - W(\tau)) \\ &\leq \varepsilon_{n_1} + \frac{\varepsilon_{n_2}}{2\varepsilon_{n_1}}(t-\tau) + \frac{\sigma\Lambda}{2}(t-\tau)^\lambda \\ &\leq \varepsilon_{n_1} + \delta_1 + \frac{\sigma\Lambda}{2}\delta_1^\lambda < y, \end{aligned}$$

which contradicts the assumption that $Y_{\varepsilon_{n_2}}(t) \geq y$. This contradiction together with all of the above implies that Y (and thus L) is continuous at each point $t \geq 0$.

Step 4. Points of growth. Now, let us prove that the points of growth of L may occur only at the zeros of Y. Indeed, let $t > 0$ be such that $Y(t) = y > 0$. Since Y is continuous, there exists $\delta_3 > 0$ such that for any $s \in (t-\delta_3, t+\delta_3)$

$$Y(s) > \frac{y}{2} > 0.$$

In turn, this implies that, for all $s \in (t - \delta_3, t + \delta_3)$ and $n \geq 1$,

$$Y_{\varepsilon_n}(s) \geq Y(s) > \frac{y}{2} > 0,$$

and thus, for any $\delta \in [0, \delta_3)$,

$$L(t+\delta) - L(t-\delta) = \lim_{n\to\infty} \frac{1}{2} \int_{t-\delta}^{t+\delta} \frac{\varepsilon_n}{Y_{\varepsilon_n}(s)} ds \leq \lim_{n\to\infty} \frac{2\delta}{y} \varepsilon_n = 0.$$

Therefore $L(t+\delta) - L(t-\delta) = 0$ and L does not grow in some neighbourhood of t. □

Remark 2.8 Let $B = \{B(t),\ t \geq 0\}$ be some standard Brownian motion and

$$U(t) = U(0) - \frac{b}{2} \int_0^t U(s)ds + \frac{\sigma}{2} B(t), \quad t \geq 0,$$

be a standard Ornstein-Uhlenbeck process with non-random positive initial value $U(0) > 0$. By Itô's formula,

$$\begin{aligned} U^2(t) &= U^2(0) + \int_0^t \left(\frac{\sigma^2}{4} - bU^2(s)\right) ds + \sigma \int_0^t U(s)dB(s) \\ &= U^2(0) + \int_0^t \left(\frac{\sigma^2}{4} - bU^2(s)\right) ds + \sigma \int_0^t |U(s)| \operatorname{sign}(U(s))dB(s) \\ &= U^2(0) + \int_0^t \left(\frac{\sigma^2}{4} - bU^2(s)\right) ds + \sigma \int_0^t |U(s)|dW(s), \end{aligned}$$

where $W(t) := \int_0^t \operatorname{sign}(U(s))dB(s)$ is a standard Brownian motion (this can be easily verified by Levy's characterization). Thus, the process $X(t) := U^2(t), t \geq 0$, is a CIR process w.r.t. W. By Theorem 2.2, the square root process $Y(t) := \sqrt{X(t)}, t \geq 0$, is a reflected Ornstein-Uhlenbeck process with respect to W satisfying the SDE of the form

$$Y(t) = U(0) - \frac{b}{2} \int_0^t Y(s)ds + \frac{\sigma}{2} W(t) + L(t), \quad t \geq 0, \tag{2.19}$$

with L being the reflection function for Y. Since $Y(t) = \sqrt{X(t)} = |U(t)|$, by Tanaka's formula (see Appendix A.2)

$$
\begin{aligned}
Y(t) &= U(0) + \int_0^t \operatorname{sign}(U(s))dU(s) + L^U_{\text{sem}}(t) \\
&= U(0) - \frac{b}{2}\int_0^t \operatorname{sign}(U(s))U(s)ds + \frac{\sigma}{2}\int_0^t \operatorname{sign}(U(s))dB(s) + L^U_{\text{sem}}(t) \\
&= U(0) - \frac{b}{2}\int_0^t Y(s)ds + \frac{\sigma}{2}W(t) + L^U_{\text{sem}}(t),
\end{aligned}
\tag{2.20}
$$

with L^U_{sem} being the semimartingale local time of U at zero. Comparing (2.19) and (2.20), we obtain that $L(t) = L^U_{\text{sem}}(t)$, i.e. the reflection function of the ROU process Y coincides with the semimartingale local time at zero of the OU process U.

2.4 Low-Dimensional CIR Processes

Let us now finalize this chapter with the remaining case $0 < d < 1$, i.e. $0 < a < \frac{\sigma^2}{4}$. Our strategy for the analysis of L in this situation will be as follows. Since the CIR process X in (2.1) is a non-negative regular diffusion, it can be represented (see e.g. [13, Chapter V, Sect. 7]) in the form

$$X(t) = S^{-1}\left(\widetilde{W}_{\tau_t}\right)$$

for some change of time τ and change of scale S of a reflected Brownian motion $\widetilde{W} = \{\widetilde{W}(t),\ t \ge 0\}$. Then, we re-write the integral in the limit (2.5) in terms of the local time $L^{\widetilde{W}} = L^{\widetilde{W}}(t, x)$ of $\widetilde{W}$ and exploit Hölder continuity of the latter to find an explicit representation of L in terms of $L^{\widetilde{W}}$.

2.4.1 *CIR Process as the Transformation of a Reflected Brownian Motion*

In order to implement our approach, we first need to represent the CIR process as a transformation of a reflected Brownian motion. For a given set of parameters a, b, σ of the SDE (2.1), define a *scale function* S: $[0, \infty) \to [0, \infty)$ by

$$S(x) := \int_0^x y^{-\frac{2a}{\sigma^2}} e^{\frac{2by}{\sigma^2}} dy, \tag{2.21}$$

and observe that, since S is strictly increasing and $S(\infty) = \infty$, there exists its inverse S^{-1}. Define also the *speed measure*

$$m(dx) = \rho(x)dx,$$

with

$$\rho(x) := \frac{1}{\sigma^2} x^{\frac{4a}{\sigma^2}-1} e^{-\frac{4b}{\sigma^2}x} \mathbb{1}_{x>0}. \tag{2.22}$$

Proposition 2.1 *Let X be the unique strong solution to the CIR equation* (2.1) *with* $0 < a < \frac{\sigma^2}{4}$. *Then there exists a reflected Brownian motion $\widetilde{W}$ starting at $S(x_0)$ such that*

$$X(t) := S^{-1}\left(\widetilde{W}(\tau_t)\right), \tag{2.23}$$

where

$$\tau_t := \varphi_t^{-1} \tag{2.24}$$

with

$$\varphi_t := \int_0^t \rho\left(S^{-1}\left(\widetilde{W}(s)\right)\right) ds. \tag{2.25}$$

Before moving to the proof of Proposition 2.1, let us make some remarks regarding its formulation.

Remark 2.9 1. The process φ in (2.25) is well-defined. Indeed, let $L^{\widetilde{W}} = \{L^{\widetilde{W}}(t, x),\ t \geq 0,\ x \geq 0\}$ be the local time of $\widetilde{W}$ (see Appendix A.2), i.e. for any bounded measurable f,

$$\int_0^t f(\widetilde{W}(s))ds = \int_0^\infty f(x) L^{\widetilde{W}}(t, x)dx \quad a.s.$$

Then, with probability 1,

$$\begin{aligned}\varphi_t &= \int_0^t \rho\left(S^{-1}\left(\widetilde{W}(s)\right)\right) ds = \int_0^\infty \rho\left(S^{-1}(y)\right) L^{\widetilde{W}}(t, y)dy \\ &= \int_0^\infty \rho(x) S'(x) L^{\widetilde{W}}(t, S(x))dx = \frac{1}{\sigma^2}\int_0^\infty x^{\frac{2a}{\sigma^2}-1} e^{-\frac{2b}{\sigma^2}x} L^{\widetilde{W}}(t, S(x))dx \\ &< \infty,\end{aligned}$$

because $\frac{2a}{\sigma^2} - 1 > -1$ and $L^{\widetilde{W}}(t, S(x)) = 0$ for $x > S^{-1}\left(\max_{s\in[0,T]} \widetilde{W}(s)\right)$.

2. Since φ is strictly increasing and $\varphi_\infty = \infty$ with probability 1, its inverse τ in (2.24) is well-defined a.s.

Remark 2.10 The transformation (2.23) is invertible. Indeed, it is straightforward to check that, with probability 1,

$$\tau_t = \int_0^t \frac{1}{\rho\left(X(s)\right)} ds.$$

Hence,

$$\widetilde{W}(t) = S\left(X\left(\varphi_t\right)\right),$$

where $\varphi = \tau^{-1}$ can be expressed as the inverse of the mapping $t \mapsto \int_0^t \frac{1}{\rho(X(s))} ds$. More detailed information on transformations of this type can be found in [6, Chapter IV, §7].

Proof (Proof of Proposition 2.1) We will split the proof into two steps. First, we prove that, for *some given reflected Brownian motion* $\widetilde{W}$, the process $X(t) := S^{-1}\left(\widetilde{W}(\tau_t)\right)$ is the weak solution to the SDE (2.1). Then we will utilize the invertability of transformation (2.23) outlined in Remark 2.10 to establish the existence of a reflected Brownian motion $\widetilde{W}$ together with the required representation for the given CIR process X.

Step 1. Let $\widetilde{Z} = \{\widetilde{Z}(t),\ t \ge 0\}$ be a standard Brownian motion starting at $\widetilde{Z}(0) = S(x_0)$. Consider a reflected Brownian motion

$$\widetilde{W}(t) := |\widetilde{Z}(t)| = S(x_0) + Z(t) + L^{\widetilde{Z}}(t),$$

where $Z(t) := \int_0^t \operatorname{sign} \widetilde{Z}(s) d\widetilde{Z}(s)$ is a Brownian motion and $L^{\widetilde{Z}}(t)$ is the local time of $L^{\widetilde{Z}}$ at zero. Put $V(t) := S^{-1}\left(\widetilde{W}(t)\right)$ and observe that, by the extension of Itô's formula in [13, Lemma IV.45.9],

$$\begin{aligned}
&V(t) - V(0)\\
&\quad= \int_0^t \left(S^{-1}\left(\widetilde{W}(u)\right)\right)^{\frac{2a}{\sigma^2}} e^{-\frac{2bS^{-1}(\widetilde{W}(u))}{\sigma^2}} d\widetilde{W}(u)\\
&\qquad+ \int_0^t \frac{1}{\sigma^2} \left(S^{-1}\left(\widetilde{W}(u)\right)\right)^{\frac{4a}{\sigma^2}-1} e^{-\frac{4b}{\sigma^2} S^{-1}(\widetilde{W}(u))} \left(a - bS^{-1}\left(\widetilde{W}(u)\right)\right) du\\
&\quad= \int_0^t \left(V(u)\right)^{\frac{2a}{\sigma^2}} e^{-\frac{2b}{\sigma^2} V(u)} dZ(u)\\
&\qquad+ \int_0^t \frac{1}{\sigma^2} \left(V(u)\right)^{\frac{4a}{\sigma^2}-1} e^{-\frac{4b}{\sigma^2} V(u)} \left(a - bV(u)\right) du.
\end{aligned}$$

Hence, by Itô's formula, for any infinitely differentiable function h with compact support, the process

$$C_t(h) := h(V(t)) - h(V(0)) - \int_0^t \frac{1}{\sigma^2} (V(s))^{\frac{4a}{\sigma^2}-1} e^{-\frac{4b}{\sigma^2}V(s)} \mathcal{A}h(V(s))ds \qquad (2.26)$$

is a local martingale, where

$$\mathcal{A}h(x) := (a - bx)h'(x) + \frac{\sigma^2 x}{2} h''(x)$$

is the generator of (2.1). Recall that

$$X(t) = V(\tau_t)$$

and observe that, by (2.26), a simple change of variables yields that

$$\begin{aligned} C_{\tau_t}(h) &= h(V(\tau_t)) - h(V(0)) - \int_0^{\tau_t} \frac{1}{\sigma^2} (V(s))^{\frac{4a}{\sigma^2}-1} e^{-\frac{4b}{\sigma^2}V(s)} \mathcal{A}h(V(s))ds \\ &= h(X(t)) - h(x_0) - \int_0^t \mathcal{A}h(X(s))ds, \end{aligned}$$

for any infinitely differentiable function h with compact support. Since $C_{\tau_t}(h), t \geq 0$, is a local martingale, by the optional stopping theorem, the process

$$h(X(t)) - h(x_0) - \int_0^t \mathcal{A}h(X(s))ds$$

is also a local martingale and therefore X is the weak solution to (2.1) by [13, V.19–V.20].

Step 2. Let now X be the unique strong solution to (2.1). Consider the process

$$\widetilde{W}(t) := S(X(\varphi_t)),$$

where φ is defined as the inverse of the mapping $t \mapsto \int_0^t \frac{1}{\rho(X(s))} ds$. By Remark 2.10 and Step 1, the process $\widetilde{W}$ is a reflected Brownian motion for which X admits the representation (2.23). □

2.4.2 Characterization of L in Terms of $L^{\widetilde{W}}$

Having the representation (2.23) at our disposal, we are now ready to characterize the process L from (2.5) in terms of the local time $L^{\widetilde{W}}$ of the corresponding reflected Brownian motion.

Let X be the unique strong solution to the SDE (2.1) with $0 < a < \frac{\sigma^2}{4}$ and $\widetilde{W}$ be the reflected Brownian motion such that

$$X(t) := S^{-1}\left(\widetilde{W}(\tau_t)\right).$$

Denote $L^{\widetilde{W}} = \{L^{\widetilde{W}}(t,x),\ t \ge 0,\ x \ge 0\}$ the local time of $\widetilde{W}$ so that

$$\int_0^t f(\widetilde{W}(s))ds = \int_0^\infty f(x)L^{\widetilde{W}}(t,x)dx, \quad t \ge 0,$$

with probability 1 for any bounded measurable f.

First of all, let us express the local time $L^Y = L^Y(t,y)$ of the process $Y = \sqrt{X}$ in terms of $L^{\widetilde{W}}$.

Proposition 2.2 *Let*

$$Y(t) := \sqrt{X(t)} = \sqrt{S^{-1}\left(\widetilde{W}(\tau_t)\right)}$$

be the square root of the CIR process X. Then, for any bounded measurable f,

$$\int_0^t f(Y(s))ds = \int_0^\infty f(y)L^Y(t,y)dy.$$

Here, with probability 1,

$$L^Y(t,y) = \frac{2}{\sigma^2} y^{\frac{4a}{\sigma^2}-1} e^{-\frac{2b}{\sigma^2}y^2} L^{\widetilde{W}}(\varphi_t, S(y^2)) \tag{2.27}$$

for φ given in (2.25).

Proof For any bounded measurable f, we can write

$$\begin{aligned}\int_0^t f(Y(s))ds &= \int_0^t f\left(\sqrt{S^{-1}\left(\widetilde{W}(\tau_u)\right)}\right)du \\ &= \int_0^{\varphi_t} f\left(\sqrt{S^{-1}\left(\widetilde{W}(z)\right)}\right)\rho\left(S^{-1}\left(\widetilde{W}(z)\right)\right)dz\end{aligned}$$

$$= \int_0^\infty f\left(\sqrt{S^{-1}(x)}\right) \rho\left(S^{-1}(x)\right) L^{\widetilde{W}}(\varphi_t, x)\, dx$$

$$= \int_0^\infty f(y)\, \rho\left(y^2\right) L^{\widetilde{W}}\left(\varphi_t, S(y^2)\right) 2yS'(y^2) dy$$

$$=: \int_0^\infty f(y) L^Y(t, y) dy.$$

The final result is obtained by recalling that

$$\begin{aligned} L^Y(t, y) &= \rho\left(y^2\right) L^{\widetilde{W}}\left(\varphi_t, S(y^2)\right) 2yS'(y^2) \\ &= \frac{2}{\sigma^2} y^{\frac{4a}{\sigma^2}-1} e^{-\frac{2b}{\sigma^2} y^2} L^{\widetilde{W}}(\varphi_t, S(y^2)). \end{aligned}$$

□

Define a normalized local time of the process Y as follows. Set

$$\ell(t, y) := y^{1-\frac{4a}{\sigma^2}} L^Y(t, y), \quad y > 0, \tag{2.28}$$

and

$$\ell(t, 0) := \lim_{y \to 0+} \ell(t, y).$$

In particular,

$$\ell(t, y) = \frac{2}{\sigma^2} e^{-\frac{2b}{\sigma^2} y^2} L^{\widetilde{W}}(\varphi_t, S(y^2)). \tag{2.29}$$

However, we want to stress that $\ell(t, y)$ is a function of the local time L^Y of the process Y without mentioning the auxiliary Brownian motion $\widetilde{W}$.

Theorem 2.4 *Let X be the CIR process satisfying (2.1) and $\widetilde{W}$ be the reflected Brownian motion such that $X(t) = S^{-1}(\widetilde{W}(\tau_t))$, $t \geq 0$. Then, with probability 1, the process $Y = \sqrt{X}$ satisfies the SDE of the form*

$$Y(t) = \sqrt{x_0} - \frac{b}{2} \int_0^t Y(s) ds + \frac{\sigma}{2} W(t) + L(t), \tag{2.30}$$

where

$$L(t) = -\frac{1}{2}\left(\frac{\sigma^2}{4} - a\right) \int_0^\infty y^{\frac{4a}{\sigma^2}-2} \left(\ell(t, y) - \ell(t, 0)\right) dy. \tag{2.31}$$

Moreover,

$$L(t) := \lim_{\varepsilon\downarrow 0}\frac{1}{2}\int_0^t\left(\frac{a}{\sqrt{X(s)+\varepsilon}}-\frac{\sigma^2}{4}\frac{X(s)}{(X(s)+\varepsilon)^{\frac{3}{2}}}\right)ds$$
$$= -\lim_{\varepsilon\downarrow 0}\frac{1}{2}\int_0^t\left(\frac{\frac{\sigma^2}{4}-a}{\sqrt{X(s)+\varepsilon}}-\frac{\sigma^2}{4}\frac{\varepsilon}{(X(s)+\varepsilon)^{\frac{3}{2}}}\right)ds.$$

Remark 2.11 Since the SDE (2.1) has a strong solution, Theorem 2.4 immediately yields that the SDE (2.30)–(2.31) also has a strong solution.

Remark 2.12 Despite the fact that $\frac{4a}{\sigma^2}-2\in(-2,-1)$, the integral

$$\int_0^\infty y^{\frac{4a}{\sigma^2}-2}\,|\ell(t,y)-\ell(t,0)|\,dy$$

is finite with probability 1. Indeed, recall that $d:=\frac{4a}{\sigma^2}\in(0,1)$. Fix $t\ge 0$ and observe that, with probability 1, there exists a value y_* such that $L^{\widetilde{W}}(t,y)=0$ for all $y\ge y_*$. Using this fact together with the representation (2.29), it is now easy to see that

$$\sup_{y\ge 1}|\ell(t,y)-\ell(t,0)|<\infty \tag{2.32}$$

with probability 1 for any $t\ge 0$. Moreover, since $L^{\widetilde{W}}(t,\cdot)$ is Hölder continuous of order up to $\frac{1}{2}$ a.s. (see Appendix A.2), for any $\delta\in\left(0,\frac{1}{2}\right)$ and any fixed $t>0$ there exists a random variable $C>0$ such that, with probability 1,

$$|\ell(t,y)-\ell(t,0)|\le C\cdot(S(y^2))^{\frac{1}{2}-\delta}. \tag{2.33}$$

Hence, on the one hand,

$$\int_1^\infty |\ell(t,y)-\ell(t,0)|\,y^{d-2}dy<\infty \text{ a.s.}$$

by (2.32). On the other hand, take $\delta\in\left(0,\frac{d}{2(2-d)}\right)$ and observe that (2.33) implies

$$\int_0^1 |\ell(t,y)-\ell(t,0)|\,y^{d-2}dy\le C\int_0^1\left(S(y^2)\right)^{\frac{1}{2}-\delta}y^{d-2}dy$$

$$
\begin{aligned}
&= C \int_0^1 \left(\int_0^{y^2} z^{-\frac{d}{2}} e^{\beta z} dz \right)^{\frac{1}{2}-\delta} y^{d-2} dy \\
&\leq C \int_0^1 \left(\int_0^{y^2} z^{-\frac{d}{2}} dz \right)^{\frac{1}{2}-\delta} y^{d-2} dy \\
&\leq C \int_0^1 y^{-1+\frac{d}{2}-\delta'} dy < \infty \text{ a.s.,}
\end{aligned}
$$

where $\delta' := (2-d)\delta \in \left(0, \frac{d}{2}\right)$ and C is a (random) constant that varies from line to line.

Now we are ready to proceed to the proof of Theorem 2.4.

Proof (Proof of Theorem 2.4) In Sect. 2.2, we obtained the representation (2.30) with L being a *ucp-limit* of the form

$$
\begin{aligned}
L(t) &= \lim_{\varepsilon \downarrow 0} \frac{1}{2} \int_0^t \left(\frac{a}{\sqrt{X(s)+\varepsilon}} - \frac{\sigma^2}{4} \frac{X(s)}{(X(s)+\varepsilon)^{\frac{3}{2}}} \right) ds \\
&= -\lim_{\varepsilon \downarrow 0} \frac{1}{2} \int_0^t \left(\frac{\frac{\sigma^2}{4} - a}{\sqrt{X(s)+\varepsilon}} - \frac{\sigma^2}{4} \frac{\varepsilon}{(X(s)+\varepsilon)^{\frac{3}{2}}} \right) ds.
\end{aligned}
$$

Hence, one is left to prove that this limit exists in the sense of a.s. convergence and to check that the last equality in (2.31) holds.

Recall again that $d := \frac{4a}{\sigma^2} \in (0, 1)$ denotes the dimension of the CIR process, i.e. we have to study the a.s.-limit of the form

$$
\begin{aligned}
L(t) &:= -\lim_{\varepsilon \downarrow 0} \frac{1}{2} \int_0^t \left(\frac{\frac{\sigma^2}{4} - a}{\sqrt{X(s)+\varepsilon}} - \frac{\sigma^2}{4} \frac{\varepsilon}{(X(s)+\varepsilon)^{\frac{3}{2}}} \right) ds \\
&= -\frac{\sigma^2}{8} \lim_{\varepsilon \downarrow 0} \int_0^t \left(\frac{1-d}{\sqrt{X(s)+\varepsilon}} - \frac{\varepsilon}{(X(s)+\varepsilon)^{\frac{3}{2}}} \right) ds.
\end{aligned}
$$

Observe that

$$\int_0^t \frac{1}{\sqrt{X(s)+\varepsilon}}ds = \int_0^t \frac{1}{\sqrt{Y^2(s)+\varepsilon}}ds = \int_0^\infty \frac{1}{\sqrt{y^2+\varepsilon}}L^Y(t,y)dy$$
$$= \int_0^\infty \frac{1}{\sqrt{y^2+\varepsilon}}y^{d-1}\ell(t,y)dy$$
$$= \int_0^\infty \frac{1}{\sqrt{y^2+\varepsilon}}y^{d-1}(\ell(t,y)-\ell(t,0))dy$$
$$+ \ell(t,0)\int_0^\infty \frac{1}{\sqrt{y^2+\varepsilon}}y^{d-1}dy,$$

and, similarly,

$$\int_0^t \frac{1}{(X(s)+\varepsilon)^{\frac{3}{2}}}ds = \int_0^t \frac{1}{(Y^2(s)+\varepsilon)^{\frac{3}{2}}}ds = \int_0^\infty \frac{1}{(y^2+\varepsilon)^{\frac{3}{2}}}L^Y(t,y)dy$$
$$= \int_0^\infty \frac{1}{(y^2+\varepsilon)^{\frac{3}{2}}}y^{d-1}\ell(t,y)dy$$
$$= \int_0^\infty \frac{1}{(y^2+\varepsilon)^{\frac{3}{2}}}y^{d-1}(\ell(t,y)-\ell(t,0))dy$$
$$+ \ell(t,0)\int_0^\infty \frac{1}{(y^2+\varepsilon)^{\frac{3}{2}}}y^{d-1}dy.$$

Therefore,

$$L(t) = -\frac{\sigma^2}{8}\lim_{\varepsilon\downarrow 0}\left(I_1(\varepsilon)-I_2(\varepsilon)+I_3(\varepsilon)\right),$$

where

$$I_1(\varepsilon) := \int_0^\infty \frac{1-d}{\sqrt{y^2+\varepsilon}}y^{d-1}(\ell(t,y)-\ell(t,0))dy,$$
$$I_2(\varepsilon) := \int_0^\infty \frac{\varepsilon}{(y^2+\varepsilon)^{\frac{3}{2}}}y^{d-1}(\ell(t,y)-\ell(t,0))dy,$$

$$I_3(\varepsilon) := \ell(t,0)\left(\int_0^\infty \frac{1-d}{\sqrt{y^2+\varepsilon}} y^{d-1} dy - \int_0^\infty \frac{\varepsilon}{(y^2+\varepsilon)^{\frac{3}{2}}} y^{d-1} dy\right).$$

Let us study separately the asymptotics of $I_1(\varepsilon)$, $I_2(\varepsilon)$, $I_3(\varepsilon)$ as $\varepsilon \downarrow 0$. First, observe that for any $y \geq 0$

$$\frac{1-d}{\sqrt{y^2+\varepsilon}} y^{d-1} |\ell(t,y) - \ell(t,0)| \leq y^{d-2} |\ell(t,y) - \ell(t,0)|$$

and note that by Remark 2.12,

$$\int_0^\infty y^{d-2} \, |\ell(t,y) - \ell(t,0)| \, dy < \infty \quad \text{a.s.}$$

Thus, by the dominated convergence theorem, with probability 1,

$$\begin{aligned}
\lim_{\varepsilon \downarrow 0} I_1(\varepsilon) &= \lim_{\varepsilon \downarrow 0} \int_0^\infty \frac{1-d}{\sqrt{y^2+\varepsilon}} y^{d-1} \, (\ell(t,y) - \ell(t,0)) \, dy \\
&= (1-d) \int_0^\infty y^{d-2} \, (\ell(t,y) - \ell(t,0)) \, dy.
\end{aligned}$$

Next, observe that, with probability 1,

$$\begin{aligned}
&\left| \int_1^\infty \frac{\varepsilon}{(y^2+\varepsilon)^{\frac{3}{2}}} y^{d-1} \, (\ell(t,y) - \ell(t,0)) \, dy \right| \\
&\leq \varepsilon \int_1^\infty |\ell(t,y) - \ell(t,0)| \, y^{d-4} dy \qquad (2.34) \\
&\leq \varepsilon C \int_1^\infty y^{d-4} dy \to 0, \quad \varepsilon \downarrow 0.
\end{aligned}$$

On the other hand, take an arbitrary $\delta \in \left(0, \frac{d}{2(2-d)}\right)$, denote $\delta' := (2-d)\delta$ and observe that (2.33) yields

$$\begin{aligned}
&\int_0^1 \frac{\varepsilon}{(y^2+\varepsilon)^{\frac{3}{2}}} y^{d-1} \left|\ell(t,y)-\ell(t,0)\right| dy \\
&\le C \int_0^1 \frac{\varepsilon}{(y^2+\varepsilon)^{\frac{3}{2}}} y^{d-1} (S(y^2))^{\frac{1}{2}-\delta} dy \\
&= C \int_0^1 \frac{\varepsilon}{(y^2+\varepsilon)^{\frac{3}{2}}} y^{d-1} \left(\int_0^{y^2} z^{-\frac{d}{2}} e^{\beta z} dz \right)^{\frac{1}{2}-\delta} dy \\
&\le C \int_0^1 \frac{\varepsilon}{(y^2+\varepsilon)^{\frac{3}{2}}} y^{d-1} \left(\int_0^{y^2} z^{-\frac{d}{2}} dz \right)^{\frac{1}{2}-\delta} dy \\
&\le C \int_0^1 \frac{\varepsilon}{(y^2+\varepsilon)^{\frac{3}{2}}} y^{\frac{d}{2}-\delta'} dy \\
&= C\varepsilon^{\frac{d}{4}-\frac{\delta'}{2}} \int_0^1 \frac{\varepsilon^{-\frac{1}{2}}}{\left(\left(y/\sqrt{\varepsilon}\right)^2+1\right)^{\frac{3}{2}}} \left(\frac{y}{\sqrt{\varepsilon}}\right)^{\frac{d}{2}-\delta'} dy,
\end{aligned} \tag{2.35}$$

where $\beta := \frac{2b}{\sigma^2}$. Hence, by substituting $z = y/\sqrt{\varepsilon}$ in (2.35), we can again use (2.33) and write

$$\begin{aligned}
&\int_0^1 \frac{\varepsilon}{(y^2+\varepsilon)^{\frac{3}{2}}} y^{d-1} \left|\ell(t,y)-\ell(t,0)\right| dy \\
&\le C\varepsilon^{\frac{d}{4}-\frac{\delta'}{2}} \int_0^\infty \frac{1}{\left(z^2+1\right)^{\frac{3}{2}}} z^{\frac{d}{2}-\delta'} dz \\
&\to 0
\end{aligned} \tag{2.36}$$

with probability 1 as $\varepsilon \downarrow 0$. Summarizing (2.34) and (2.36), we obtain that, with probability 1,

$$\lim_{\varepsilon \downarrow 0} I_2(\varepsilon) = 0.$$

Finally, integration by parts yields

$$\int_0^\infty \frac{y^{d-1}}{\sqrt{y^2+\varepsilon}} dy = \frac{1}{d} \int_0^\infty \frac{y^{d+1}}{(y^2+\varepsilon)^{\frac{3}{2}}} dy,$$

and the right-hand side of the last equation is equal to

$$\frac{1}{d}\int_0^\infty \frac{y^{d-1}(y^2+\varepsilon-\varepsilon)}{(y^2+\varepsilon)^{\frac{3}{2}}}dy$$
$$=\frac{1}{d}\int_0^\infty \frac{y^{d-1}}{\sqrt{y^2+\varepsilon}}dy-\frac{1}{d}\int_0^\infty \frac{\varepsilon y^{d-1}}{(y^2+\varepsilon)^{\frac{3}{2}}}dy.$$

Therefore

$$\int_0^\infty \frac{\varepsilon}{(y^2+\varepsilon)^{\frac{3}{2}}}y^{d-1}dy=\int_0^\infty \frac{1-d}{\sqrt{y^2+\varepsilon}}y^{d-1}dy,$$

and

$$I_3(\varepsilon)=\ell(t,0)\left(\int_0^\infty \frac{1-d}{\sqrt{y^2+\varepsilon}}y^{d-1}dy-\int_0^\infty \frac{\varepsilon}{(y^2+\varepsilon)^{\frac{3}{2}}}y^{d-1}dy\right)=0.$$

Summarizing all of the above and recalling that $d=\frac{4a}{\sigma^2}$, we finally obtain that with probability 1

$$\begin{aligned}L(t)&=-\frac{\sigma^2}{8}\lim_{\varepsilon\downarrow 0}\left(I_1(\varepsilon)-I_2(\varepsilon)+I_3(\varepsilon)\right)\\&=-\frac{\sigma^2}{8}(1-d)\int_0^\infty y^{d-2}\left(\ell(t,y)-\ell(t,0)\right)dy\\&=-\frac{1}{2}\left(\frac{\sigma^2}{4}-a\right)\int_0^\infty y^{\frac{4a}{\sigma^2}-2}\left(\ell(t,y)-\ell(t,0)\right)dy\end{aligned}$$

which ends the proof. □

Remark 2.13 Theorem 2.4 implies that the limit

$$L(t)=\lim_{n\to\infty}\frac{1}{2}\int_0^t\left(\frac{a}{\sqrt{X(s)+\varepsilon_n}}-\frac{\sigma^2}{4}\frac{X(s)}{(X(s)+\varepsilon_n)^{\frac{3}{2}}}\right)ds \tag{2.37}$$

exists a.s. for any sequence $\{\varepsilon_n,\ n\geq 1\}$ such that $\varepsilon_n\downarrow 0$ and does not depend on the particular choice of the sequence. Moreover, the proof of Theorem 2.4 yields that the existence of the limit (2.37) is ensured for all ω such that $L^{\widetilde{W}}(\omega;t,\cdot)$ is Hölder continuous. In other words, the set of full probability where (2.37) holds can be chosen independently of a particular sequence $\{\varepsilon_n,\ n\geq 1\}$, as anticipated in Remark 2.4.

References

1. Blei, S.: On symmetric and skew Bessel processes. Stoch. Process. Appl. **122**(9), 3262–3287 (2012). https://doi.org/10.1016/j.spa.2012.05.008
2. Cherny, A.S.: On the strong and weak solutions of stochastic differential equations governing Bessel processes. Stoch. Rep. **70**(3–4), 213–219 (2000). https://doi.org/10.1080/17442500008834252
3. Cherny, A.S., Engelbert, H.J.: Singular Stochastic Differential Equations. Springer, Berlin (2005)
4. Cohen, S.N., Elliott, R.J.: Stochastic calculus and applications, 2nd edn. Springer, Cambridge, MA (2015)
5. Ikeda, N., Watanabe, S.: A comparison theorem for solutions of stochastic differential equations and its applications. Osaka J. Math. **14**(3), 619–633 (1977). https://doi.org/10.18910/7664
6. Ikeda, N., Watanabe, S.: Stochastic Differential Equations and Diffusion Processes, Elsevier (1981)
7. Jeanblanc, M., Yor, M., Chesney, M.: Mathematical methods for financial markets. Springer, London (2009)
8. Mishura, Y., Pilipenko, A., Yurchenko-Tytarenko, A.: Low-dimensional Cox-Ingersoll-Ross process. Stochastics **96**(5), 1530–1550 (2024). https://doi.org/10.1080/17442508.2023.2300291
9. Mishura, Y., Yurchenko-Tytarenko, A.: Standard and fractional reflected Ornstein–Uhlenbeck processes as the limits of square roots of Cox–Ingersoll–ross processes. Stochastics 1–19 (2022). https://doi.org/10.1080/17442508.2022.2047188
10. Pilipenko, A.: An introduction to stochastic differential equations with reflection. Universitätsverlag Potsdam, Potsdam (2014)
11. Piterbarg, V.I.: Twenty Lectures About Gaussian Processes. Atlantic Financial Press (2015)
12. Revuz, D., Yor, M.: Continuous Martingales and Brownian Motion. Springer, Berlin, Heidelberg (1999)
13. Rogers, L.C.G., Williams, D.: Diffusions, Markov Processes and Martingales. Cambridge University Press (2000)
14. Skorokhod, A.V.: Stochastic equations for diffusion processes in a bounded region. Theory Probab. Appl. **6**(3), 264–274 (1961). https://doi.org/10.1137/1106035
15. Skorokhod, A.V.: Stochastic equations for diffusion processes in a bounded region. II. Theory Prob. Appl. **7**(1), 3–23 (1962). https://doi.org/10.1137/1107002
16. Ward, A.R., Glynn, P.W.: Properties of the reflected Ornstein-Uhlenbeck process. Queueing Syst. **44**, 109–123 (2003)
17. Yamada, T., Watanabe, S.: On the uniqueness of solutions of stochastic differential equations. Kyoto J. Math. **11**(1), 155–167 (1971)
18. Ylvisaker, D.: A note on the absence of tangencies in Gaussian sample paths. Ann. Math. Stat. **39**(1), 261–262 (1968)

Chapter 3
Sandwiched Processes Driven by Hölder Noises

Abstract This chapter develops the theoretical foundation for a new class of volatility models inspired by the repelling drift behavior of the CIR process discussed in Chap. 2. We introduce a general stochastic integro-differential equation driven by a general locally Hölder-continuous noise. The defining feature of the model is an explosive drift term that enforces lower (and potentially upper) bounds on the solution—hence the name *sandwiched* process.

3.1 Introduction

The main goal of this chapter is to provide theoretical foundations for the volatility model that will be used later on. Namely, we aim to discuss a stochastic integrodifferential equation of the form

$$Y(t) = y_0 + \int_0^t b(s, Y(s))ds + \int_0^t \mathcal{G}(t, s)a(s, Y(s))ds + Z(t) \qquad (3.1)$$

driven by a general locally Hölder continuous random process $Z = \{Z(t),\ t \geq 0\}$ that can potentially carry nontrivial dependence on the past. In order to make sure that the solution to (3.1) is bounded from below, we will use the observation from the previous chapter and take the function b with an explosive growth to $+\infty$ in the proximity of the desirable bound. As we will also see later, it is possible to make the solution Y bounded from above by letting b have an explosive decay to $-\infty$ near the upper bound. This chapter draws on the material originally published in [4].

3.2 Assumptions and Examples

We start this chapter by specifying the detailed assumptions on the coefficients of the Eq. (3.1) and illustrating each assumption with specific examples.

G. Di Nunno et al., *Volterra Volatility Models*, Springer Finance,
https://doi.org/10.1007/978-3-032-26576-0_3

3.2.1 Noise Z

Fix a probability space $(\Omega, \mathcal{F}, \mathbb{P})$ and let $T > 0$ be a finite time horizon. Consider a stochastic process $Z = \{Z(t),\ t \in [0, T]\}$ satisfying the following assumption.

Assumption 3.1 $Z = \{Z(t),\ t \in [0, T]\}$ is a stochastic process on $(\Omega, \mathcal{F}, \mathbb{P})$ such that

(Z1) $Z(0) = 0$ a.s.;
(Z2) Z has a.s. Hölder continuous paths of order $\lambda \in (0, 1)$, i.e. there exists a random variable $\Lambda = \Lambda_\lambda(\omega) > 0$ such that, with probability 1,

$$|Z(t) - Z(s)| \le \Lambda |t - s|^\lambda, \quad t, s \in [0, T]. \tag{3.2}$$

Note that, for now, we do not require any particular assumptions on the distribution of the noise Z. However, in what follows, we will need the random variable Λ from (3.2) to have moments of sufficiently high orders. For the analysis of the latter, we will use a corollary from the well-known Garsia-Rodemich-Rumsey inequality (for more details, see [6] and Appendix A.1).

Lemma 3.1 *Let $f\colon [0, T] \to \mathbb{R}$ be a continuous function, $p \ge 1$ and $a > \frac{1}{p}$. Then for all $t, s \in [0, T]$ one has*

$$|f(t) - f(s)| \le A_{a,p} |t - s|^{\alpha - \frac{1}{p}} \left(\int_0^T \int_0^T \frac{|f(x) - f(y)|^p}{|x - y|^{\alpha p + 1}} dx dy \right)^{\frac{1}{p}},$$

with the convention $0/0 = 0$, where

$$A_{\alpha,p} = 2^{3+\frac{2}{p}} \left(\frac{\alpha p + 1}{\alpha p - 1} \right). \tag{3.3}$$

Let us list several examples of admissible noises satisfying Assumption 3.1 together with the analysis of the corresponding random variable Λ.

Example 3.1 (*degenerate noise*) The process Z with $Z(t) = 0$ for all $t \in [0, T]$ obviously satisfies conditions **(Z1)** and **(Z2)**.

Example 3.2 (*Hölder continuous Gaussian processes*) Let $Z = \{Z(t),\ t \ge 0\}$ be a centered Gaussian process with $Z(0) = 0$ and $H \in (0, 1)$ be a given constant. Then, by [1], Z has a modification with Hölder continuous paths of any order $\lambda \in (0, H)$ if and only if for any $\lambda \in (0, H)$ there exists a constant $C_\lambda > 0$ such that

$$\left(\mathbb{E}|Z(t) - Z(s)|^2\right)^{\frac{1}{2}} \le C_\lambda |t - s|^\lambda, \quad s, t \in [0, T]. \tag{3.4}$$

Furthermore, using Lemma 3.1, one can prove that the corresponding random variable Λ can be chosen to have moments of all positive orders. Namely, assume that $\lambda \in (0, H)$ and take $p \geq 1$ such that $\frac{1}{p} < H - \lambda$. If we take

$$\Lambda = A_{\lambda+\frac{1}{p},p} \left(\int_0^T \int_0^T \frac{|Z(x) - Z(y)|^p}{|x - y|^{\lambda p+2}} dxdy \right)^{\frac{1}{p}}, \tag{3.5}$$

then, for any $r \geq 1$, $\mathbb{E}\Lambda^r < \infty$ and for all $s, t \in [0, T]$:

$$|Z(t) - Z(s)| \leq \Lambda |t - s|^\lambda,$$

see e.g. Appendix A.1 as well as [1, Theorem 1].

An important example of Gaussian processes considered in this book is the class of *Gaussian Volterra processes* described in the example below.

Example 3.3 (*Hölder continuous Gaussian Volterra processes*) Let $B = \{B(t),\ t \in [0, T]\}$ be a Brownian motion. Also, let $\mathcal{K}$: $[0, T]^2 \to \mathbb{R}$ be a function such that $\mathcal{K}(t, s) = 0$ for all $s \geq t$, and $\mathcal{K}$ satisfies two assumptions:

(i) $\int_0^T \int_0^T \mathcal{K}^2(t, s) ds dt < \infty$;
(ii) there exists $H \in (0, 1)$ such that for all $\lambda \in (0, H)$ and $0 \leq t_1, t_2 \leq T$

$$\int_0^T (\mathcal{K}(t_2, s) - \mathcal{K}(t_1, s))^2 ds \leq C_\lambda^2 |t_2 - t_1|^{2\lambda},$$

where $C_\lambda > 0$ is some constant depending on λ.

Define a Gaussian Volterra process $Z = \{Z(t),\ t \in [0, T]\}$ as

$$Z(t) = \int_0^t \mathcal{K}(t, s) dB(s), \quad t \in [0, T]. \tag{3.6}$$

Then, for any $\lambda \in (0, H)$ and $0 \leq t_1, t_2 \leq T$,

$$\left(\mathbb{E}|Z(t_2) - Z(t_1)|^2\right)^{\frac{1}{2}} = \left(\int_0^T (\mathcal{K}(t_2, s) - \mathcal{K}(t_1, s))^2 ds \right)^{\frac{1}{2}} \leq C_\lambda |t_2 - t_1|^\lambda,$$

i.e. Z has a modification with Hölder continuous trajectories of any order $\lambda \in (0, H)$.

Let us now present some concrete examples of Gaussian Volterra processes with kernels satisfying conditions (i) and (ii) from Example 3.3.

Example 3.4 (*Riemann-Liouville fractional Brownian motion*) Let $H \in (0, 1)$ and $B = \{B(t),\ t \in [0, T]\}$ be a Brownian motion. Define

$$\mathcal{K}(t, s) = \frac{1}{\Gamma\left(H + \frac{1}{2}\right)}(t - s)^{H-\frac{1}{2}}\mathbb{1}_{s<t} \tag{3.7}$$

and put

$$Z(t) = \int_0^t \mathcal{K}(t, s)dB(s) = \frac{1}{\Gamma\left(H + \frac{1}{2}\right)}\int_0^t (t - s)^{H-\frac{1}{2}}dB(s), \quad t \in [0, T]. \tag{3.8}$$

The Gaussian Volterra process Z defined by (3.8) is called the *Riemann-Liouville fractional Brownian motion with Hurst index H* and the kernel $\mathcal{K}$ given by (3.7) is called the *fractional kernel*. Note that there is a constant C such that

$$\begin{aligned}
&\int_0^T (\mathcal{K}(t_2, s) - \mathcal{K}(t_1, s))^2 ds \\
&= \frac{1}{\Gamma\left(H + \frac{1}{2}\right)}\left(\int_0^{t_1} ((t_2 - s)^{H-\frac{1}{2}} - (t_1 - s)^{H-\frac{1}{2}})^2 ds + \int_{t_1}^{t_2} (t_2 - s)^{2H-1} ds\right) \\
&\leq C(t_2 - t_1)^{2H}
\end{aligned}$$

for any $0 \leq t_1 < t_2 \leq T$. Indeed,

$$\int_{t_1}^{t_2} (t_2 - s)^{2H-1} ds = \frac{1}{2H}(t_2 - t_1)^{2H}.$$

Further, making a change of variable $u = t_1 - s$, we can write

$$\begin{aligned}
&\int_0^{t_1} ((t_2 - s)^{H-\frac{1}{2}} - (t_1 - s)^{H-\frac{1}{2}})^2 ds \\
&= \int_0^{t_1} \left((t_2 - t_1 + u)^{H-\frac{1}{2}} - u^{H-\frac{1}{2}}\right)^2 du \\
&= \int_0^{t_1} u^{2H-1}\left(\left(\frac{t_2 - t_1}{u} + 1\right)^{H-\frac{1}{2}} - 1\right)^2 du,
\end{aligned}$$

and, with another change of variable $v = \frac{t_2 - t_1}{u}$, we obtain

$$\int_0^{t_1} u^{2H-1}\left(\left(\frac{t_2-t_1}{u}+1\right)^{H-\frac{1}{2}}-1\right)^2 du$$
$$= (t_2-t_1)^{2H}\int_{\frac{t_2-t_1}{t_1}}^{\infty}\left((v+1)^{H-\frac{1}{2}}-1\right)^2 v^{-1-2H}dv$$
$$\le (t_2-t_1)^{2H}\int_0^{\infty}\left((v+1)^{H-\frac{1}{2}}-1\right)^2 v^{-1-2H}dv.$$

It remains to notice that

$$\int_0^{\infty}\left((v+1)^{H-\frac{1}{2}}-1\right)^2 v^{-1-2H}dv < \infty,$$

because

$$\left((v+1)^{H-\frac{1}{2}}-1\right)^2 v^{-1-2H} = O(v^{1-2H}), \quad v \to 0+,$$

and

$$\left((v+1)^{H-\frac{1}{2}}-1\right)^2 v^{-1-2H} = O(v^{-2}) + O(v^{-1-2H}), \quad v \to +\infty.$$

In other words, the fractional kernel satisfies the conditions from Example 3.3 and hence the Riemann-Liouville fractional Brownian motion Z has a modification that is Hölder continuous of any order $\lambda \in (0, H)$.

Example 3.5 *(fractional Brownian motion)* Let $Z = B^H = \{B^H(t),\ t \in [0, T]\}$ be a fractional Brownian motion with Hurst index $H \in (0, 1)$, i.e. a centered Gaussian process with covariance function

$$\mathbb{E}\left[B^H(t)B^H(s)\right] = \frac{1}{2}\left(t^{2H} + s^{2H} - |t-s|^{2H}\right), \quad s, t \in [0, T].$$

Then Z satisfies (3.4) since, for any $\lambda \in (0, H)$,

$$\left(\mathbb{E}|Z(t_2) - Z(t_1)|^2\right)^{\frac{1}{2}} = \left(\mathbb{E}|B^H(t_2) - B^H(t_1)|^2\right)^{\frac{1}{2}} = |t_2 - t_1|^H, \tag{3.9}$$

i.e. $Z = B^H$ has a modification with Hölder continuous paths of any order $\lambda \in (0, H)$. According to [8], B^H admits a Volterra-type representation (3.6) with the kernel

$$\mathcal{K}(t,s) = a_H\left(\left(\frac{t}{s}\right)^{H-\frac{1}{2}}(t-s)^{H-\frac{1}{2}} - \left(H-\frac{1}{2}\right)s^{\frac{1}{2}-H}\int_s^t (u-s)^{H-\frac{1}{2}}u^{H-\frac{3}{2}}du\right)\mathbb{1}_{s<t}, \tag{3.10}$$

$$a_H = \left(\frac{2H\Gamma\left(\frac{3}{2}-H\right)}{\Gamma\left(H+\frac{1}{2}\right)\Gamma(2-2H)}\right)^{1/2},$$

with Γ denoting the gamma-function. If $H \in \left(\frac{1}{2}, 1\right)$, the shape of $\mathcal{K}$ in (3.10) can be simplified to the form

$$\mathcal{K}(t,s) = c_H s^{\frac{1}{2}-H}\int_s^t u^{H-\frac{1}{2}}(u-s)^{H-\frac{3}{2}}du\,\mathbb{1}_{s<t}, \qquad c_H = \left(H-\frac{1}{2}\right)a_H. \tag{3.11}$$

Note that the calculation given in (3.9) immediately yields that $\mathcal{K}$ from (3.10) automatically satisfies condition (ii) from Example 3.3.

Example 3.6 (*Gaussian Volterra processes with fBm-type kernel*) Consider a natural generalization of the kernel (3.11) of the form

$$\mathcal{K}(t,s) = f_1(s)\int_s^t f_2(u)f_3(u-s)du\,\mathbb{1}_{s<t},$$

where $f_1 \in L^p[0,T]$, $f_2 \in L^q[0,T]$ and $f_3 \in L^r[0,T]$ with p, q, r such that

(1) $p \in [2,\infty]$, $q \in (1,\infty]$, $r \in [1,\infty]$,
(2) $\frac{1}{p}+\frac{1}{r} \geq \frac{1}{2}$,
(3) $\frac{1}{p}+\frac{1}{q}+\frac{1}{r} < \frac{3}{2}$.

Then, by [7, Lemma 1], for any $0 \leq t_1 \leq t_2 \leq T$,

$$\left(\mathbb{E}|Z(t_2)-Z(t_1)|^2\right)^{\frac{1}{2}} = \left(\int_0^T (\mathcal{K}(t_2,s)-\mathcal{K}(t_1,s))^2 ds\right)^{\frac{1}{2}} \leq \|f_1\|_p\|f_2\|_q\|f_3\|_r|t-s|^{\frac{3}{2}-\frac{1}{p}-\frac{1}{q}-\frac{1}{r}},$$

and therefore $\mathcal{K}$ satisfies condition (ii) from Example 3.3 and Z has a modification with Hölder continuous paths of all orders $\lambda \in \left(0, \frac{3}{2}-\frac{1}{p}-\frac{1}{q}-\frac{1}{r}\right)$.

Example 3.7 (*multifractional Brownian motion*) Consider a β-Hölder continuous function $H\colon [0, T] \to (0, 1)$ and let $B = \{B(t),\ t \in [0, T]\}$ be a Brownian motion. A Gaussian process $Z = \{Z(t),\ t \in [0, T]\}$ defined by

$$Z(t) := \frac{1}{\Gamma\left(H(t) + \frac{1}{2}\right)} \int_{-\infty}^{0} \left((t - s)^{H(t)-\frac{1}{2}} - (-s)^{H(t)-\frac{1}{2}}\right) dB(s)$$
$$+ \frac{1}{\Gamma\left(H(t) + \frac{1}{2}\right)} \int_{0}^{t} (t - s)^{H(t)-\frac{1}{2}} dB(s)$$

is called the *multifractional Brownian motion* with functional Hurst parameter H. For more detail on this process and its variations, see e.g. [2, 5, 9, 12] and references therein. Here, we note that Z has a modification with Hölder continuous trajectories of any order $\lambda < \min\{\beta, \min_{t\in[0,T]} H(t)\}$. Indeed, put

$$Z_1(t) := \int_{-\infty}^{0} \left((t - s)^{H(t)-\frac{1}{2}} - (-s)^{H(t)-\frac{1}{2}}\right) dB(s),$$

$$Z_2(t) := \int_{0}^{t} (t - s)^{H(t)-\frac{1}{2}} dB(s)$$

By [9, Proposition 3, Step 1], Z_1 has a modification that is λ-Hölder continuous of any order $\lambda < \beta$. Next, for any $0 \le t_1 < t_2 \le T$,

$$\begin{aligned}\mathbb{E}[(Z_2(t_1) - Z_2(t_2))^2] &= \mathbb{E}[(N_1(t_1, t_2) + N_2(t_1, t_2) + N_3(t_1, t_2))^2]\\ &\le 3(\mathbb{E}[N_1^2(t_1, t_2)] + \mathbb{E}[N_2^2(t_1, t_2)] + \mathbb{E}[N_3^2(t_1, t_2)]),\end{aligned}$$

where

$$N_1(t_1, t_2) = \int_{0}^{t_2} ((t_2 - s)^{H(t_2)-\frac{1}{2}} - (t_2 - s)^{H(t_1)-\frac{1}{2}}) dB(s),$$

$$N_2(t_1, t_2) = \int_{t_1}^{t_2} (t_2 - s)^{H(t_1)-\frac{1}{2}} dB(s),$$

$$N_3(t_1, t_2) = \int_{0}^{t_1} ((t_2 - s)^{H(t_1)-\frac{1}{2}} - (t_1 - s)^{H(t_1)-\frac{1}{2}}) dB(s).$$

By [9, Proposition 3, Step 2], there is constant $C > 0$ that does not depend on t_1, t_2, such that

$$\mathbb{E}[N_1^2(t_1, t_2)] \le C|t_2 - t_1|^{2\beta},$$

$$\mathbb{E}[N_2^2(t_1, t_2)] \le C|t_2 - t_1|^{2\min_{t\in[0,T]} H(t)}.$$

In addition, similarly to calculation in Example 3.4, one can write:

$$\begin{aligned}\mathbb{E}[N_3^2(t_1, t_2)]) &= \int_0^{t_1} \left((t_2 - s)^{H(t_1)-\frac{1}{2}} - (t_1 - s)^{H(t_1)-\frac{1}{2}}\right)^2 ds \\ &= (t_2 - t_1)^{2H(t_1)} \int_{\frac{t_2-t_1}{t_1}}^{\infty} \left((v+1)^{H(t_1)-\frac{1}{2}} - 1\right)^2 v^{-1-2H(t_1)} dv \\ &\le (t_2 - t_1)^{2H(t_1)} \int_0^{\infty} \left((v+1)^{H(t_1)-\frac{1}{2}} - 1\right)^2 v^{-1-2H(t_1)} dv.\end{aligned}$$

Note that

$$(t_2 - t_1)^{2H(t_1)} \le (t_2 - t_1)^{2\min_{t\in[0,T]} H(t)} (1+T)^{2(\max_{t\in[0,T]} H(t) - \min_{t\in[0,T]} H(t))}$$

and

$$\begin{aligned}&\int_0^{\infty} \left((v+1)^{H(t_1)-\frac{1}{2}} - 1\right)^2 v^{-1-2H(t_1)} dv \\ &\quad\le \int_0^{1} \left((v+1)^{\max_{t\in[0,T]} H(t)-\frac{1}{2}} - 1\right)^2 v^{-1-2\max_{t\in[0,T]} H(t)} dv \\ &\quad\quad + \int_1^{\infty} \left((v+1)^{\max_{t\in[0,T]} H(t)-\frac{1}{2}} - 1\right)^2 v^{-1-2\min_{t\in[0,T]} H(t)} dv \\ &\quad< \infty,\end{aligned}$$

so, in summary, there exists a deterministic constant $C > 0$ that does not depend on t_1, t_2 such that

$$\mathbb{E}[N_3^2(t_1, t_2)]) \le C(t_2 - t_1)^{2\min_{t\in[0,T]} H(t)},$$

Therefore,

$$\mathbb{E}[(Z_2(t_1) - Z_2(t_2))^2] \le C(t_2 - t_1)^{2\min\{\beta, \min_{t\in[0,T]} H(t)\}},$$

which yields the existence of the required Hölder continuous modification by [1] (see also Appendix A.2).

Let us finish this subsection with an example of non-Gaussian noise Z satisfying Assumption 3.1.

Example 3.8 (*non-Gaussian continuous martingales*) Denote $B = \{B(t),\ t \in [0, T]\}$ a standard Brownian motion and $\sigma = \{\sigma(t),\ t \in [0, T]\}$ an Itô integrable process such that, for all $\beta > 0$,

$$\sup_{u \in [0,T]} \mathbb{E}\sigma^{2+2\beta}(u) < \infty. \tag{3.12}$$

Define

$$Z(t) := \int_0^t \sigma(u) dB(u), \quad t \in [0, T].$$

Then, by the Burkholder-Davis-Gundy inequality, for any $0 \le s < t \le T$ and any $\beta > 0$:

$$\begin{aligned} \mathbb{E}|Z(t) - Z(s)|^{2+2\beta} &\le C_\beta \mathbb{E}\left[\left(\int_s^t \sigma^2(u)du\right)^{1+\beta}\right] \le C_\beta (t-s)^\beta \int_s^t \mathbb{E}\sigma^{2+2\beta}(u)du \\ &\le C_\beta \sup_{u \in [0,T]} \mathbb{E}\sigma^{2+2\beta}(u)(t-s)^{1+\beta}. \end{aligned} \tag{3.13}$$

Therefore, by the Kolmogorov continuity theorem and an arbitrary choice of β, Z has a modification that is λ-Hölder continuous of any order $\lambda \in \left(0, \frac{1}{2}\right)$.

Next, for an arbitrary $\lambda \in \left(0, \frac{1}{2}\right)$, choose $p \ge 1$ such that $\lambda + \frac{1}{p} < \frac{1}{2}$ and put

$$\Lambda := A_{\lambda + \frac{1}{p}, p} \left(\int_0^T \int_0^T \frac{|Z(x) - Z(y)|^p}{|x-y|^{\lambda p + 2}} dx dy \right)^{\frac{1}{p}},$$

where $A_{\lambda+\frac{1}{p},p}$ is defined by (3.3). By the the Burkholder-Davis-Gundy inequality, for any $r > p$, we obtain

$$\mathbb{E}|Z(t) - Z(s)|^r \le |t-s|^{\frac{r}{2}} C_r \sup_{u \in [0,T]} \mathbb{E}\sigma^r(u), \quad s, t \in [0, T].$$

Hence, using Lemma 3.1 and the Minkowski integral inequality, we have:

$$
\begin{aligned}
\left(\mathbb{E}\Lambda^r\right)^{\frac{p}{r}} &= A^p_{\lambda+\frac{1}{p},p}\left(\mathbb{E}\left[\left(\int_0^T\int_0^T \frac{|Z(u)-Z(v)|^p}{|u-v|^{\lambda p+2}}dudv\right)^{\frac{r}{p}}\right]\right)^{\frac{p}{r}} \\
&\le A^p_{\lambda+\frac{1}{p},p}\int_0^T\int_0^T \frac{\left(\mathbb{E}[|Z(u)-Z(v)|^r]\right)^{\frac{p}{r}}}{|u-v|^{\lambda p+2}}dudv \\
&\le A^p_{\lambda+\frac{1}{p},p}C_r^{\frac{p}{r}}\left(\sup_{t\in[0,T]}\mathbb{E}\sigma^r(t)\right)^{\frac{p}{r}}\int_0^T\int_0^T |u-v|^{\frac{p}{2}-\lambda p-2}dudv < \infty,
\end{aligned}
$$

since $\frac{p}{2} - \lambda p - 2 > -1$, i.e. $\mathbb{E}\Lambda^r < \infty$ for all $r > 0$. Note that condition (3.12) can actually be relaxed (see e.g. [3, Lemma 14.2]).

3.2.2 *The Drift*

Let $Z = \{Z(t),\ t \in [0, t]\}$ be a stochastic process satisfying Assumption 3.1 with $\lambda \in (0, 1)$ being the corresponding order of Hölder continuity. Fix some deterministic value $y_0 \in \mathbb{R}$ and a function $\varphi\colon [0, T] \to \mathbb{R}$ such that $\varphi(0) < y_0$ and

$$|\varphi(t) - \varphi(s)| \le L_\varphi |t - s|^\lambda, \quad t, s \in [0, T]. \tag{3.14}$$

For an arbitrary $x \ge 0$, denote

$$
\begin{aligned}
\mathcal{D}_x &:= \{(t, y) \mid t \in [0, T],\ y \in (\varphi(t) + x, \infty)\}, \\
\overline{\mathcal{D}}_x &:= \{(t, y) \mid t \in [0, T],\ y \in [\varphi(t) + x, \infty)\}.
\end{aligned}
\tag{3.15}
$$

Next, consider a function $b\colon \mathcal{D}_0 \to \mathbb{R}$ satisfying the following assumption.

Assumption 3.2 (**A1**) $b \in C(\mathcal{D}_0)$;

(**A2**) there exists a Borel-measurable mapping $\varepsilon \in (0, \infty) \mapsto c_\varepsilon \in (0, \infty)$ such that, for any $\varepsilon > 0$,

$$|b(t, y_1) - b(t, y_2)| \le c_\varepsilon |y_1 - y_2|, \quad (t, y_1), (t, y_2) \in \mathcal{D}_\varepsilon;$$

(**A3**) there are positive constants y_*, c and γ such that for all $(t, y) \in \mathcal{D}_0 \setminus \mathcal{D}_{y_*}$:

$$b(t, y) \ge \frac{c}{(y - \varphi(t))^\gamma};$$

(**A4**) the constant γ from assumption (**A3**) satisfies the condition

$$\gamma > \frac{1-\lambda}{\lambda}$$

with λ being the order of Hölder continuity of φ and the paths of Z.

Example 3.9 Let $\alpha_1\colon [0,T] \to (0,\infty)$ be an arbitrary continuous function and $\alpha_2\colon \mathcal{D}_0 \to \mathbb{R}$ be such that

$$|\alpha_2(t,y_1) - \alpha_2(t,y_2)| \le C|y_1 - y_2|, \quad (t,y_1),(t,y_2) \in \mathcal{D}_0,$$

for some constant $C > 0$. Then, if one chooses $\gamma > \frac{1-\lambda}{\lambda}$, the function

$$b(t,y) := \frac{\alpha_1(t)}{(y-\varphi(t))^\gamma} + \alpha_2(t,y), \quad (t,y) \in \mathcal{D}_0, \tag{3.16}$$

satisfies assumptions **(A1)**–**(A4)**.

Example 3.10 A particular example of the function (3.16) with $\varphi \equiv 0$ is the drift coefficient of the square root of the CIR process (2.10)

$$b(y) = \frac{1}{2}\left(a - \frac{\sigma^2}{4}\right)\frac{1}{y}, \quad y > 0, \tag{3.17}$$

where $a > \frac{\sigma^2}{4}$.

Next, we assume that the function $a\colon [0,T] \times \overline{\mathcal{D}_0}$ from (3.1) satisfies the following conditions.

Assumption 3.3 The function a is continuous on $\overline{\mathcal{D}_0}$ and, for any $(t,y_1),(t,y_2) \in \overline{\mathcal{D}_0}$,

$$|a(t,y_2) - a(t,y_1)| \le L_a|y_2 - y_1|.$$

Finally, the Volterra kernel $\mathcal{G}$ in (3.1) is assumed to satisfy similar conditions as the kernel $\mathcal{K}$ in Example 3.3 as summarized below.

Assumption 3.4 $\mathcal{G}\colon [0,T]^2 \to \mathbb{R}$ is a Volterra kernel, i.e. $\mathcal{G}(t,s) = 0$ if $s \ge t$, and

- **(G1)** $\int_0^T \int_0^T \mathcal{G}^2(t,s)ds dt < \infty$ and
- **(G2)** for any $0 \le t_1 \le t_2 \le T$

$$\int_0^T (\mathcal{G}(t_2,s) - \mathcal{G}(t_1,s))^2 ds \le \ell_\lambda^2 |t_2 - t_1|^{2\lambda},$$

where $\lambda \in (0,1)$ is the order of Hölder continuity of φ and the paths of Z and $\ell_\lambda > 0$ is some constant depending on λ.

Remark 3.1 Note that, if $t_1 = 0$, the inequality from Assumption 3.4(**G2**) reads as

$$\int_0^T (\mathcal{G}(t_2, s) - \mathcal{G}(t_1, s))^2 ds = \int_0^T \mathcal{G}^2(t_2, s) ds \le \ell_\lambda^2 t_2^{2\lambda},$$

since $\mathcal{G}(0, s) = 0$.

Example 3.11 Let $\lambda \in (0, 1)$ and

$$\mathcal{G}(t, s) = (t - s)^{\lambda - \frac{1}{2}} \mathbb{1}_{s<t}.$$

Then Assumption 3.4(G1) automatically holds and, moreover, with a computation identical to the one in Example 3.4, for any $0 \le t_1 < t_2 \le T$,

$$\int_0^T (\mathcal{G}(t_2, s) - \mathcal{G}(t_1, s))^2 ds \le \ell_\lambda^2 (t_2 - t_1)^{2\lambda},$$

for some constant $\ell_\lambda > 0$, i.e. Assumption 3.4(**G2**) also holds.

3.3 Existence and Uniqueness of the Solution

Having all the assumptions on the coefficients of (3.1) in place, we now want to prove the corresponding existence and uniqueness result. Our strategy will be as follows: we will show that the solution exists and is unique until the first moment of hitting the lower bound φ and then prove that this moment actually never happens, i.e. the solution exists and is unique globally.

Let us start with an auxiliary result for deterministic equations.

Theorem 3.1 *Let $z = \{z(t),\ t \in [0, T]\}$ be a fixed deterministic continuous function, $\mathcal{G}$ satisfy Assumption 3.4 and $f, g \in C([0, T] \times \mathbb{R})$ be such that*

$$|f(t, y_2) - f(t, y_1)| + |g(t, y_2) - g(t, y_1)| \le L|y_2 - y_1|, \quad t \in [0, T],\ y_1, y_2 \in \mathbb{R}, \tag{3.18}$$

for some positive constant $L > 0$. Then the (deterministic) integrodifferential equation of the form

$$y(t) = y_0 + \int_0^t f(s, y(s)) ds + \int_0^t \mathcal{G}(t, s) g(s, y(s)) ds + z(t), \quad t \in [0, T], \tag{3.19}$$

has a unique continuous solution.

Proof **Existence**. In order to prove the existence of a continuous solution, we will use the standard Picard iteration argument. Denote

$$
\begin{aligned}
y_0(t) &= y_0 + z(t),\\
y_{n+1}(t) &= y_0 + \int_0^t f(s, y_n(s))ds + \int_0^t \mathcal{G}(t, s)g(s, y_n(s))ds + z(t), \quad n \geq 0,
\end{aligned}
\tag{3.20}
$$

and observe that, by induction, $y_n \in C([0, T])$ for each $n \geq 0$. Indeed, $y_0 \in C([0, T])$ by definition and, if $y_n \in C([0, T])$, then, for any $0 \leq t_1 \leq t_2 \leq T$,

$$
\begin{aligned}
|y_{n+1}(t_2) - y_{n+1}(t_1)| &\\
&\leq \int_{t_1}^{t_2} |f(s, y_n(s))|ds + |z(t_2) - z(t_1)|\\
&\quad + \int_0^{t_2} |\mathcal{G}(t_2, s) - \mathcal{G}(t_1, s)||g(s, y_n(s))|ds\\
&\leq \max_{s\in[0,T]} |f(s, y_n(s))||t_2 - t_1| + |z(t_2) - z(t_1)|\\
&\quad + \max_{s\in[0,T]} |g(s, y_n(s))|\sqrt{T}\sqrt{\int_0^{t_2} |\mathcal{G}(t_2, s) - \mathcal{G}(t_1, s)|^2 ds}\\
&\leq |z(t_2) - z(t_1)| + \max_{s\in[0,T]} |f(s, y_n(s))||t_2 - t_1|\\
&\quad + \max_{s\in[0,T]} |g(s, y_n(s))|\sqrt{T}\ell_\lambda |t_2 - t_1|^\lambda,
\end{aligned}
$$

which immediately yields that $y_{n+1} \in C([0, T])$ thanks to the continuity of z. Next, put

$$
M := \max_{s\in[0,T]} |f(s, y_0(s))| + \max_{s\in[0,T]} |g(s, y_0(s))|,
$$

and notice that, by Remark 3.1,

$$
\begin{aligned}
|y_1(t) - y_0(t)| &\leq \int_0^t |f(s, y_0(s))|\, ds + \int_0^t |\mathcal{G}(t, s)g(s, y_0(s))|\, ds\\
&\leq M\left(t + \int_0^t |\mathcal{G}(t, s)|ds\right) \leq M\left(\sqrt{T} + \sqrt{\int_0^t |\mathcal{G}(t, s)|^2\, ds}\right) t^{\frac{1}{2}}\\
&\leq M\left(\sqrt{T} + \ell_\lambda T^\lambda\right) t^{\frac{1}{2}}
\end{aligned}
$$

$$=: MKt^{\frac{1}{2}},$$

with $K := \sqrt{T} + \ell_\lambda T^\lambda$,

$$\begin{aligned}
|y_2(t) - y_1(t)| &\le \int_0^t |f(s, y_1(s)) - f(s, y_0(s))|\, ds \\
&\quad + \int_0^t |\mathcal{G}(t,s)|\, |g(s, y_1(s)) - g(s, y_0(s))|\, ds \\
&\le L\left(\int_0^t |y_1(s) - y_0(s)| ds + \int_0^t |\mathcal{G}(t,s)|\, |y_1(s) - y_0(s)|\, ds\right) \\
&\le LMK\left(\int_0^t s^{\frac{1}{2}} ds + \int_0^t |\mathcal{G}(t,s)|\, s^{\frac{1}{2}} ds\right) \\
&\le LMK\left(\sqrt{T}\sqrt{\int_0^t s ds} + \sqrt{\int_0^t |\mathcal{G}(t,s)|^2 ds}\sqrt{\int_0^t s ds}\right) \\
&\le LMK\left(\frac{\sqrt{T} + \ell_\lambda T^\lambda}{\sqrt{2}}\right) t \\
&=: \frac{LMK^2}{\sqrt{2}} t,
\end{aligned}$$

and, in general,

$$|y_{n+1}(t) - y_n(t)| \le \frac{ML^n K^{n+1}}{\sqrt{(n+1)!}} (\sqrt{t})^{n+1}.$$

Therefore, for any $n \ge 0$ and $m \ge 1$

$$\begin{aligned}
|y_{n+m}(t) - y_n(t)| &\le \sum_{k=1}^{m} |y_{n+k}(t) - y_{n+k-1}(t)| \le \sum_{k=1}^{m} \frac{ML^{n+k-1} K^{n+k}}{\sqrt{(n+k)!}} (\sqrt{t})^{n+k} \\
&\le \sum_{k=1}^{m} \frac{ML^{n+k-1} (K\sqrt{T})^{n+k}}{\sqrt{(n+k)!}} = \frac{M}{L} \sum_{p=n+1}^{n+m} \frac{(KL\sqrt{T})^p}{\sqrt{p!}} \\
&\to 0, \quad m, n \to \infty,
\end{aligned}$$

since the series

$$\sum_{p=0}^{\infty} \frac{|x|^p}{\sqrt{p!}}$$

converges for any $x \in \mathbb{R}$. Thus, the sequence of continuous functions $\{y_n,\ n \geq 0\}$ converges in the sup-norm to some continuous function $y = \{y(t),\ t \in [0, T]\}$. Moreover, since

$$\sup_{s\in[0,T]} |f(s, y(s)) - f(s, y_n(s))| \leq L \sup_{s\in[0,T]} |y(s) - y_n(s)| \to 0, \quad n \to \infty,$$

it is clear that, for any $t \in [0, T]$,

$$\int_0^t f(s, y_n(s))ds \to \int_0^t f(s, y(s))ds, \quad n \to \infty,$$

and, since

$$\sup_{s\in[0,T]} |g(s, y(s)) - g(s, y_n(s))| \leq L \sup_{s\in[0,T]} |y(s) - y_n(s)| \to 0, \quad n \to \infty,$$

it follows that, for any $t \in [0, T]$,

$$\begin{aligned}
\left| \int_0^t \mathcal{G}(t, s)g(s, y(s))ds - \int_0^t \mathcal{G}(t, s)g(s, y_n(s))ds \right| \\
&\leq \int_0^t |\mathcal{G}(t, s)||g(s, y(s)) - g(s, y_n(s))|ds \\
&\leq L \sup_{s\in[0,T]} |y(s) - y_n(s)| \int_0^t |\mathcal{G}(t, s)|ds \\
&\leq L \sup_{s\in[0,T]} |y(s) - y_n(s)| \sqrt{T} \sqrt{\int_0^t |\mathcal{G}(t, s)|^2 ds} \\
&\leq L\sqrt{T}\ell_\lambda T^\lambda \sup_{s\in[0,T]} |y(s) - y_n(s)| \\
&\to 0, \quad n \to \infty.
\end{aligned}$$

Therefore, by moving $n \to \infty$ in (3.20) for each $t \in [0, T]$, we immediately get that y is the solution to (3.19).

Uniqueness. Let y and $\widetilde{y}$ be two continuous solutions to (3.19). Then, for any $t \in [0, T]$,

$$
\begin{aligned}
|y(t)-\widetilde{y}(t)|^2 &\le 2\left(\int_0^t (f(s,y(s))-f(s,\widetilde{y}(s)))ds\right)^2 \\
&\quad + 2\left(\int_0^t \mathcal{G}(t,s)(g(s,y(s))-g(s,\widetilde{y}(s)))ds\right)^2 \\
&\le 2T\int_0^t (f(s,y(s))-f(s,\widetilde{y}(s)))^2 ds \\
&\quad + 2\int_0^t \mathcal{G}^2(t,s)ds\int_0^t (g(s,y(s))-g(s,\widetilde{y}(s)))^2 ds \\
&\le \left(2TL^2+2\ell_\lambda^2 L^2 T^{1+2\lambda}\right)\int_0^t |y(s)-\widetilde{y}(s)|^2 ds,
\end{aligned}
$$

and the uniqueness follows from the Gronwall inequality. □

For any $\varepsilon \in (0, y_0 - \varphi(0))$ with y_0 being the initial value in (3.1), denote

$$
b_\varepsilon(t,y) := \begin{cases} b(t,y), & (t,y)\in\mathcal{D}_\varepsilon, \\ b(t,\varphi(t)+\varepsilon), & (t,y)\notin\mathcal{D}_\varepsilon, \end{cases}
$$

where $\mathcal{D}_\varepsilon$ is defined in (3.15), and consider a stochastic integrodifferential equation of the form

$$
Y_\varepsilon(t) = y_0 + \int_0^t b_\varepsilon(s, Y_\varepsilon(s))ds + \int_0^t \mathcal{G}(t,s)a(s,Y_\varepsilon(s))ds + Z(t). \tag{3.21}
$$

Observe that $b_\varepsilon \in C([0,T]\times\mathbb{R})$ and, moreover, there exists a constant $L_\varepsilon > 0$ such that for any $t\in[0,T]$ and $y_1, y_2 \in \mathbb{R}$

$$
|b_\varepsilon(t,y_2) - b_\varepsilon(t,y_1)| \le L_\varepsilon |y_2 - y_1|.
$$

Therefore, by treating the Eq. (3.21) pathwisely, one can deduce using Theorem 3.1 that (3.21) has a unique strong a.s. continuous solution $Y_\varepsilon = \{Y_\varepsilon(t),\ t\in[0,T]\}$.

The idea of the existence and uniqueness proof for the equation (3.1) is as follows. Let $\Omega' \subset \Omega$, $\mathbb{P}(\Omega') = 1$, be such that for all $\omega\in\Omega'$ and $0\le t_1\le t_2\le T$

$$
|Z(\omega,t_2) - Z(\omega,t_1)| \le \Lambda(\omega)|t_2-t_1|^\lambda.
$$

By the definition of b_ε, the process Y_ε also satisfies the original equation (3.1) up until the first moment of crossing the line $\{\varphi(t) + \varepsilon,\ t \in [0, T]\}$. We want to prove that, for any *fixed* $\omega \in \Omega'$, there exists $\varepsilon = \varepsilon(\omega)$ such that Y_ε *always* stays above the level $\{\varphi(t) + \varepsilon,\ t \in [0, T]\}$ and hence it is the unique continuous solution to (3.1) on this fixed $\omega \in \Omega'$. Before implementing this plan, let us prove the following auxiliary result.

Proposition 3.1 *Let Assumptions 3.1, 3.2, 3.3 and 3.4 hold and $\Omega' \subset \Omega$, $\mathbb{P}(\Omega') = 1$, be such that the paths of Z are λ-Hölder continuous on Ω'. Fix $\varepsilon \in (0, y_0 - \varphi(0))$, take the corresponding unique continuous solution Y_ε to (3.21) and denote*

$$T_\varepsilon := \sup\{t \in [0, T] \mid \forall s \in [0, t] : \ Y_\varepsilon(s) > \varphi(s)\}.$$

Then there exist deterministic constants M_1, $M_2 > 0$ that do not depend on ε such that, for all $\omega \in \Omega'$ and $t \in [0, T_\varepsilon]$,

$$Y_\varepsilon(t) - \varphi(t) \le M_1 + M_2 \Lambda. \tag{3.22}$$

Proof Fix $\omega \in \Omega'$ (in what follows, we will not write ω in brackets for the notational simplicity). Take an arbitrary $t \in [0, T_\varepsilon]$ and assume that $Y_\varepsilon(t) - \varphi(t) > y_0 - \varphi(0)$ (otherwise, (3.22) holds automatically). Since $Y_\varepsilon - \varphi$ is continuous and $Y_\varepsilon(0) - \varphi(0) = y_0 - \varphi(0)$, one can define

$$\tau(t) := \sup\{s \in [0, t] \mid Y_\varepsilon(s) - \varphi(s) = y_0 - \varphi(0)\}.$$

It is clear that $\tau(t) < t$, $Y_\varepsilon(\tau(t)) - \varphi(\tau(t)) = y_0 - \varphi(0)$ and, for all $s \in [\tau(t), t]$, $Y_\varepsilon(s) - \varphi(s) \ge y_0 - \varphi(0) > \varepsilon$. Additionally, for any $s \in [0, T_\varepsilon]$,

$$Y_\varepsilon(s) - \varphi(s) > 0,$$

therefore, by Assumption 3.2(**A2**), for any $s \in [\tau(t), t]$,

$$\begin{aligned}
|b_\varepsilon(s, Y_\varepsilon(s))| &= |b(s, Y_\varepsilon(s))| \\
&\le |b(s, Y_\varepsilon(s)) - b(s, \varphi(s) + y_0 - \varphi(0))| + |b(s, \varphi(s) + y_0 - \varphi(0))| \\
&\le c_{y_0 - \varphi(0)} |Y_\varepsilon(s) - \varphi(s) - y_0 + \varphi(0)| + \max_{s \in [0,T]} |b(s, \varphi(s) + y_0 - \varphi(0))| \\
&\le c_{y_0 - \varphi(0)} (y_0 - \varphi(0)) + \max_{s \in [0,T]} |b(s, \varphi(s) + y_0 - \varphi(0))| \\
&\quad + c_{y_0 - \varphi(0)} |Y_\varepsilon(s) - \varphi(s)| \\
&=: L_{b,\varphi,y_0} + c_{y_0 - \varphi(0)} (Y_\varepsilon(s) - \varphi(s))
\end{aligned}$$

with $c_{y_0 - \varphi(0)}$ being the constant from Assumption 3.2(**A2**) corresponding to $\varepsilon = y_0 - \varphi(0)$ and

$$L_{b,\varphi,y_0} := c_{y_0 - \varphi(0)} (y_0 - \varphi(0)) + \max_{s \in [0,T]} |b(s, \varphi(s) + y_0 - \varphi(0))|.$$

Furthermore, by Assumption 3.3,

$$\begin{aligned}|a(s, Y_\varepsilon(s))| &\le |a(s, Y_\varepsilon(s)) - a(s, \varphi(s))| + |a(s, \varphi(s))| \\ &\le L_a(Y_\varepsilon(s) - \varphi(s)) + \max_{s\in[0,T]} |a(s, \varphi(s))| \\ &=: L_{a,\varphi,y_0} + L_a(Y_\varepsilon(s) - \varphi(s))\end{aligned}$$

with $L_{a,\varphi} := \max_{s\in[0,T]} |a(s, \varphi(s))|$.

Next, write

$$\begin{aligned}Y_\varepsilon(t) - \varphi(t) &= y_0 - \varphi(t) + \int_0^t b_\varepsilon(s, Y_\varepsilon(s))ds + \int_0^t \mathcal{G}(t, s)a(s, Y_\varepsilon(s))ds + Z(t) \\ &= Y_\varepsilon(\tau(t)) - \varphi(t) + \int_{\tau(t)}^t b_\varepsilon(s, Y_\varepsilon(s))ds \\ &\quad + \int_0^t (\mathcal{G}(t, s) - \mathcal{G}(\tau(t), s))a(s, Y_\varepsilon(s))ds + Z(t) - Z(\tau(t)) \\ &= \varphi(\tau(t)) + y_0 - \varphi(0) - \varphi(t) + \int_{\tau(t)}^t b_\varepsilon(s, Y_\varepsilon(s))ds \\ &\quad + \int_0^t (\mathcal{G}(t, s) - \mathcal{G}(\tau(t), s))a(s, Y_\varepsilon(s))ds + Z(t) - Z(\tau(t)),\end{aligned} \tag{3.23}$$

and observe that

$$\varphi(\tau(t)) - \varphi(t) \le L_\varphi(t - \tau(t))^\lambda \le L_\varphi T^\lambda, \tag{3.24}$$

$$Z(t) - Z(\tau(t)) \le \Lambda(t - \tau(t))^\lambda \le \Lambda T^\lambda, \tag{3.25}$$

$$\begin{aligned}\int_{\tau(t)}^t b_\varepsilon(s, Y_\varepsilon(s))ds &\le L_{b,\varphi,y_0}(t - \tau(t)) + c_{y_0-\varphi(0)} \int_{\tau(t)}^t (Y_\varepsilon(s) - \varphi(s))ds \\ &\le L_{b,\varphi,y_0} T + c_{y_0-\varphi(0)} \int_0^t (Y_\varepsilon(s) - \varphi(s))ds,\end{aligned} \tag{3.26}$$

and, by Assumption 3.4,

$$
\begin{aligned}
&\int_0^t (\mathcal{G}(t,s) - \mathcal{G}(\tau(t),s))a(s,Y_\varepsilon(s))ds \\
&\qquad \le \sqrt{\int_0^t (\mathcal{G}(t,s) - \mathcal{G}(\tau(t),s))^2 ds}\sqrt{\int_0^t a^2(s,Y_\varepsilon(s))ds} \\
&\qquad \le \ell_\lambda (t-\tau(t))^\lambda \sqrt{\int_0^t (L_{a,\varphi,0} + L_a(Y_\varepsilon(s) - \varphi(s)))^2 ds} \\
&\qquad \le \ell_\lambda T^\lambda \sqrt{2L_{a,\varphi,0}^2 T + 2L_a^2 \int_0^t (Y_\varepsilon(s) - \varphi(s))^2 ds}.
\end{aligned}
\tag{3.27}
$$

Using (3.24)–(3.27) as well as (3.23), we can write

$$
\begin{aligned}
Y_\varepsilon(t) - \varphi(t) \le y_0 - \varphi(0) &+ (L_\varphi + \Lambda)T^\lambda + L_{b,\varphi,y_0} T \\
&+ c_{y_0 - \varphi(0)} \int_0^t (Y_\varepsilon(s) - \varphi(s))ds \\
&+ \ell_\lambda T^\lambda \sqrt{2L_{a,\varphi,0}^2 T + 2L_a^2 \int_0^t (Y_\varepsilon(s) - \varphi(s))^2 ds}
\end{aligned}
$$

and hence

$$
\begin{aligned}
&(Y_\varepsilon(t) - \varphi(t))^2 \\
&\quad \le 4\,(y_0 - \varphi(0))^2 + 4\left((L_\varphi + \Lambda)T^\lambda + L_{b,\varphi,y_0} T\right)^2 \\
&\qquad + 4c_{y_0-\varphi(0)}^2 \left(\int_0^t (Y_\varepsilon(s) - \varphi(s))ds\right)^2 \\
&\qquad + 4\ell_\lambda^2 T^{2\lambda}\left(2L_{a,\varphi,0}^2 T + 2L_a^2 \int_0^t (Y_\varepsilon(s) - \varphi(s))^2 ds\right) \\
&\quad \le 4\,(y_0 - \varphi(0))^2 + 4\left((L_\varphi + \Lambda)T^\lambda + L_{b,\varphi,y_0} T\right)^2 + 8\ell_\lambda^2 T^{2\lambda+1} L_{a,\varphi,0}^2 \\
&\qquad + \left(4c_{y_0-\varphi(0)}^2 T + 8\ell_\lambda^2 T^{2\lambda} L_a^2\right) \int_0^t (Y_\varepsilon(s) - \varphi(s))^2 ds.
\end{aligned}
\tag{3.28}
$$

Note that (3.28) also holds automatically if $Y_\varepsilon(t) - \varphi(t) < y_0 - \varphi(0)$, i.e. for any $t \in [0, T_\varepsilon]$,

$$
\begin{aligned}
&(Y_\varepsilon(t) - \varphi(t))^2 \\
&\quad\le 4\,(y_0 - \varphi(0))^2 + 4\left(L_\varphi T^\lambda + L_{b,\varphi,y_0} T + \Lambda T^\lambda\right)^2 + 8\ell_\lambda^2 T^{2\lambda+1} L_{a,\varphi,0}^2 \\
&\quad+ \left(4c_{y_0-\varphi(0)}^2 T + 8\ell_\lambda^2 T^{2\lambda} L_a^2\right) \int_0^t (Y_\varepsilon(s) - \varphi(s))^2 ds \\
&\quad\le 4\,(y_0 - \varphi(0))^2 + 8\left(L_\varphi T^\lambda + L_{b,\varphi,y_0} T\right)^2 + 8\ell_\lambda^2 T^{2\lambda+1} L_{a,\varphi,0}^2 + 8T^{2\lambda}\Lambda^2 \\
&\quad+ \left(4c_{y_0-\varphi(0)}^2 T + 8\ell_\lambda^2 T^{2\lambda} L_a^2\right) \int_0^t (Y_\varepsilon(s) - \varphi(s))^2 ds.
\end{aligned}
$$

Hence, by Gronwall inequality, for any $t \in [0, T_\varepsilon]$,

$$
\begin{aligned}
(Y_\varepsilon(t) - \varphi(t))^2 \le 8T^{2\lambda} \exp\left\{4c_{y_0-\varphi(0)}^2 T^2 + 8\ell_\lambda^2 T^{2\lambda+1} L_a^2\right\} \Lambda^2 \\
+ \left(4\,(y_0 - \varphi(0))^2 + 8\left(L_\varphi T^\lambda + L_{b,\varphi,y_0} T\right)^2 + 8\ell_\lambda^2 T^{2\lambda+1} L_{a,\varphi,0}^2\right) \\
\times \exp\left\{4c_{y_0-\varphi(0)}^2 T^2 + 8\ell_\lambda^2 T^{2\lambda+1} L_a^2\right\},
\end{aligned}
$$

which immediately implies the claim of this proposition. □

We are now ready to prove the main result of this section.

Theorem 3.2 *Let Assumptions 3.1, 3.2, 3.3 and 3.4 hold and $\Omega' \subset \Omega$, $\mathbb{P}(\Omega') = 1$, be such that the paths of Z are λ-Hölder continuous on Ω'. Then, for all $\omega \in \Omega'$, the stochastic integro-differential equation* (3.1) *has a unique solution Y with continuous paths. Moreover, on this Ω',*

$$
Y(t) > \varphi(t), \quad t \in [0, T]. \tag{3.29}
$$

Proof Choose $\omega \in \Omega'$, i.e. the corresponding path $Z(\omega, t)$, $t \in [0, T]$, is λ-Hölder continuous. In what follows, we will omit ω in brackets for notational convenience, but all equations below should be treated as deterministic for the fixed ω.

We shall split the proof into several steps.

Step 1. Denote

$$
D := \frac{c^{\frac{\lambda}{\gamma\lambda+\lambda-1}}}{2^{\frac{\gamma\lambda-\lambda+1}{\gamma\lambda+\lambda-1}}\left(\lambda^{\frac{\lambda}{1-\lambda}} - \lambda^{\frac{1}{1-\lambda}}\right)} > 0,
$$

where c is the constant from Assumption 3.2(A3) and λ, γ are from Assumption 3.2(A4), and put

$$
C_1 := \left(\frac{\min\{y_*, y_0 - \varphi(0)\}}{4D}\right)^{\gamma\lambda+\lambda-1} + L_\varphi + \ell_\lambda \sqrt{2L_{a,\varphi}^2 T + 4TM_1^2 L_a^2}, \tag{3.30}
$$

and

$$C_2 := 2\sqrt{T}\ell_\lambda L_a M_2 + 1, \tag{3.31}$$

with

- $y_0 > \varphi(0)$ being the initial value in the equation (3.1),
- y_* being from Assumption 3.2(A3),
- $L_\varphi > 0$ being from (3.14),
- $L_{a,\varphi} := \max_{s\in[0,T]} |a(s, \varphi(s))|$,
- $L_a > 0$ being from Assumption 3.3,
- ℓ_λ being from Assumption 3.1 and
- constants M_1, M_2 being from Proposition 3.1.

Finally, denote

$$\xi = \xi(\omega) := \frac{D}{(C_1 + C_2\Lambda(\omega))^{\frac{1}{\gamma\lambda+\lambda-1}}}. \tag{3.32}$$

Using the fact that $\gamma\lambda + \lambda - 1 > 0$ by Assumption 3.2(A4), it is straightforward to check that

$$\xi \le \frac{1}{4}\min\{y_*, y_0 - \varphi(0)\}.$$

For the $\omega \in \Omega'$ fixed above, take the corresponding continuous path $Y_\xi(\omega)$ solving the Eq. (3.21) with $\varepsilon = \xi(\omega)$. As noted previously, Y_ξ satisfies the original equation (3.1) up until the moment τ_ξ with

$$\tau_\xi := \sup\{t \in [0, T] \mid \forall s \in [0, t] : Y_\xi(s) - \varphi(s) \ge \xi\} \in [0, T],$$

(note that $\tau_\xi > 0$ since $\xi < y_0 - \varphi(0)$). If $\widetilde{Y}_\xi$ is another continuous path satisfying (3.1) on $[0, \tau_\xi]$, then it must satisfy (3.21) for all $t \in [0, \widetilde{\tau}_\xi]$ with

$$\widetilde{\tau}_\xi := \sup\{t \in [0, T] \mid \forall s \in [0, t] : \widetilde{Y}_\xi(s) - \varphi(s) \ge \xi\} \in [0, T].$$

However, since there is only one continuous solution to (3.21), $\widetilde{\tau}_\xi = \tau_\xi$ and $\widetilde{Y}_\xi(t) = Y_\xi(t)$ for all $t \in [0, \tau_\xi]$ and (3.29) holds.

Our ultimate goal is to prove that $Y_\xi(t) \ge \varphi(t) + \xi$ *for all* $t \in [0, T]$, i.e. that $\tau_\xi = T$ and Y_ξ is the unique continuous solution to (3.1) on the entire $[0, T]$ for the fixed $\omega \in \Omega'$.

Step 2. Assume that $\tau_\xi < T$. Since $2\xi < y_0 - \varphi(0)$, the path Y_ξ must cross the line $\varphi + 2\xi$ before hitting $\varphi + \xi$, i.e. one can define

$$\tau_{2\xi} := \sup\{t \in [0, \tau_\xi] \mid Y_\xi(t) - \varphi(t) = 2\xi\}.$$

It is clear that $0 < \tau_{2\xi} < \tau_\xi$ and, by continuity,

$$Y_\xi(\tau_\xi) - \varphi(\tau_\xi) = \xi, \quad Y_\xi(\tau_{2\xi}) - \varphi(\tau_{2\xi}) = 2\xi.$$

Therefore

$$
\begin{aligned}
\xi &= Y_\xi(\tau_\xi) - \varphi(\tau_\xi) \\
&= y_0 - \varphi(\tau_\xi) + \int_0^{\tau_\xi} b_\xi(s, Y_\xi(s))ds \\
&\quad + \int_0^{\tau_\xi} \mathcal{G}(\tau_\xi, s)a(s, Y_\xi(s))ds + Z(\tau_\xi) \\
&= -\varphi(\tau_\xi) + Y_\xi(\tau_{2\xi}) + \int_{\tau_{2\xi}}^{\tau_\xi} b_\xi(s, Y_\xi(s))ds \\
&\quad + \int_0^{\tau_\xi} (\mathcal{G}(\tau_\xi, s) - \mathcal{G}(\tau_{2\xi}, s))a(s, Y_\xi(s))ds + Z(\tau_\xi) - Z(\tau_{2\xi}) \\
&= 2\xi + \varphi(\tau_{2\xi}) - \varphi(\tau_\xi) + \int_{\tau_{2\xi}}^{\tau_\xi} b_\xi(s, Y_\xi(s))ds \\
&\quad + \int_0^{\tau_\xi} (\mathcal{G}(\tau_\xi, s) - \mathcal{G}(\tau_{2\xi}, s))a(s, Y_\xi(s))ds + Z(\tau_\xi) - Z(\tau_{2\xi}).
\end{aligned}
\tag{3.33}
$$

By (3.14) and Assumption 3.1(Z2),

$$
\varphi(\tau_{2\xi}) - \varphi(\tau_\xi) \geq -L_\varphi(\tau_\xi - \tau_{2\xi})^\lambda \tag{3.34}
$$

and

$$
Z(\tau_\xi) - Z(\tau_{2\xi}) \geq -\Lambda(\tau_\xi - \tau_{2\xi})^\lambda. \tag{3.35}
$$

Moreover, since $2\xi < y_*$ and $Y_\xi(t) - \varphi(t) \in [\xi, 2\xi]$ for all $t \in [\tau_{2\xi}, \tau_\xi]$, Assumption 3.2(A3) implies that

$$
\int_{\tau_{2\xi}}^{\tau_\xi} b_\xi(s, Y_\xi(s))ds \geq \frac{c}{(2\xi)^\gamma}(\tau_\xi - \tau_{2\xi}). \tag{3.36}
$$

Finally, using the fact that

$$
\begin{aligned}
|a(s, Y_\xi(s))| &\leq |a(s, Y_\xi(s)) - a(s, \varphi(s))| + |a(s, \varphi(s))| \\
&\leq L_a|Y_\xi(s) - \varphi(s)| + \max_{s\in[0,T]} |a(s, \varphi(s))| \\
&=: L_{a,\varphi} + L_a|Y_\xi(s) - \varphi(s)|,
\end{aligned}
$$

where, as before, L_a is from Assumption 3.3 and $L_{a,\varphi} := \max_{s\in[0,T]} |a(s,\varphi(s))|$, one can use Proposition 3.1 and Assumption 3.4 to deduce that

$$\begin{aligned}
&\int_0^{\tau_\xi} (\mathcal{G}(\tau_\xi, s) - \mathcal{G}(\tau_{2\xi}, s))a(s, Y_\xi(s))ds \\
&\geq - \sqrt{\int_0^{\tau_\xi} (\mathcal{G}(\tau_\xi, s) - \mathcal{G}(\tau_{2\xi}, s))^2 ds} \sqrt{\int_0^{\tau_\xi} a^2(s, Y_\xi(s))ds} \\
&\geq -\ell_\lambda(\tau_\xi - \tau_{2\xi})^\lambda \sqrt{\int_0^{\tau_\xi} (L_{a,\varphi} + L_a|Y_\xi(s) - \varphi(s)|)^2 ds} \\
&\geq -\ell_\lambda(\tau_\xi - \tau_{2\xi})^\lambda \sqrt{2L_{a,\varphi}^2 T + 2L_a^2 \int_0^{\tau_\xi} |Y_\xi(s) - \varphi(s)|^2 ds} \\
&\geq -\ell_\lambda(\tau_\xi - \tau_{2\xi})^\lambda \sqrt{2L_{a,\varphi}^2 T + 2L_a^2 \int_0^{\tau_\xi} |M_1 + M_2\Lambda|^2 ds} \\
&\geq -\ell_\lambda(\tau_\xi - \tau_{2\xi})^\lambda \sqrt{2L_{a,\varphi}^2 T + 4TM_1^2 L_a^2 + 4TL_a^2 M_2^2 \Lambda^2} \\
&\geq -(\ell_\lambda \sqrt{2L_{a,\varphi}^2 T + 4TM_1^2 L_a^2} + 2\sqrt{T}\ell_\lambda L_a M_2 \Lambda)(\tau_\xi - \tau_{2\xi})^\lambda.
\end{aligned} \tag{3.37}$$

Taking into account (3.33) together with (3.34)–(3.37), we deduce that, if it is indeed true that $\tau_\xi < T$, one must have

$$\begin{aligned}
\xi &\geq 2\xi + \frac{c}{(2\xi)^\gamma}(\tau_\xi - \tau_{2\xi}) \\
&\quad - \left(\left(L_\varphi + \ell_\lambda \sqrt{2L_{a,\varphi}^2 T + 4TM_1^2 L_a^2}\right) + \left(2\sqrt{T}\ell_\lambda L_a M_2 + 1\right)\Lambda\right)(\tau_\xi - \tau_{2\xi})^\lambda \\
&\geq 2\xi + \frac{c}{(2\xi)^\gamma}(\tau_\xi - \tau_{2\xi}) - (C_1 + C_2\Lambda)(\tau_\xi - \tau_{2\xi})^\lambda
\end{aligned}$$

with C_1 and C_2 being defined by (3.30) and (3.31) respectively. In other words, if $\tau_\xi < T$, one must have

$$\xi + \frac{c}{(2\xi)^\gamma}(\tau_\xi - \tau_{2\xi}) - (C_1 + C_2\Lambda)(\tau_\xi - \tau_{2\xi})^\lambda \leq 0. \tag{3.38}$$

Now, consider a function

$$F_\xi(x) := \xi + \frac{c}{(2\xi)^\gamma} x - (C_1 + C_2\Lambda)\, x^\lambda, \quad x \ge 0.$$

It is straightforward to check that F_ξ attains its minimal value at

$$x_* := \left(\frac{\lambda 2^\gamma (C_1 + C_2\Lambda)}{c} \right)^{\frac{1}{1-\lambda}} \xi^{\frac{\gamma}{1-\lambda}}$$

and, taking into account (3.32),

$$F_\xi(x_*) = \xi - \left(\frac{2^{\gamma\lambda}(C_1 + C_2\Lambda)}{c^\lambda} \right)^{\frac{1}{1-\lambda}} \left(\lambda^{\frac{\lambda}{1-\lambda}} - \lambda^{\frac{1}{1-\lambda}} \right) \xi^{\frac{\gamma\lambda}{1-\lambda}} = \frac{\xi}{2} > 0.$$

Thus we obtain that

$$\xi + \frac{c}{(2\xi)^\gamma}(\tau_\xi - \tau_{2\xi}) - (C_1 + C_2\Lambda)\,(\tau_\xi - \tau_{2\xi})^\lambda \ge F_\xi(x_*) = \frac{\xi}{2} > 0,$$

which contradicts (3.38). Therefore, $\tau_\xi = T$ and Y_ξ is the unique continuous solution to (3.1) for the fixed $\omega \in \Omega'$. □

Definition 3.1 Given the property (3.29) under Assumptions 3.1, 3.2, 3.3 and 3.4, we will call the solution to the stochastic integro-differential equation (3.1) a **one-sided sandwiched process**.

In fact, the proof of Theorem 3.2 establishes more than just existence and uniqueness of the continuous solution: it shows that, for all $\omega \in \Omega$ such that the corresponding path $\{Z(\omega, t),\ t \in [0, T]\}$ is λ-Hölder continuous, the solution $Y = \{Y(t),\ t \in [0, T]\}$ to (3.1) stays above the level $\{\varphi(t) + \xi,\ t \in [0, T]\}$, where the random variable $\xi < y_0 - \varphi(0)$ and has the form (3.32). Moreover, since $\{Y(\omega, t),\ t \in [0, T]\}$ coincides with $\{Y_{\xi(\omega)}(\omega, t),\ t \in [0, T]\}$ for $\mathbb{P}$-almost all $\omega \in \Omega$, the upper bound 3.22 holds for Y with probability 1. These two observations are summarized in the following theorem.

Theorem 3.3 *Let Assumptions 3.1, 3.2, 3.3 and 3.4 hold and let* $Y = \{Y(t),\ t \in [0, T]\}$ *be the unique continuous solution to stochastic integrodifferential equation* (3.1). *Then, for all* $\omega \in \Omega$ *such that the corresponding path* $\{Z(\omega, t),\ t \in [0, T]\}$ *is* λ*-Hölder continuous, there exist deterministic constants* C_1, $C_2 > 0$ *that do not depend on the noise* Z *such that*

$$\varphi(t) + \frac{C_1}{(1 + \Lambda)^{\frac{1}{\gamma\lambda + \lambda - 1}}} \le Y(t) \le \varphi(t) + C_2(1 + \Lambda), \quad t \in [0, T]. \tag{3.39}$$

The bounds in (3.39) immediately allow to establish the existence of moments for Y, as summarized in the following corollary.

Corollary 3.1 *Let Assumptions 3.1, 3.2, 3.3 and 3.4 hold, let* $Y = \{Y(t),\ t \in [0, T]\}$ *be the unique continuous solution to stochastic integro-differential equation* (3.1) *and fix* $r > 0$.

(1) If Λ *in* (3.2) *can be chosen such that* $\mathbb{E}[\Lambda^{\frac{r}{\gamma\lambda+\lambda-1}}] < \infty$, *then*

$$\mathbb{E}\left[\sup_{t\in[0,T]} \frac{1}{(Y(t)-\varphi(t))^r}\right] < \infty.$$

(2) If Λ *in* (3.2) *can be chosen such that* $\mathbb{E}[\Lambda^r] < \infty$, *then*

$$\mathbb{E}\left[\sup_{t\in[0,T]} |Y(t)|^r\right] < \infty.$$

Finally, let us discuss the regularity of paths for Y.

Theorem 3.4 *Let Assumptions 3.1, 3.2, 3.3 and 3.4 hold and* $\Omega' \subset \Omega$, $\mathbb{P}(\Omega') = 1$, *be such that the paths of* Z *are* λ*-Hölder continuous on* Ω'. *Then, on* Ω', *the continuous solution* $Y = \{Y(t),\ t \in [0, T]\}$ *to the stochastic integro-differential equation* (3.1) *is* λ*-Hölder continuous, i.e. there is a random variable* $\Upsilon > 0$ *such that for all* $0 \le t_1 \le t_2 \le T$

$$|Y(t_2) - Y(t_1)| \le \Upsilon |t_2 - t_1|^{\lambda}. \tag{3.40}$$

Moreover, it is possible to take Υ *of the form*

$$\Upsilon = C(1 + \Lambda + c_\xi + c_\xi \Lambda),$$

where $C > 0$ *is some deterministic constant,* Λ *is from* (3.2),

$$\xi := \frac{C_1}{2(1+\Lambda)^{\frac{1}{\gamma\lambda+\lambda-1}}},$$

with C_1 *being from Theorem 3.3, and* c_ξ *being the evaluation in the (random) point* ξ *of the mapping* $\varepsilon \in (0, \infty) \mapsto c_\varepsilon \in (0, \infty)$ *from Assumption 3.2(A2).*

Remark 3.2 By Assumption 3.2(A2), the mapping $\varepsilon \in (0, \infty) \mapsto c_\varepsilon \in (0, \infty)$ is Borel-measurable, hence c_ξ is a random variable.

Proof Fix $\omega \in \Omega'$, i.e. the corresponding path $Z(\omega, t)$, $t \in [0, T]$, is λ-Hölder continuous. By the proof of Theorem 3.2, one can choose C_1 such that $\xi < y_0 - \varphi(0)$. By Theorem 3.3, $(s, Y(s)) \in \mathcal{D}_\xi$ for any $s \in [0, T]$, so, by Assumption 3.2(A2), for all $s \in [0, T]$,

$$
\begin{aligned}
|b(s, Y(s))| &\le |b(s, Y(s)) - b(s, \varphi(s) + y_0 - \varphi(0))| + |b(s, \varphi(s) + y_0 - \varphi(0))| \\
&\le c_\xi |Y(s) - \varphi(s)| + c_\xi |y_0 - \varphi(0)| + \max_{s\in[0,T]} |b(s, \varphi(s) + y_0 - \varphi(0))| \\
&\le c_\xi (M_1 + M_2\Lambda) + c_\xi |y_0 - \varphi(0)| + \max_{s\in[0,T]} |b(s, \varphi(s) + y_0 - \varphi(0))| \\
&\le C(1 + c_\xi(1 + \Lambda))
\end{aligned}
$$

with $C > 0$ being some deterministic constant. Additionally, using Theorem 3.3, one can see that for any $s \in [0, T]$

$$
\begin{aligned}
|a(s, Y(s))| &\le |a(s, Y(s)) - a(s, \varphi(s)| + |a(s, \varphi(s))| \\
&\le L_a |Y(s) - \varphi(s)| + \max_{s\in[0,T]} |a(s, \varphi(s))| \\
&\le L_a C_2(1 + \Lambda) + \max_{s\in[0,T]} |a(s, \varphi(s))| \\
&\le C(1 + \Lambda),
\end{aligned}
$$

where $C > 0$ is again a deterministic constant. Therefore, for any $0 \le t_1 \le t_2 \le T$, there exists a positive constant $C > 0$ such that

$$
\begin{aligned}
|Y(t_2) - Y(t_1)| &\le \int_{t_1}^{t_2} |b(s, Y(s))|ds + \int_0^{t_2} |\mathcal{G}(t_2, s) - \mathcal{G}(t_1, s)||a(s, Y(s))|ds \\
&\quad + |Z(t_2) - Z(t_1)| \\
&\le C(1 + c_\xi + c_\xi \Lambda)(t_2 - t_1) \\
&\quad + \sqrt{\int_0^{t_2} |\mathcal{G}(t_2, s) - \mathcal{G}(t_1, s)|^2 ds} \sqrt{\int_0^{t_2} a^2(s, Y(s))ds} \\
&\quad + \Lambda |t_2 - t_1|^\lambda \\
&\le C(1 + c_\xi(1 + \Lambda))(t_2 - t_1) + \ell_\lambda \sqrt{T} C(1 + \Lambda)|t_2 - t_1|^\lambda \\
&\quad + \Lambda |t_2 - t_1|^\lambda \\
&\le \Upsilon |t_2 - t_1|^\lambda,
\end{aligned}
$$

with $\Upsilon := CT^{1-\lambda}(1 + c_\xi(1 + \Lambda)) + \ell_\lambda \sqrt{T} C(1 + \Lambda) + \Lambda$, which ends the proof. □

Remark 3.3 Note that arguments in this subsection are pathwise: in all proofs, we fix $\omega \in \Omega'$ and use the tools of the "deterministic" calculus. This leads to an interesting implication: all the results here hold without any changes, if the lower boundary φ is itself a stochastic process with λ-Hölder continuous trajectories.

3.4 Connection to Skorokhod Reflections

We have seen that, under Assumptions 3.1, 3.2, 3.3 and 3.4, the solution Y to (3.1) exceeds φ. Note that the inequality in (3.29) is strict: Y never reaches the lower bound φ and hence cannot be called a *reflected* process. In the same time, it is still possible to establish a connection between (3.1) and reflected processes introduced by A. Skorokhod in his seminal papers [10, 11].

For any $\varepsilon > 0$, consider a particular case of (3.1) of the form

$$Y_\varepsilon(t) = y_0 + \int_0^t \frac{\varepsilon}{(Y_\varepsilon(s) - \varphi(s))^\gamma} ds - \int_0^t a(s, Y_\varepsilon(s))ds + Z(t), \tag{3.41}$$

where $Z = \{Z(t),\ t \in [0, T]\}$ is a stochastic process satisfying Assumption 3.1, $\gamma > \frac{1-\lambda}{\lambda}$, where λ is the joint order of Hölder continuity of φ and the paths of Z, $y_0 > \varphi(0)$ and $a\colon [0, T] \times \mathbb{R} \to \mathbb{R}$ is a continuous function such that

$$|a(t, y_1) - a(t, y_2)| \le c|y_1 - y_2|, \quad y_1, y_2 \in \mathbb{R}, \quad t \in [0, T], \tag{3.42}$$

for some constant $c > 0$. It is clear that the drift $b_\varepsilon(t, y) := \frac{\varepsilon}{(y-\varphi(t))^\gamma} - a(t, y)$ satisfies Assumption 3.2 and hence there exists a unique solution Y_ε to (3.41) and $Y_\varepsilon(t) > \varphi(t)$, $t \in [0, T]$.

Next, consider an equation of the form

$$Y_0(t) = y_0 - \int_0^t a(s, Y_0(s))ds + Z(t) + L_0(t), \tag{3.43}$$

which is considered as an equation w.r.t. both Y_0 and L_0 simultaneously and $L_0 = \{L_0(t),\ t \in [0, T]\}$ is searched among the processes satisfying the following conditions:

(i) $L_0(0) = 0$ a.s.,
(ii) L_0 is non-decreasing a.s.,
(iii) L_0 is continuous a.s.,
(iv) the points of growth for L_0 occur a.s. only at the points where $Y_0(t) - \varphi(t) = 0$,
(v) $Y_0(t) \ge \varphi(t)$, $t \in [0, T]$, a.s.

Remark 3.4 Equations of the type (3.43) are called *Skorokhod problems*. In what follows, we will also call the process L_0 the *φ-reflection* function (or φ-regulator) for Y_0.

Let us start with the proof that the problem (3.43) cannot have multiple solutions.

Theorem 3.5 *If the Skorokhod reflection problem (3.43) has a solution (Y_0, L_0), it is unique.*

Proof First, note that, without loss of generality, we can put $\varphi \equiv 0$. Indeed, let (Y_0, L_0) be a solution to the Skorokhod reflection problem (3.43) with the lower boundary φ. Then the process $Y_0^\varphi(t) := Y_0(t) - \varphi(t)$, $t \in [0, T]$, satisfies

$$Y_0^\varphi(t) = Y^\varphi(0) - \int_0^t a^\varphi(s, Y_0^\varphi(s))ds + Z^\varphi(t) + L_0(t), \tag{3.44}$$

where $Y^\varphi(0) := y_0 - \varphi(0)$, $a^\varphi(t, y) := a(t, y + \varphi(t))$, $Z^\varphi(t) := Z(t) - (\varphi(t) - \varphi(0))$. It is easy to check that L_0 is a 0-reflection function for Y_0^φ, i.e. (Y_0^φ, L_0) is a solution to the Skorokhod reflection problem (3.44) with the lower boundary 0. Similar reasoning allows to establish that the opposite is also true: if (Y_0^φ, L_0) is a solution to (3.44), then $(Y_0 = Y_0^\varphi + \varphi, L_0)$ is a solution to (3.43) and hence (3.43) has a solution if and only if (3.44) does, and uniqueness of solution of one Skorokhod problem implies uniqueness of the other. Therefore, in this proof we assume that $\varphi \equiv 0$.

The further arguments essentially follow the arguments of Skorokhod's seminal papers [10, 11]. The only difference is that we have a general Hölder continuous noise Z instead of a classical Brownian motion, but the additive form of Z in (3.43) makes the arguments shorter.

Let (Y_0, L_0) and (Y_0', L_0') be two solutions to (3.43). Denote

$$\Delta^+(t) := \begin{cases} Y_0(t) - Y_0'(t), & \text{if } Y_0(t) - Y_0'(t) > 0, \\ 0, & \text{otherwise,} \end{cases}$$

$$\Delta^-(t) := \begin{cases} Y_0'(t) - Y_0(t), & \text{if } Y_0'(t) - Y_0(t) > 0, \\ 0, & \text{otherwise.} \end{cases}$$

By definition of a solution to Skorokhod reflection problem, both Δ^+ and Δ^- are continuous with probability 1. Let

$$\tau(t) := \sup\{s \in [0, t] \mid \Delta^+(s) = 0\}.$$

If $\tau(t) < t$, we have that for all $s \in (\tau(t), t]$

$$\Delta^+(s) > 0$$

and therefore $Y_0(s) > Y_0'(s) \geq 0$. This means that Y_0 does not hit zero on $(\tau(t), t]$, so $L_0(t) = L_0(\tau(t))$ by definition of the reflection function. Moreover, since $Y_0 - Y_0'$ is continuous, $Y_0(\tau(t)) - Y_0'(\tau(t)) = 0$ and hence

$$Y_0(t) - Y_0'(t) = -\int_{\tau(t)}^t \left(a(s, Y_0(s)) - a(s, Y_0'(s))\right) ds + L_0'(\tau(t)) - L_0'(t).$$

However, $Y_0(t) - Y_0'(t) > 0$ and $L_0'(\tau(t)) - L_0'(t) \le 0$, therefore

$$\Delta^+(t) \le \left| \int_{\tau(t)}^{t} \big(a(s, Y_0(s)) - a(s, Y_0'(s))\big)\, ds \right|$$
$$\le \int_0^t \left| a(s, Y_0(s)) - a(s, Y_0'(s)) \right| ds$$
$$\le c \int_0^t \left| Y_0(s) - Y_0'(s) \right| ds,$$

which also holds true if $\tau(t) = t$ (i.e. if $\Delta^+(t) = 0$). Similarly,

$$\Delta^-(t) \le c \int_0^t \left| Y_0(s) - Y_0'(s) \right| ds,$$

and hence, for all $t \in [0, T]$,

$$|Y_0(t) - Y_0'(t)| \le c \int_0^t \left| Y_0(s) - Y_0'(s) \right| ds. \tag{3.45}$$

The equality of $Y_0(t)$ and $Y_0'(t)$ with probability 1 now follows immediately from Gronwall lemma and (3.45) which, in turn, immediately implies that $L_0(t) = L_0'(t)$ a.s. □

Note that Theorem 3.5 does not clarify whether the solution to (3.43) exists – the existence will be obtained automatically in Theorem 3.6 below. However, before stating it, we will need a simple yet useful comparison-type result.

Lemma 3.2 *Assume that continuous random processes* $\{X_1(t),\ t \ge 0\}$ *and* $\{X_2(t), t \ge 0\}$ *satisfy a.s. the equations of the form*

$$X_i(t) = X(0) + \int_0^t f_i(s, X_i(s))ds + Z(t), \quad t \ge 0, \quad i = 1, 2,$$

where $X(0)$ *is a constant and* $f_1,\ f_2$: $[0, \infty) \times \mathbb{R} \to \mathbb{R}$ *are continuous functions such that for any* $(t, x) \in [0, \infty) \times \mathbb{R}$:

$$f_1(t, x) < f_2(t, x).$$

Then $X_1(t) < X_2(t)$ *a.s. for any* $t > 0$.

Proof The proof is straightforward. Denote

$$\Delta(t) := X_2(t) - X_1(t) = \int_0^t (f_2(s, X_2(s)) - f_1(s, X_1(s)))\,ds, \quad t \geq 0,$$

and observe that $\Delta(0) = 0$ and that the function Δ is differentiable with

$$\Delta'_+(0) = f_2(0, X(0)) - f_1(0, X(0)) > 0.$$

It is clear that $\Delta(t) = \Delta'_+(0)t + o(t)$, $t \to 0+$, whence there exists the maximal interval $(0, t^*) \subset (0, \infty)$ such that $\Delta(t) > 0$ for all $t \in (0, t^*)$. Assume that $t^* < \infty$. By the definition of t^* and continuity of Δ, $\Delta(t^*) = 0$. Hence $X_1(t^*) = X_2(t^*) = X^*$ and

$$\Delta'(t^*) = f_2(t^*, X^*) - f_1(t^*, X^*) > 0.$$

As $\Delta(t) = \Delta'(t^*)(t - t^*) + o(t - t^*)$, $t \to t^*$, there exists such $\varepsilon > 0$ that $\Delta(t) < 0$ for all $t \in (t^* - \varepsilon, t^*)$ which contradicts the definition of t^*. Therefore $t^* = \infty$ and for all $t > 0$:

$$X_1(t) < X_2(t).$$

Theorem 3.6 *The solution (Y_0, L_0) to the Skorokhod reflection problem (3.43) exists and, with probability 1,*

$$\sup_{t\in[0,T]} |Y_\varepsilon(t) - Y_0(t)| \to 0, \quad \sup_{t\in[0,T]} \left| \int_0^t \frac{\varepsilon}{(Y_\varepsilon(s) - \varphi(s))^\gamma} ds - L_0(t) \right| \to 0 \quad (3.46)$$

as $\varepsilon \downarrow 0$, where Y_ε is the solution to (3.41).

Proof Fix an arbitrary path $\omega \in \Omega$ such that $Z(\omega, t)$ is λ-Hölder continuous with respect to t (in what follows, ω in brackets will be omitted). For any fixed t, $Y_\varepsilon(t)$ is non-increasing w.r.t. ε by Lemma 3.2 and hence the limit

$$Y_0(t) := \lim_{\varepsilon\downarrow 0} Y_\varepsilon(t)$$

is well-defined. Since a is continuous,

$$a(s, Y_\varepsilon(s)) \to a(s, Y_0(s)), \quad \varepsilon \downarrow 0.$$

Moreover, (3.42) implies that there exists a constant $C > 0$ such that

$$|a(t, y)| \leq C(1 + |y|), \quad t \in [0, T], \ y \in \mathbb{R},$$

hence, by Lemma 3.2 and Theorem 3.3, for any $\varepsilon \in (0, 1]$ and $s \in [0, T]$,

$$|a(s, Y_\varepsilon(s))| \le C(1 + |Y_\varepsilon(s)|) \le C(1 + |Y_1(s)|) \le C(1 + \Lambda),$$

where $C > 0$, as usually, denotes a constant whose exact value is not important and can vary within the line. Therefore, by the dominated convergence theorem, for any $t \in [0, T]$

$$\int_0^t a(s, Y_\varepsilon(s))ds \to \int_0^t a(s, Y_0(s))ds, \quad \varepsilon \downarrow 0.$$

In particular, this means that the left-hand side of

$$Y_\varepsilon(t) - y_0 + \int_0^t a(s, Y_\varepsilon(s))ds - Z(t) = \int_0^t \frac{\varepsilon}{(Y_\varepsilon(s) - \varphi(s))^\gamma} ds$$

converges for any $t \in [0, 1]$ and hence there exists the limit

$$L_0(t) := \lim_{\varepsilon \downarrow 0} \int_0^t \frac{\varepsilon}{(Y_\varepsilon(s) - \varphi(s))^\gamma} ds.$$

It remains to prove that L_0 is the φ-reflection function for Y_0. For reader's convenience, the further proof will be split into 4 steps.

Step 1. It is easy to see by definition that $L_0(0) = 0$, $L_0(\cdot)$ is non-decreasing and $Y_0(t) \ge \varphi(t)$, $t \in [0, T]$.

Step 2. Let us prove the continuity of L_0 on $(0, T)$. Take $t \in (0, T)$ and assume that $L_0(t+) - L_0(t-) = \ell > 0$ (one-sided limits of L_0, and hence of Y_0, exist due to monotonicity of L_0). Since

$$Y_0(t) = y_0 - \int_0^t a(s, Y_0(s))ds + Z(t) + L_0(t),$$

this implies that $Y_0(t+) - Y_0(t-) = \ell$. Moreover, since L_0 is non-decreasing,

$$L_0(t-) \le L_0(t) \le L_0(t+)$$

which in turn implies that $Y_0(t-) \le Y_0(t) \le Y_0(t+)$.

Now, there are only two possible cases.

Case 1: $Y_0(t-) - \varphi(t-) = Y_0(t-) - \varphi(t) = y > 0$. Since the left-sided limit $Y_0(t-)$ exists and φ is continuous, there exists $\delta > 0$ such that for all $s \in [t - \delta, t]$ $Y_0(s) - \varphi(s) > \frac{y}{2} > 0$. Moreover, since Y_0 is assumed to have a positive jump in t and $Y_0(t+)$ exists, one can choose $\delta > 0$ such that $Y_0(s) - \varphi(s) > \frac{y}{2} > 0$ for all $s \in [t - \delta, t + \delta]$. Thus for any $t_1, t_2 \in [t - \delta, t + \delta]$

$$L_0(t_2) - L_0(t_1) = \lim_{\varepsilon \downarrow 0} \int_{t_1}^{t_2} \frac{\varepsilon}{(Y_\varepsilon(s) - \varphi(s))^\gamma} ds \le \lim_{\varepsilon \downarrow 0} \int_{t_1}^{t_2} \frac{\varepsilon}{(Y_0(s) - \varphi(s))^\gamma} ds = 0,$$

and hence, in such case, $L_0(t-) = L_0(t+) = L_0(t)$ that contradicts to the assumption $L_0(t+) - L_0(t-) = \ell > 0$.

Case 2: $Y_0(t-) - \varphi(t) = 0$, $Y_0(t+) - \varphi(t) = \ell > 0$. Choose ε_1, $\delta_1 > 0$ such that $\varepsilon_1 < 1, t + \delta_1 < T$ and

$$\begin{aligned} &\varepsilon_1 + 2^\gamma (K + \Lambda)\delta_1^\gamma + 2\delta_1 \\ &\quad + 2C \left(1 + M_1(1, T) + M_2(1, T)\Lambda + 2 \max_{s \in [0,T]} |\varphi(s)| \right) \delta_1 < \frac{\ell}{2}, \end{aligned} \tag{3.47}$$

where K is such that $|\varphi(s_1) - \varphi(s_2)| \le K|s_1 - s_2|^\lambda$, $s_1, s_2 \in [0, T]$, Λ is from **(Z2)**, C is such that $|a(s, y)| \le C(1 + |y|)$ and $M_1(1, T)$, $M_2(1, T)$ are such that

$$\sup_{s \in [0,T]} |Y_1(s)| \le M_1(1, T) + M_2(1, T)\Lambda.$$

Next, note that there exists $\delta_2 < \delta_1$ such that $Y_0(t - \delta_2) - \varphi(t - \delta_2) < \varepsilon_1$ and $Y_0(t + \delta_2) - \varphi(t + \delta_2) > \frac{\ell}{2}$. Moreover, there exists $\varepsilon_2 < \varepsilon_1^\gamma$ such that $Y_{\varepsilon_2}(t - \delta_2) - \varphi(t - \delta_2) < \varepsilon_1$ and, since

$$Y_{\varepsilon_2}(t + \delta_2) - \varphi(t + \delta_2) \ge Y_0(t + \delta_2) - \varphi(t + \delta_2) > \frac{\ell}{2}, \tag{3.48}$$

one can define

$$\tau := \sup\{s \in (t - \delta_2, t + \delta_2) \mid Y_{\varepsilon_2}(s) - \varphi(s) = \varepsilon_1\}.$$

Due to continuity, $Y_{\varepsilon_2}(\tau) - \varphi(\tau) = \varepsilon_1$ and, by the definition of τ, $Y_{\varepsilon_2}(s) - \varphi(s) \ge \varepsilon_1$ for all $s \in [\tau, t + \delta_2)$. Therefore

$$\begin{aligned} Y_{\varepsilon_2}(t + \delta_2) &= Y_{\varepsilon_2}(\tau) + \int_\tau^{t+\delta_2} \frac{\varepsilon_2}{(Y_{\varepsilon_2}(s) - \varphi(s))^\gamma} ds \\ &\quad - \int_\tau^{t+\delta_2} a(s, Y_{\varepsilon_2}(s)) ds + Z(t + \delta_2) - Z(\tau) \\ &= \varphi(t + \delta_2) + \big(Y_{\varepsilon_2}(\tau) - \varphi(\tau)\big) + (\varphi(\tau) - \varphi(t + \delta_2)) \\ &\quad + \int_\tau^{t+\delta_2} \frac{\varepsilon_2}{(Y_{\varepsilon_2}(s) - \varphi(s))^\gamma} ds \end{aligned}$$

$$
\begin{aligned}
&- \int\limits_{\tau}^{t+\delta_2} a(s, Y_{\varepsilon_2}(s))ds + Z(t+\delta_2) - Z(\tau) \\
&\le \varphi(t+\delta_2) + \varepsilon_1 + K(t+\delta_2-\tau)^\gamma + \frac{\varepsilon_2}{\varepsilon_1^\gamma}(t+\delta_2-\tau) \\
&\quad + C\left(1 + \sup_{s\in[0,T]} |Y_{\varepsilon_2}(s)|\right)(t+\delta_2-\tau) + \Lambda(t+\delta_2-\tau)^\gamma .
\end{aligned}
$$

Note that

$$
\begin{aligned}
\sup_{s\in[0,T]} |Y_{\varepsilon_2}(s)| &\le \sup_{s\in[0,T]} |Y_1(s)| + 2\max_{s\in[0,T]} |\varphi(s)| \\
&\le M_1(1,T) + M_2(1,T)\Lambda + 2\max_{s\in[0,T]} |\varphi(s)|
\end{aligned}
$$

and, consequently, by (3.47) and the fact that $\frac{\varepsilon_2}{\varepsilon_1^\gamma} \le 1$,

$$
\begin{aligned}
Y_{\varepsilon_2}(t+\delta_2) - \varphi(t+\delta_2) &\le \varepsilon_1 + 2^\gamma(K+\Lambda)\delta_1^\gamma + 2\delta_1 \\
&\quad + 2C\left(1 + M_1(1,T) + M_2(1,T)\Lambda + 2\max_{s\in[0,T]} |\varphi(s)|\right)\delta_1 \\
&< \frac{\ell}{2},
\end{aligned}
$$

which contradicts to (3.48).

Contradictions in both cases above prove that L_0 is continuous at any point $t \in (0,T)$.

Step 3. Let us show that L_0 is continuous in 0 and in T.

Left-continuity in T. Let $\widetilde{T} > T$. Denote

$$
\widetilde{\varphi}(t) = \begin{cases} \varphi(t), & t \in [0,T], \\ \varphi(T), & t \in [T,\widetilde{T}], \end{cases} \qquad \widetilde{Z}(t) = \begin{cases} Z(t), & t \in [0,T], \\ Z(T), & t \in [T,\widetilde{T}], \end{cases}
$$

$$
\widetilde{a}(t,y) = \begin{cases} a(t,y), & t \in [0,T], \\ a(T,y), & t \in [T,\widetilde{T}], \end{cases}
$$

and consider

$$
\widetilde{Y}_\varepsilon(t) = y_0 + \int\limits_0^t \frac{\varepsilon}{\left(\widetilde{Y}_\varepsilon(s) - \widetilde{\varphi}(s)\right)^\gamma} ds - \int\limits_0^t \widetilde{a}(s, \widetilde{Y}_\varepsilon(s))ds + \widetilde{Z}(t).
$$

Using similar arguments as above, one can prove that $\widetilde{Y}_0(t) := \lim_{\varepsilon\downarrow 0} \widetilde{Y}_\varepsilon(t)$ and $\widetilde{L}_0(t) := \lim_{\varepsilon\downarrow 0} \int_0^t \frac{\varepsilon}{(\widetilde{Y}_\varepsilon(s)-\widetilde{\varphi}(s))^\gamma} ds$ are well-defined and continuous at any point $t \in (0, \widetilde{T})$. Moreover, $\widetilde{Y}_\varepsilon$, $\widetilde{L}_0$ and $\widetilde{Y}_0(t)$ coincide with Y_ε, L_0 and Y_0 respectively on $[0, T]$ and whence L_0 and Y_0 are left-continuous in $T \in (0, \widetilde{T})$.

Right-continuity in 0. By Lemma 3.2, each Y_ε exceeds the process U defined by

$$U(t) = y_0 - \int_0^t a(s, U(s))ds + Z(t).$$

Define $\tau := \inf\{t \in [0, T] \mid U(t) - \varphi(t) = y_0/2\}$. Then, for any $t \in [0, \tau]$ $Y_\varepsilon(t) - \varphi(t) \geq U(t) - \varphi(t) \geq \frac{y_0}{2}$ and hence

$$L_0(t) = \lim_{\varepsilon\downarrow 0} \int_0^t \frac{\varepsilon}{(Y_\varepsilon(s) - \varphi(s))^\gamma} ds \leq \lim_{\varepsilon\downarrow 0} \frac{2^\gamma \tau}{y_0^\gamma} \varepsilon = 0,$$

i.e. $L_0(0+) = 0 = L_0(0)$.

Step 4. It remains to prove that L_0 has points of growth only in such $t \in [0, T]$ where $Y_0(t) = \varphi(t)$. Let t be such that $Y_0(t) - \varphi(t) = y > 0$. Then, by continuity of Y_0, there exists an interval $[t - \delta, t + \delta]$ such that $Y_0(s) - \varphi(s) > \frac{y}{2}$ for all $s \in [t - \delta, t + \delta]$ and hence, for any $\varepsilon > 0$, $Y_\varepsilon(s) - \varphi(s) > \frac{y}{2}$. Therefore,

$$L(t + \delta) - L(t - \delta) = \lim_{\varepsilon\downarrow 0} \int_{t-\delta}^{t+\delta} \frac{\varepsilon}{(Y_\varepsilon(s) - \varphi(s))^\gamma} ds \leq \lim_{\varepsilon\downarrow 0} \frac{2^{1+\gamma}\delta}{y^\gamma} \varepsilon = 0,$$

i.e. t is not a point of growth for L_0.

Therefore, L_0 is a φ-reflection function for Y_0 and the pair (Y_0, L_0) is the unique solution to the Skorokhod reflection problem (3.43) as required. Note that the uniform convergence from (3.46) immediately follows from the continuity of L_0 (and hence Y_0) and the pointwise convergence established above.

3.5 Two-Sided Sandwiched Process

Previously, we discussed the situation when the drift exploded near the line $\{\varphi(t), t \in [0, T]\}$: this structure ensured that the solution of (3.1) is "repelled" from φ and stays above it. It is possible to exploit such an effect even further: if we assume that b has an explosive decrease to $-\infty$ in the proximity of some other function ψ such that $\psi(t) > \varphi(t)$, $t \in [0, T]$, then we can expect that the solution will be "*sandwiched*" between φ and ψ.

In this section, we formalize this idea and get the bounds of the type (3.39). In some sense, such a "sandwiched" process may be preferable for modeling volatility: the lower bound can be used to ensure the positivity of paths whereas the upper bound would guarantee the sufficient integrability of the price.

Let the noise Z satisfy Assumption 3.1 for some $\lambda \in (0, 1)$. Fix two continuous functions $\varphi, \psi\colon [0, T] \to \mathbb{R}$ such that $\varphi(t) < \psi(t)$, $t \in [0, T]$, and

$$|\varphi(t_2) - \varphi(t_1)| + |\psi(t_2) - \psi(t_1)| \le L_{\varphi,\psi}|t_2 - t_1|^{\lambda}, \quad t_1, t_2 \in [0, T], \tag{3.49}$$

take a value $y_0 \in (\varphi(0), \psi(0))$ and, for arbitrary $x_1, x_2 \in \mathbb{R}$ such that

$$\max_{t\in[0,T]} \varphi(t) + x_1 < \min_{t\in[0,T]} \psi(t) - x_2,$$

denote

$$\begin{aligned} \mathcal{D}_{x_1,x_2} &:= \{(t, y) \mid t \in [0, T], y \in (\varphi(t) + x_1, \psi(t) - x_2)\}, \\ \overline{\mathcal{D}_{x_1,x_2}} &:= \{(t, y) \mid t \in [0, T], y \in [\varphi(t) + x_1, \psi(t) - x_2]\}. \end{aligned} \tag{3.50}$$

Finally, consider the integro-differential equation (3.1) with the Volterra kernel $\mathcal{G}$: $[0, T]^2 \to \mathbb{R}$ satisfying Assumption 3.4 and functions $b\colon \mathcal{D}_{0,0} \to \mathbb{R}$ and $a\colon \overline{\mathcal{D}_{0,0}} \to \mathbb{R}$ satisfying the following assumptions.

Assumption 3.5

- **(B1)** $b \in C(\mathcal{D}_{0,0})$;
- **(B2)** there exists a Borel-measurable mapping

$$\varepsilon \in \left(0, \frac{1}{2} \min_{t\in[0,T]} |\psi(t) - \varphi(t)|\right) \mapsto c_\varepsilon \in (0, \infty)$$

  such that, for any $\varepsilon \in (0, \frac{1}{2} \min_{t\in[0,T]} |\psi(t) - \varphi(t)|)$,

$$|b(t, y_1) - b(t, y_2)| \le c_\varepsilon |y_1 - y_2|, \quad (t, y_1), (t, y_2) \in \mathcal{D}_{\varepsilon,\varepsilon};$$

- **(B3)** there are constants $y_* \in (0, \frac{1}{2} \min_{t\in[0,T]} |\psi(t) - \varphi(t)|)$, $c > 0$ and $\gamma > 0$ that for all $(t, y) \in \mathcal{D}_{0,0} \setminus \mathcal{D}_{y_*,0}$:

$$b(t, y) \ge \frac{c}{(y - \varphi(t))^\gamma},$$

  and for all $(t, y) \in \mathcal{D}_{0,0} \setminus \mathcal{D}_{0,y_*}$:

$$b(t, y) \le -\frac{c}{(\psi(t) - y)^\gamma}.$$

- **(B4)** the constant γ from assumption **(B3)** satisfies the condition

$$\gamma > \frac{1-\lambda}{\lambda}$$

with λ being the order of Hölder continuity of φ, ψ and the paths of Z.

Assumption 3.6 The function a is continuous on $\overline{\mathcal{D}_{0,0}}$ and, for any $(t, y_1), (t, y_2) \in \overline{\mathcal{D}_{0,0}}$,

$$|a(t, y_2) - a(t, y_1)| \le L_a |y_2 - y_1|.$$

Example 3.12 Let $\alpha_1\colon [0, T] \to (0, \infty)$, $\alpha_2\colon [0, T] \to (0, \infty)$ and $\alpha_3\colon \mathcal{D}_{0,0} \to \mathbb{R}$ be continuous and

$$|\alpha_3(t, y_1) - \alpha_3(t, y_2)| \le C|y_1 - y_2|, \quad (t, y_1), (t, y_2) \in \mathcal{D}_{0,0},$$

for some constant $C > 0$. Then

$$b(t, y) := \frac{\alpha_1(t)}{(y - \varphi(t))^\gamma} - \frac{\alpha_2(t)}{(\psi(t) - y)^\gamma} - \alpha_3(t, y), \quad t \in [0, T], y \in \mathcal{D}_{0,0},$$

satisfies Assumption 3.5 provided that $\gamma > \frac{1-\lambda}{\lambda}$.

Theorem 3.7 *Let Assumptions 3.1, 3.4, 3.5 and 3.6 hold. Then the stochastic integro-differential equation* (3.1) *has a unique solution Y with a.s. continuous paths. Moreover, with probability 1,*

$$\varphi(t) < Y(t) < \psi(t), \quad t \in [0, T]. \tag{3.51}$$

Proof For any $\varepsilon \in (0, \min\{y_0 - \varphi(0), \psi(0) - y_0\}$ with y_0 being the initial value in (3.1), denote

$$b_\varepsilon(t, y) := \begin{cases} b(t, y), & (t, y) \in \mathcal{D}_{\varepsilon,\varepsilon}, \\ b(t, \varphi(t) + \varepsilon), & (t, y) \in \mathcal{D}_{0,0} \setminus \mathcal{D}_{\varepsilon,0}, \\ b(t, \psi(t) - \varepsilon), & (t, y) \in \mathcal{D}_{0,0} \setminus \mathcal{D}_{0,\varepsilon}. \end{cases}$$

Consider a stochastic integrodifferential equation of the form

$$Y_\varepsilon(t) = y_0 + \int_0^t b_\varepsilon(s, Y_\varepsilon(s))ds + \int_0^t \mathcal{G}(t, s)a(s, Y_\varepsilon(s))ds + Z(t). \tag{3.52}$$

Observe that $b_\varepsilon \in C([0, T] \times \mathbb{R})$ and, moreover, there exists a constant $L_\varepsilon > 0$ such that for any $t \in [0, T]$ and $y_1, y_2 \in \mathbb{R}$

$$|b_\varepsilon(t, y_2) - b_\varepsilon(t, y_1)| \le L_\varepsilon |y_2 - y_1|.$$

Therefore, by treating the Eq. (3.21) pathwisely, one can deduce using Theorem 3.1 that (3.21) has a unique strong a.s. continuous solution $Y_\varepsilon = \{Y_\varepsilon(t),\ t \in [0, T]\}$.

The rest of the proof is similar to Theorem 3.2. Fix $\omega \in \Omega$ such that $Z(\omega, t)$ is λ-Hölder continuous w.r.t. t, denote

$$M_1 := 2^{-\frac{\gamma\lambda-\lambda+1}{\gamma\lambda+\lambda-1}} \left(\frac{c^{\frac{\lambda}{1-\lambda}}}{\lambda^{\frac{\lambda}{1-\lambda}} - \lambda^{\frac{1}{1-\lambda}}} \right)^{\frac{\gamma\lambda+\lambda-1}{1-\lambda}} > 0,$$

where c is the constant from Assumption 3.5(B3) and λ, γ are from Assumption 3.5(B4) and put

$$M_2 := \left(\frac{2D}{\min\{y_*, y_0 - \varphi(0), \psi(0) - y_0\}} \right)^{\gamma\lambda+\lambda-1} + L_{\varphi,\psi} + T^{\frac{1}{2}} \ell_\lambda \sup_{(s,y)\in\overline{\mathcal{D}_{0,0}}} |a(s, y)| \tag{3.53}$$

with

- $y_0 \in (\varphi(0), \psi(0))$ being the initial value in the equation (3.1),
- y_* being from Assumption 3.5(B3),
- $L_{\varphi,\psi} > 0$ being from (3.49) and
- ℓ_λ being from Assumption 3.1.

Finally, define

$$\xi := \frac{M_1}{(M_2 + \Lambda)^{\frac{1}{\gamma\lambda+\lambda-1}}}$$

and observe that $2\xi < \min\{y_*, y_0 - \varphi(0), \psi(0) - y_0\}$. In particular, this means that $y_0 \in \mathcal{D}_{\xi,\xi}$ and one can define

$$\tau_\xi := \sup\{t \in [0, T] \mid (t, Y_\xi(t)) \in \overline{\mathcal{D}_{\xi,\xi}}\} \in [0, T].$$

Note that Y coincides with Y_ξ on $[0, \tau_\xi]$, so our goal is to prove that $\tau_\xi = T$. Assume that is is not true, i.e. $\tau_\xi < T$. Since $2\xi < \min\{y_0 - \varphi(0), \psi(0) - y_0\}$, the path of Y_ξ must cross either $\varphi + 2\xi$ or $\psi - 2\xi$ before the moment τ_ξ and one can define

$$\tau_{2\xi} := \sup\{t \in [0, \tau_\xi] \mid (t, Y_\xi(t)) \in \overline{\mathcal{D}_{2\xi,2\xi}}\}.$$

Clearly, $0 < \tau_{2\xi} < \tau_\xi$ and, for any $t \in [0, \tau_\xi]$

$$|a(t, Y_\xi(t))| \le \sup_{(s,y)\in\overline{\mathcal{D}_{0,0}}} |a(s, y)|.$$

Moreover, for any $t \in [\tau_{2\xi}, \tau_\xi]$, either

$$Y_\xi(t) \in [\varphi(t) + \xi, \varphi(t) + 2\xi], \quad Y_\xi(\tau_{2\xi}) = \varphi(\tau_{2\xi}) + 2\xi, \quad Y_\xi(\tau_\xi) = \varphi(\tau_\xi) + \xi, \tag{3.54}$$

or

$$Y_\xi(t) \in [\psi(t) - 2\xi, \psi(t) - \xi], \quad Y_\xi(\tau_{2\xi}) = \psi(\tau_{2\xi}) - 2\xi, \quad Y_\xi(\tau_\xi) = \psi(\tau_\xi) - \xi. \tag{3.55}$$

If (3.54) is true, then, just like in Theorem 3.2,

$$\begin{aligned}
\xi &= Y_\xi(\tau_\xi) - \varphi(\tau_\xi) \\
&= y_0 - \varphi(\tau_\xi) + \int_0^{\tau_\xi} b_\xi(s, Y_\xi(s))ds \\
&\quad + \int_0^{\tau_\xi} \mathcal{G}(\tau_\xi, s)a(s, Y_\xi(s))ds + Z(\tau_\xi) \\
&= -\varphi(\tau_\xi) + Y_\xi(\tau_{2\xi}) + \int_{\tau_{2\xi}}^{\tau_\xi} b_\xi(s, Y_\xi(s))ds \\
&\quad + \int_0^{\tau_\xi} (\mathcal{G}(\tau_\xi, s) - \mathcal{G}(\tau_{2\xi}, s))a(s, Y_\xi(s))ds + Z(\tau_\xi) - Z(\tau_{2\xi}) \\
&= 2\xi + \varphi(\tau_{2\xi}) - \varphi(\tau_\xi) + \int_{\tau_{2\xi}}^{\tau_\xi} b_\xi(s, Y_\xi(s))ds \\
&\quad + \int_0^{\tau_\xi} (\mathcal{G}(\tau_\xi, s) - \mathcal{G}(\tau_{2\xi}, s))a(s, Y_\xi(s))ds + Z(\tau_\xi) - Z(\tau_{2\xi}) \\
&\geq 2\xi - L_{\varphi,\psi}(\tau_\xi - \tau_{2\xi})^\lambda + \frac{c}{(2\xi)^\gamma}(\tau_\xi - \tau_{2\xi}) - \Lambda(\tau_\xi - \tau_{2\xi})^\lambda \\
&\quad - \left(\int_0^{\tau_\xi} (\mathcal{G}(\tau_\xi, s) - \mathcal{G}(\tau_{2\xi}, s))^2 ds \right)^{\frac{1}{2}} \left(\int_0^{\tau_\xi} |a(s, Y_\xi(s))|^2 ds \right)^{\frac{1}{2}} \\
&\geq 2\xi - L_{\varphi,\psi}(\tau_\xi - \tau_{2\xi})^\lambda + \frac{c}{(2\xi)^\gamma}(\tau_\xi - \tau_{2\xi}) - \Lambda(\tau_\xi - \tau_{2\xi})^\lambda \\
&\quad - T^{\frac{1}{2}} \ell_\lambda \sup_{(s,y)\in\overline{\mathcal{D}_{0,0}}} |a(s, y)|(\tau_\xi - \tau_{2\xi})^\lambda \\
&\geq 2\xi + \frac{c}{(2\xi)^\gamma}(\tau_\xi - \tau_{2\xi}) - (M_2 + \Lambda)(\tau_\xi - \tau_{2\xi})^\lambda.
\end{aligned}$$

Therefore, if (3.54) holds,

$$\xi \geq 2\xi + \frac{c}{(2\xi)^\gamma}(\tau_\xi - \tau_{2\xi}) - (M_2 + \Lambda)(\tau_\xi - \tau_{2\xi})^\lambda,$$

i.e.

$$\xi + \frac{c}{(2\xi)^\gamma}(\tau_\xi - \tau_{2\xi}) - (M_2 + \Lambda)(\tau_\xi - \tau_{2\xi})^\lambda \le 0. \tag{3.56}$$

Similarly, if (3.55) is true, we have

$$\begin{aligned}
-\xi &= Y_\xi(\tau_\xi) - \psi(\tau_\xi) \\
&= y_0 - \psi(\tau_\xi) + \int_0^{\tau_\xi} b_\xi(s, Y_\xi(s))ds \\
&\quad + \int_0^{\tau_\xi} \mathcal{G}(\tau_\xi, s)a(s, Y_\xi(s))ds + Z(\tau_\xi) \\
&= -\psi(\tau_\xi) + Y_\xi(\tau_{2\xi}) + \int_{\tau_{2\xi}}^{\tau_\xi} b_\xi(s, Y_\xi(s))ds \\
&\quad + \int_0^{\tau_\xi} (\mathcal{G}(\tau_\xi, s) - \mathcal{G}(\tau_{2\xi}, s))a(s, Y_\xi(s))ds + Z(\tau_\xi) - Z(\tau_{2\xi}) \\
&= -\psi(\tau_\xi) + \psi(\tau_{2\xi}) - 2\xi + \int_{\tau_{2\xi}}^{\tau_\xi} b_\xi(s, Y_\xi(s))ds \\
&\quad + \int_0^{\tau_\xi} (\mathcal{G}(\tau_\xi, s) - \mathcal{G}(\tau_{2\xi}, s))a(s, Y_\xi(s))ds + Z(\tau_\xi) - Z(\tau_{2\xi}) \\
&\le -2\xi + L_{\varphi,\psi}(\tau_\xi - \tau_{2\xi})^\lambda - \frac{c}{(2\xi)^\gamma}(\tau_\xi - \tau_{2\xi}) + \Lambda(\tau_\xi - \tau_{2\xi})^\lambda \\
&\quad + \left(\int_0^{\tau_\xi} (\mathcal{G}(\tau_\xi, s) - \mathcal{G}(\tau_{2\xi}, s))^2 ds\right)^{\frac{1}{2}} \left(\int_0^{\tau_\xi} |a(s, Y_\xi(s))|^2 ds\right)^{\frac{1}{2}} \\
&\le 2\xi + L_{\varphi,\psi}(\tau_\xi - \tau_{2\xi})^\lambda - \frac{c}{(2\xi)^\gamma}(\tau_\xi - \tau_{2\xi}) + \Lambda(\tau_\xi - \tau_{2\xi})^\lambda \\
&\quad + T^{\frac{1}{2}}\ell_\lambda \sup_{(s,y)\in\overline{\mathcal{D}_{0,0}}} |a(s, y)|(\tau_\xi - \tau_{2\xi})^\lambda \\
&\le 2\xi - \frac{c}{(2\xi)^\gamma}(\tau_\xi - \tau_{2\xi}) + (M_2 + \Lambda)(\tau_\xi - \tau_{2\xi})^\lambda.
\end{aligned}$$

and hence (3.56) also holds.

Next, consider a function

$$F_\xi(x) := \xi + \frac{c}{(2\xi)^\gamma}x - (M_2 + \Lambda)\,x^\lambda, \quad x \ge 0.$$

It is easy to check that F_ξ attains its minimal value at

$$x_* := \left(\frac{\lambda 2^\gamma (M_2 + \Lambda)}{c}\right)^{\frac{1}{1-\lambda}} \xi^{\frac{\gamma}{1-\lambda}}$$

and, taking into account the definition of ξ,

$$F_\xi(x_*) = \xi - \left(\frac{2^{\gamma\lambda}(C + \Lambda)}{c^\lambda}\right)^{\frac{1}{1-\lambda}} \left(\lambda^{\frac{\lambda}{1-\lambda}} - \lambda^{\frac{1}{1-\lambda}}\right) \xi^{\frac{\gamma\lambda}{1-\lambda}} = \frac{\xi}{2} > 0,$$

which contradicts (3.56). Therefore, $(t, Y_\xi(t)) \in \overline{\mathcal{D}_{\xi,\xi}}$ for all $t \in [0, T]$ and (3.1) has a unique solution Y which coincides with Y_ξ. In particular, (3.51) holds.

Just like in the one-sided case described in Section 3.3, the proof of Theorem 3.7 reveals a refinement of (3.51).

Theorem 3.8 *Let Assumptions 3.1, 3.4, 3.5 and 3.6 hold and let* $Y = \{Y(t),\ t \in [0, T]\}$ *be the unique continuous solution to stochastic integrodifferential equation* (3.1). *Then, for all* $\omega \in \Omega$ *such that the corresponding path* $\{Z(\omega, t),\ t \in [0, T]\}$ *is* λ*-Hölder continuous, there exists a deterministic constant* C_1 *that does not depend on the noise* Z *such that*

$$\varphi(t) + \frac{C_1}{(1 + \Lambda)^{\frac{1}{\gamma\lambda+\lambda-1}}} \le Y(t) \le \psi(t) - \frac{C_1}{(1 + \Lambda)^{\frac{1}{\gamma\lambda+\lambda-1}}}, \quad t \in [0, T]. \tag{3.57}$$

Corollary 3.2 *Let Assumptions 3.1, 3.4, 3.5 and 3.6 hold, let* $Y = \{Y(t),\ t \in [0, T]\}$ *be the unique continuous solution to stochastic integro-differential equation* (3.1) *and fix* $r > 0$. *If* Λ *in* (3.2) *can be chosen such that* $\mathbb{E}[\Lambda^{\frac{r}{\gamma\lambda+\lambda-1}}] < \infty$, *then*

$$\mathbb{E}\left[\sup_{t\in[0,T]} \frac{1}{(Y(t) - \varphi(t))^r}\right] < \infty, \quad \mathbb{E}\left[\sup_{t\in[0,T]} \frac{1}{(\psi(t) - Y(t))^r}\right] < \infty.$$

Definition 3.2 Given the property (3.51) under Assumptions 3.1, 3.4, 3.5 and 3.6, we will call the solution to the stochastic integro-differential equation (3.1) a **two-sided sandwiched process**.

Finally, we state the two-sided version of Theorem 3.4. The proof of is almost identical to the one-sided version, so we will leave it as an exercise.

Theorem 3.9 *Let Assumptions 3.1, 3.5, 3.6 and 3.4 hold. Then, with probability 1, the continuous solution* $Y = \{Y(t),\ t \in [0, T]\}$ *to the stochastic integro-differential equation* (3.1) *is* λ*-Hölder continuous, i.e. for all* $0 \le t_1 \le t_2 \le T$ *there is a random variable* $\Upsilon > 0$ *such that*

$$|Y(t_2) - Y(t_1)| \le \Upsilon |t_2 - t_1|^\lambda. \tag{3.58}$$

Moreover, it is possible to take Υ *of the form*

$$\Upsilon = C(1 + \Lambda + c_\xi + c_\xi \Lambda),$$

where $C > 0$ *is some deterministic constant,* Λ *is from* (3.2),

$$\xi := \frac{C_1}{(1+\Lambda)^{\frac{1}{\gamma\lambda+\lambda-1}}},$$

with C_1 *being from Theorem 3.8, and* c_ξ *being the evaluation in the (random) point* ξ *of the mapping* $\varepsilon \in\mapsto c_\varepsilon$ *from Assumption 3.5(B2).*

References

1. Azmoodeh, E., Sottinen, T., Viitasaari, L., Yazigi, A.: Necessary and sufficient conditions for Hölder continuity of Gaussian processes. Statist. Probab. Lett. **94**, 230–235 (2014). https://doi.org/10.1016/j.spl.2014.07.030
2. Benassi, A., Jaffard, S., Roux, D.: Elliptic Gaussian random processes. Rev. Matemát. Iberoam. **13**(1), 19–90 (1997). http://eudml.org/doc/39530
3. Boguslavskaya, E., Mishura, Y., Shevchenko, G.: Replication of Wiener-transformable stochastic processes with application to financial markets with memory. In: Silvestrov, S., Malyarenko, A., Rančić, M. (eds.) Stochastic Processes and Applications, pp. 335–361. Springer International Publishing, Cham (2018)
4. Di Nunno, G., Mishura, Y., Yurchenko-Tytarenko, A.: Sandwiched SDEs with unbounded drift driven by Hölder noises. Adv. Appl. Probab. **55**(3) (2023)
5. Dozzi, M., Kozachenko, Y., Mishura, Y., Ralchenko, K.: Asymptotic growth of trajectories of multifractional Brownian motion, with statistical applications to drift parameter estimation. Stat. Inference Stoch. Process. **21**(1), 21–52 (2018). https://doi.org/10.1007/s11203-016-9147-z
6. Garsia, A., Rodemich, E., Rumsey, H.: A real variable lemma and the continuity of paths of some Gaussian processes. Indiana Univ. Math. J. **20**, 565–578 (1970)
7. Mishura, Y., Shevchenko, G., Shklyar, S.: Gaussian processes with Volterra kernels. In: Malyarenko, A., Ni, Y., Rančić, M., Silvestrov, S. (eds.) Stochastic Processes, Statistical Methods, and Engineering Mathematics, pp. 249–276. Springer International Publishing, Cham (2022)
8. Norros, I., Valkeila, E., Virtamo, J.: An elementary approach to a Girsanov formula and other analytical results on fractional Brownian motions. Bernoulli **5**(4), 571–587 (1999). https://doi.org/10.2307/3318691
9. Peltier, R.F., Lévy Véhel, J.: Multifractional Brownian motion: definition and preliminary results. INRIA **RR-2645** (1995). https://inria.hal.science/inria-00074045v1
10. Skorokhod, A.V.: Stochastic equations for diffusion processes in a bounded region. Theory Prob. Appl. **6**(3), 264–274 (1961). https://doi.org/10.1137/1106035
11. Skorokhod, A.V.: Stochastic equations for diffusion processes in a bounded region. II. Theory Prob. Appl. **7**(1), 3–23 (1962). https://doi.org/10.1137/1107002
12. Stoev, S.A., Taqqu, M.S.: How rich is the class of multifractional Brownian motions? Stoch. Process. Appl. **116**(2), 200–221 (2006). https://doi.org/10.1016/j.spa.2005.09.007

Chapter 4
Volatility Models with Explosive Drifts

Abstract In this chapter, we build on the explosive drift mechanism introduced earlier and define a class of generalized Cox–Ingersoll–Ross (GCIR) and Chan–Karolyi–Longstaff–Sanders (GCKLS) processes driven by Hölder-continuous noises. We discuss existence, uniqueness, and moment bounds for these processes as well as examine the behavior of these models under extremely rough driving noises.

4.1 CIR and CKLS Processes Driven by Hölder Continuous Noises

The structure of the Eq. (3.1) is quite similar to the Eq. (2.10) for the square root of the CIR process: both (2.10) and (3.1) have an additive noise and explosive drift. In fact, this similarity with the CIR dynamics can be exploited to a greater extent allowing to generalize the CIR process for a much broader set of noises. This Chapter is based on [2, 5, 6].

Consider a stochastic differential equation

$$Y(t) = y_0 + \int_0^t \left(\frac{\kappa}{Y^{\frac{\alpha}{1-\alpha}}(s)} - \theta Y(s) \right) ds + Z(t), \quad y_0 > 0, \tag{4.1}$$

where $\kappa, \theta > 0$ are positive constants, $\alpha \in (0, 1)$, and the process Z starts at zero and is λ-Hölder continuous with $\alpha + \lambda > 1$. It is easy to verify that Assumptions 3.2, 3.3 and 3.4 hold with $\gamma = \frac{\alpha}{1-\alpha}$ and $\varphi \equiv 0$, so (4.1) has a unique positive solution. Furthermore, as it is noted in Corollary 3.1, if the corresponding Hölder continuity constant Λ can be chosen to have all positive moments, Y will have moments of all real orders, including the negative ones.

Definition 4.1 We call the process $X = \{X(t),\ t \in [0, T]\}$ of the form

$$X(t) = Y^{\frac{1}{1-\alpha}}(t), \quad t \in [0, T],$$

G. Di Nunno et al., *Volterra Volatility Models*, Springer Finance,
https://doi.org/10.1007/978-3-032-26576-0_4

- a generalized Cox-Ingersoll-Ross (GCIR) process, if $\alpha = \frac{1}{2}$;
- a generalized Chan-Karolyi-Longstaff-Sanders (GCKLS) process, if $\alpha \in \left(0, \frac{1}{2}\right) \cup \left(\frac{1}{2}, 1\right)$.

The intuition behind this definition can be explained in the theorem below.

Theorem 4.1 *Assume that $\lambda > \frac{1}{2}$ and $\alpha \in (0, 1)$ are such that $\alpha + \lambda > 1$. Then, with probability 1, the process X from Definition 4.1 satisfies the equation*

$$X(t) = x_0 + \frac{1}{1-\alpha}\int_0^t (\kappa - \theta X(s))\,ds + \frac{1}{1-\alpha}\int_0^t X^{\alpha}(s)dZ(s), \tag{4.2}$$

where $x_0 := y_0^{\frac{1}{1-\alpha}}$ the integral $\int_0^t X^{\alpha}(s)dZ(s)$ exists as the pathwise limit of Riemann-Stieltjes integral sums.

Proof Fix the partition $0 = t_0 < t_1 < t_2 < \cdots < t_n = t$ where $t \in [0, T]$, $|\Delta t| := \max_{k=1,\dots,n}(t_k - t_{k-1})$. It is clear that

$$X(t) = x_0 + \sum_{k=1}^{n}(X(t_k) - X(t_{k-1})) = x_0 + \sum_{k=1}^{n}(Y^{\frac{1}{1-\alpha}}(t_k) - Y^{\frac{1}{1-\alpha}}(t_{k-1})),$$

so, using the Taylor's expansion, we obtain that

$$\begin{aligned} X(t) = x_0 &+ \sum_{k=1}^{n} \frac{1}{1-\alpha} Y^{\frac{\alpha}{1-\alpha}}(t_{k-1})(Y(t_k) - Y(t_{k-1})) \\ &+ \sum_{k=1}^{n} \frac{\alpha \Theta_{n,k}^{\frac{2\alpha-1}{1-\alpha}}}{2(1-\alpha)^2}(Y(t_k) - Y(t_{k-1}))^2 \end{aligned}$$

with $\Theta_{n,k}$ being a real value between $Y(t_k)$ and $Y(t_{k-1})$. By Theorem 3.3, there are constants C_1, C_2 which do not depend on the partition such that for all n and k

$$\frac{C_1}{(1+\Lambda)^{\frac{1}{\gamma\lambda+\lambda-1}}} \le \Theta_{n,k} \le C_2(1+\Lambda) \tag{4.3}$$

with $\gamma = \frac{\alpha}{1-\alpha}$. In particular, this implies that

$$\sup_{\substack{n\ge 1,\\ k=0,1,\dots,n}} |\Theta_{n,k}|^{\frac{2\alpha-1}{1-\alpha}} < \infty \quad a.s.$$

Moreover, by Theorem 3.4, Y a.s. has λ-Hölder continuous paths, i.e. there is a random variable $\Upsilon > 0$ such that, with probability 1,

$$|Y(t_2) - Y(t_1)| \le \Upsilon |t_2 - t_1|^{\lambda}, \quad t_1, t_2 \in [0, T].$$

In addition, since Y has λ-Hölder continuous trajectories a.s., $\lambda > \frac{1}{2}$, and it is positive a.s., the trajectories of $X^{\alpha} = Y^{\frac{\alpha}{1-\alpha}}$ are also λ-Hölder continuous. Therefore, as $\lambda > \frac{1}{2}$ and Z is λ-Hölder continuous as well, the classical result of Young [7] (see also [3, Section 4.1] or [8]) yields that the integral

$$\int_0^t X^{\alpha}(s) dZ(s)$$

exists as a pathwise limit of Riemann-Stieltjes integral sums. Therefore, with probability 1,

$$\begin{aligned}
&\sum_{k=1}^{n} \frac{\lambda \Theta_{n,k}^{\frac{2\alpha-1}{1-\alpha}}}{2(1-\alpha)^2} (Y(t_k) - Y(t_{k-1}))^2 \\
&\le \frac{\lambda \Upsilon^2}{2(1-\alpha)^2} \sup_{\substack{n \ge 1, \\ k=0,1,\dots,n}} |\Theta_{n,k}|^{\frac{2\alpha-1}{1-\alpha}} \sum_{k=1}^{n} (t_k - t_{k-1})^{2\lambda} \\
&\le \frac{\lambda \Upsilon^2}{2(1-\alpha)^2} \sup_{\substack{n \ge 1, \\ k=0,1,\dots,n}} |\Theta_{n,k}|^{\frac{2\alpha-1}{1-\alpha}} T |\Delta t|^{2\lambda - 1} \\
&\to 0, \quad |\Delta t| \to 0,
\end{aligned}$$

and

$$\begin{aligned}
&\sum_{k=1}^{n} \frac{1}{1-\alpha} Y^{\frac{\alpha}{1-\alpha}}(t_{k-1})(Y(t_k) - Y(t_{k-1})) \\
&= \frac{1}{1-\alpha} \sum_{k=1}^{n} X^{\alpha}(t_{k-1})(Y(t_k) - Y(t_{k-1})) \\
&= \frac{1}{1-\alpha} \sum_{k=1}^{n} X^{\alpha}(t_{k-1}) \int_{t_{k-1}}^{t_k} \left(\frac{\kappa}{Y(s)^{\frac{\alpha}{1-\alpha}}} - \theta Y(s) \right) ds \\
&\quad + \frac{1}{1-\alpha} \sum_{k=1}^{n} X^{\alpha}(t_{k-1})(Z(t_k) - Z(t_{k-1})) \qquad (4.4) \\
&= \frac{1}{1-\alpha} \sum_{k=1}^{n} X^{\alpha}(t_{k-1}) \int_{t_{k-1}}^{t_k} \left(\frac{\kappa}{X^{\alpha}(s)} - \theta X^{1-\alpha}(s) \right) ds \\
&\quad + \frac{1}{1-\alpha} \sum_{k=1}^{n} X^{\alpha}(t_{k-1})(Z(t_k) - Z(t_{k-1})) \\
&\to \frac{1}{1-\alpha} \int_0^t (\kappa - \theta X(s)) ds + \frac{1}{1-\alpha} \int_0^t X^{\alpha}(s) dZ(s), \quad |\Delta t| \to 0.
\end{aligned}$$

Taking into account all of the above, the X satisfies the stochastic differential equation (4.2) which concludes the proof. □

Remark 4.1 Note that the reasoning described above also implies that, for $\alpha \in \left[\frac{1}{2}, 1\right)$ and $\lambda > \frac{1}{2}$, the SDE (4.2), where the integral w.r.t. Z is understood pathwisely, has a unique strong solution in the class of non-negative stochastic processes with λ-Hölder continuous trajectories. Indeed, $\{Y^{\frac{1}{1-\alpha}}(t),\ t \in [0, T]\}$ with Y defined by (4.1) is a solution to (4.2). Moreover, if X is another solution to (4.2), then by the chain rule [8, Theorem 4.3.1], the process $\{X^{1-\alpha}(t),\ t \in [0, T]\}$ must satisfy the SDE (4.1) until the first moment of zero hitting. However, the SDE (4.1) has a unique continuous solution that never hits zero and thus $X^{1-\alpha}$ coincides with Y.

4.2 Mixed-Fractional CKLS-Process

Assume that $y_0, \kappa, \theta, \nu_1, \nu_2$ are positive constants, $B = \{B(t),\ t \in [0, T]\}$ is an a.s. continuous modification of a standard Wiener process, $B^H = \{B^H(t),\ t \in [0, T]\}$ is an a.s. continuous modification of a fractional Brownian motion independent of B with $H \in (0, 1)$, $Z = \nu_1 B + \nu_2 B^H$, $\alpha \in \left(\frac{1}{2}, 1\right)$ is such that $H \wedge \frac{1}{2} + \alpha > 1$ and the function b has the form

$$b(t, y) = b(y) = \frac{\kappa}{y^{\frac{\alpha}{1-\alpha}}} - \frac{\alpha\nu_1^2}{2y} - \theta y.$$

Then, by Theorem 3.2, the process Y defined by the equation

$$Y(t) = y_0 + \int_0^t \left(\frac{\kappa}{Y^{\frac{\alpha}{1-\alpha}}(s)} - \frac{\alpha\nu_1^2}{2Y(s)} - \theta Y(s)\right) ds + \nu_1 B(t) + \nu_2 B^H(t) \tag{4.5}$$

exists, is unique, positive and has all the moments of real orders. Define the process $X(t) := Y^{\frac{1}{1-\alpha}}(t)$, $t \in [0, T]$, and observe that the integral $\int_0^t X^\alpha(s) dB(s)$ is well-defined in the Itô sense. Indeed, B is a martingale w.r.t. the filtration generated jointly by (B, B^H) and the process X^α is adapted to this filtration. Moreover, by Theorem 3.3,

$$\sup_{t\in[0,T]} X^{2\alpha}(t) \le C_2^{2\alpha}(1 + \Lambda)^{2\alpha},$$

and, since the random variable Λ can be chosen to have moments of all orders (see e.g. Appendix A.1 or [1]),

$$\int_0^t \mathbb{E}[X^{2\alpha}(s)] ds < \infty.$$

Like in Sect. 4.1, if $H > \frac{1}{2}$, X can be interpreted as a generalization of the CKLS-process as summarized in the following theorem.

Theorem 4.2 *Let $H > \frac{1}{2}$, $\alpha > \frac{1}{2}$ and $H \wedge \frac{1}{2} + \alpha > 1$. Then the process $X(t) := Y^{\frac{1}{1-\alpha}}(t)$, $t \in [0, T]$, satisfies the SDE of the form*

$$\begin{aligned} X(t) = x_0 + \frac{1}{1-\alpha}\Bigg(\int_0^t (\kappa - \theta X(s))\,ds + \nu_1 \int_0^t X^\alpha(s)dB(s) \\ + \nu_2 \int_0^t X^\alpha(s)dB^H(s)\Bigg), \end{aligned} \tag{4.6}$$

where $x_0 := y_0^{\frac{1}{1-\alpha}}$, the integral with respect to B is the regular Itô integral (w.r.t. the filtration generated jointly by (B, B^H)) and the integral with respect to B^H is understood as the L^2-limit of Riemann-Stieltjes integral sums.

Proof First of all, choose $\lambda \in \left(\frac{1}{2}, H\right)$, $\lambda' \in \left[\frac{1}{3}, \frac{1}{2}\right)$ and $\varepsilon > 0$ such that $\lambda + \lambda' > 1$ and $\lambda + \varepsilon < H, \lambda' + \varepsilon < \frac{1}{2}$. Note that the noise Z is a.s. $(\lambda' + \varepsilon)$-Hölder continuous, i.e., with probability 1,

$$|Z(t_2) - Z(t_1)| \le \Lambda_{\lambda'+\varepsilon}|t_2 - t_1|^{\lambda'+\varepsilon}, \quad t_1, t_2 \in [0, T],$$

where $\Lambda_{\lambda'+\varepsilon}$ can be chosen to have moments of all orders (see [1] or Appendix A.1).

Our goal is to use the argument that is similar to the one from Sect. 4.1 with one main difference: since we are going to treat the integral w.r.t. the Brownian motion B as a regular Itô integral, all the convergences (including convergence of integral sums w.r.t. B^H) must be considered in L^2-sense. For reader's convenience, we split the further proof into several steps.

Step 1. By Theorem 3.3, there is a constant $C_1 > 0$ (that depends on $\lambda' + \varepsilon$) such that

$$Y(t) \ge \frac{C_1}{(1 + \Lambda_{\lambda'+\varepsilon})^{\frac{1}{\gamma(\lambda'+\varepsilon)+\lambda'+\varepsilon-1}}}.$$

Next, observe that, for any $\varepsilon > 0$ and $y_1, y_2 > \varepsilon$,

$$\begin{aligned} &\left|\left(\frac{\kappa}{y_1^{\frac{\alpha}{1-\alpha}}} - \frac{\alpha\nu_1^2}{2y_1} - \theta y_1\right) - \left(\frac{\kappa}{y_2^{\frac{\alpha}{1-\alpha}}} - \frac{\alpha\nu_1^2}{2y_2} - \theta y_2\right)\right| \\ &\le \kappa\left|y_1^{-\frac{\alpha}{1-\alpha}} - y_2^{-\frac{\alpha}{1-\alpha}}\right| + \frac{\alpha\nu_1^2}{2}\left|y_1^{-1} - y_2^{-1}\right| + \theta\,|y_1 - y_2| \\ &\le \left(\frac{\alpha\kappa}{(1-\alpha)\varepsilon^{\frac{1}{1-\alpha}}} + \frac{\alpha\nu_1^2}{2\varepsilon^2} + \theta\right)|y_1 - y_2|, \end{aligned}$$

so one can put the value of c_ε from Assumption 3.2 equal to

$$c_\varepsilon = \frac{\alpha\kappa}{(1-\alpha)\varepsilon^{\frac{1}{1-\alpha}}} + \frac{\alpha\nu_1^2}{2\varepsilon^2} + \theta.$$

Then, by Theorem 3.4, with probability 1,

$$|Y(t_2) - Y(t_1)| \le \Upsilon_{\lambda'+\varepsilon}|t_2 - t_1|^{\lambda'+\varepsilon}, \quad t_1, t_2 \in [0, T],$$

where

$$\Upsilon_{\lambda'+\varepsilon} = C(1 + \Lambda_{\lambda'+\varepsilon} + c_\xi + c_\xi \Lambda_{\lambda'+\varepsilon})$$

with

$$c_\xi = \frac{\alpha\kappa}{(1-\alpha)C_1^{\frac{1}{1-\alpha}}}(1+\Lambda_{\lambda'+\varepsilon})^{\frac{1}{(1-\alpha)(\gamma(\lambda'+\varepsilon)+\lambda'+\varepsilon-1)}} + \frac{\alpha\nu_1^2}{2C_1^2}(1+\Lambda)^{\frac{2}{\gamma(\lambda'+\varepsilon)+\lambda'+\varepsilon-1}} + \theta.$$

In particular, since $\Lambda_{\lambda'+\varepsilon}$ has moments of all orders, so does $\Upsilon_{\lambda'+\varepsilon}$.

Next, observe that

$$\begin{aligned} |X(t_2) - X(t_1)| &= |Y^{\frac{\alpha}{1-\alpha}}(t_2) - Y^{\frac{\alpha}{1-\alpha}}(t_1)| \\ &\le \frac{\alpha}{1-\alpha} \max_{s\in[0,T]} Y^{\frac{2\alpha-1}{1-\alpha}}(s)|Y(t_2) - Y(t_1)| \\ &\le \frac{\alpha}{1-\alpha} \max_{s\in[0,T]} Y^{\frac{2\alpha-1}{1-\alpha}}(s)\Upsilon_{\lambda'+\varepsilon}|t_2 - t_1|^{\lambda'+\varepsilon}. \end{aligned} \tag{4.7}$$

Since, by Theorem 3.3, for some constant $C > 0$,

$$Y^{\frac{2\alpha-1}{1-\alpha}}(s) \le C(1+\Lambda)^{\frac{2\alpha-1}{1-\alpha}},$$

we can re-write (4.7) as

$$|X(t_2) - X(t_1)| \le C(1+\Lambda_{\lambda'+\varepsilon})^{\frac{2\alpha-1}{1-\alpha}}\Upsilon_{\lambda'+\varepsilon}|t_2 - t_1|^{\lambda'+\varepsilon} =: \Upsilon_X|t_2 - t_1|^{\lambda'+\varepsilon}, \tag{4.8}$$

where Υ_X also has moments of all orders.

Step 2. Let us now show that the integral $\int_0^t X^\alpha(s)dB^H(s)$ is indeed well-defined as the L^2-limit of Riemann-Stieltjes integral sums. Let $0 = t_0 < t_1 < t_2 < \cdots < t_n = t$ be a partition of $[0, t]$ with the mesh $|\Delta t| := \max_{k=0,\dots,n-1}(t_{k+1} - t_k)$. By the Young-Lóeve inequality (see e.g. [4, Theorem 6.8]), it holds a.s. that

$$\left|\int_0^t X^\alpha(s)dB^H(s) - \sum_{k=0}^{n-1} X^\alpha(t_k)(B^H(t_{k+1}) - B^H(t_k))\right|$$

$$\leq \sum_{k=0}^{n-1} \left| \int_{t_k}^{t_{k+1}} X^\alpha(s) dB^H(s) - X^\alpha(t_k)(B^H(t_{k+1}) - B^H(t_k)) \right|$$
$$\leq \frac{1}{2^{1-(\lambda+\lambda')}} \sum_{k=0}^{n-1} [X^\alpha]_{\lambda';[t_k,t_{k+1}]} [B^H]_{\lambda;[t_k,t_{k+1}]},$$

where

$$[f]_{\lambda;[t,t']} := \left(\sup_{\Pi[t,t']} \sum_{l=0}^{m-1} |f(s_{l+1}) - f(s_l)|^{\frac{1}{\lambda}} \right)^\lambda,$$

with supremum taken over all partitions $\Pi[t, t'] = \{t = s_0 < \cdots < s_m = t'\}$ of $[t, t']$.

Using the definition of X, the lower bound for Y as well as (4.8), we can write that, for any $t, t' \in [0, T]$,

$$\begin{aligned} |X^\alpha(t) - X^\alpha(t')| &\leq \frac{\alpha}{\min_{s\in[0,T]} X^{1-\alpha}(s)} |X(t) - X(t')| \\ &= \frac{\alpha}{\min_{s\in[0,T]} Y(s)} |X(t) - X(t')| \\ &\leq C(1 + \Lambda_{\lambda'+\varepsilon})^{\frac{1}{\gamma(\lambda'+\varepsilon)+\lambda'+\varepsilon-1}} |X(t) - X(t')| \\ &\leq C\Upsilon_X(1 + \Lambda_{\lambda'+\varepsilon})^{\frac{1}{\gamma(\lambda'+\varepsilon)+\lambda'+\varepsilon-1}} |t_2 - t_1|^{\lambda'+\varepsilon} \\ &=: \Upsilon_{X^\alpha} |t_2 - t_1|^{\lambda'+\varepsilon}, \end{aligned}$$

where the random variable $\Upsilon_{X^\alpha} > 0$ has moments of all orders. Therefore, with probability 1,

$$\begin{aligned} [X^\alpha]_{\lambda';[t_k,t_{k+1}]} &= \left(\sup_{\Pi[t_k,t_{k+1}]} \sum_{l=0}^{m-1} |X^\alpha(s_{l+1}) - X^\alpha(s_l)|^{\frac{1}{\lambda'}} \right)^{\lambda'} \\ &\leq \Upsilon_{X^\alpha} \left(\sup_{\Pi[t_k,t_{k+1}]} \sum_{k=0}^{m-1} (s_{l+1} - s_l)^{1+\frac{\varepsilon}{\lambda'}} \right)^{\lambda'} \leq \Upsilon_{X^\alpha} |\Delta t|^{\lambda'+\varepsilon} \end{aligned}$$

and, similarly, since $\lambda + \varepsilon < H$,

$$[B^H]_{\lambda;[t_k,t_{k+1}]} \leq \Lambda_{B^H} |\Delta t|^{\lambda+\varepsilon},$$

where Λ_{B^H} has moments of all orders and

$$|B^H(t) - B^H(s)| \leq \Lambda_{B^H} |t - s|^{\lambda+\varepsilon}.$$

Therefore,

$$\mathbb{E}\left|\int_0^t X^\alpha(s)dB^H(s) - \sum_{k=0}^{n-1} X^\alpha(t_k)(B^H(t_{k+1}) - B^H(t_k))\right|^2$$

$$\leq \mathbb{E}\left[\left(\frac{1}{2^{1-(\lambda+\lambda')}}\sum_{k=0}^{n-1}[X^\alpha]_{\lambda';[t_k,t_{k+1}]}[B^H]_{\lambda;[t_k,t_{k+1}]}\right)^2\right]$$

$$\leq \mathbb{E}\left[\Lambda^2_{B^H}\Upsilon^2_{X^\alpha}\frac{1}{2^{2-2(\lambda+\lambda')}}\left(\sum_{k=0}^{n-1}|\Delta t|^{\lambda+\lambda'+2\varepsilon}\right)^2\right] \to 0,$$

as $|\Delta t| \to 0$. In other words, the integral $\int_0^t X^\alpha(s)dB^H(s)$ is indeed well-defined as the L^2-limit of Riemann-Stieltjes integral sums.

Step 3. Finally, we would like to get representation (4.6). For a partition $0 = t_0 < t_1 < t_2 < \cdots < t_n = t$, one can write

$$X(t) = x_0 + \sum_{k=1}^{n}\left(Y^{\frac{1}{1-\alpha}}(t_k) - Y^{\frac{1}{1-\alpha}}(t_{k-1})\right)$$

$$= x_0 + \frac{1}{1-\alpha}\sum_{k=0}^{n-1}\left(Y^{\frac{\alpha}{1-\alpha}}(t_{k-1})(Y(t_k) - Y(t_{k-1}))\right)$$

$$+ \frac{1}{2}\frac{\alpha}{(1-\alpha)^2}\sum_{k=0}^{n-1}\left(Y^{\frac{2\alpha-1}{1-\alpha}}(t_{k-1})(Y(t_k) - Y(t_{k-1}))^2\right)$$

$$+ \frac{1}{6}\frac{\alpha(2\alpha-1)}{(1-\alpha)^3}\sum_{k=1}^{n}\left(\Theta_{n,k}^{\frac{3\alpha-2}{1-\alpha}}(Y(t_k) - Y(t_{k-1}))^3\right),$$

where $\Theta_{n,k}$ is a value between $Y(t_{k-1})$ and $Y(t_k)$. Using Theorem 3.3, it is easy to check that there are constants $C_1, C_2 > 0$ which do not depend on the partition such that for all n and k

$$\frac{C_1}{(1+\Lambda_{\lambda'+\varepsilon})^{\frac{1}{\gamma\lambda+\lambda-1}}} \leq \Theta_{n,k} \leq C_2(1+\Lambda_{\lambda'+\varepsilon}) \tag{4.9}$$

with $\gamma = \frac{\alpha}{1-\alpha}$. In particular, this implies that there exists a random variable $\Theta_\alpha > 0$ that does not depend on the partition and has moments of all orders (including the negative ones) such that

$$\sup_{\substack{n\geq 1,\\ k=0,1,\ldots,n}} \Theta_{n,k}^{\frac{3\alpha-2}{1-\alpha}} < \Theta_\alpha.$$

Recall that $\lambda' \in \left[\frac{1}{3}, \frac{1}{2}\right)$, whence, taking into account that

$$\mathbb{E}\left[\Theta_\alpha^2\Upsilon^6_{\lambda'+\varepsilon}\right] < \infty,$$

one can see that

$$\left|\sum_{k=1}^{n}\left(\Theta_{n,k}^{\frac{3\alpha-2}{1-\alpha}}(Y(t_k)-Y(t_{k-1}))^3\right)\right|$$
$$\leq \Theta_\alpha \Upsilon^3_{\lambda'+\varepsilon}\sum_{k=1}^{n}(t_k-t_{k-1})^{3\lambda'+3\varepsilon}\xrightarrow{L^2}0,\quad |\Delta t|\to 0.$$

Finally, using Step 1, it remains to notice that, as $|\Delta t|\to 0$,

$$\frac{1}{1-\alpha}\sum_{k=0}^{n-1}\left(Y^{\frac{\alpha}{1-\alpha}}(t_{k-1})(Y(t_k)-Y(t_{k-1}))\right)$$
$$\xrightarrow{L^2}\frac{1}{1-\alpha}\left(\int_0^t(\kappa-\theta X(s))\,ds+\nu_1\int_0^t X^\alpha(s)dB(s)+\nu_2\int_0^t X^\alpha(s)dB^H(s)\right)$$
$$-\frac{\alpha\nu_1^2}{2(1-\alpha)^2}\int_0^t Y^{\frac{2\alpha-1}{1-\alpha}}(s)ds$$

and

$$\frac{1}{2}\frac{\alpha}{(1-\alpha)^2}\sum_{k=0}^{n-1}\left(Y^{\frac{2\alpha-1}{1-\alpha}}(t_{k-1})(Y(t_k)-Y(t_{k-1}))^2\right)\xrightarrow{L^2}\frac{\alpha\nu_1^2}{2(1-\alpha)^2}\int_0^t Y^{\frac{2\alpha-1}{1-\alpha}}(s)ds,$$

which concludes the proof. □

4.3 CIR and CKLS Driven by Very Rough Noises

In the previous sections, in order to ensure existence and uniqueness of the continuous solution to (4.1), we had to make sure that Assumption 3.2(A4) holds, i.e. that

$$\gamma=\frac{\alpha}{1-\alpha}>\frac{1-\lambda}{\lambda}$$

or, equivalently,

$$\lambda>\frac{1}{1+\gamma}.$$

This condition is equivalent to assuming $\alpha+\lambda>1$; in particular, if $\alpha=\frac{1}{2}$ as in the CIR case, Hölder regularity of the driving noise λ must exceed $\frac{1}{2}$. In other words, we imposed a restriction on the regularity of the driving process Z, since the existence-uniqueness results of Chap. 2 do not allow the GCIR process to be driven by truly *rough* noises.

However, a natural question arises:

what happens, if this condition does not hold?

For example, one may be interested in a process of the type (4.1) with $\alpha = \frac{1}{2}$ (corresponding to $\gamma = 1$) driven by a "very rough" Z with $\lambda \approx 0$ – but such a case has not yet been covered. The goal of this Section is to fill this gap and extend our analysis to this situation.

Namely, let $Z = Z(t), t \in [0, T]$, be a λ-Hölder continuous noise satisfying Assumption 3.1 and let λ and $\gamma > 0$ be such that the "opposite" of Assumption 3.2(A4) holds, i.e.

$$0 < \gamma \leq \frac{1-\lambda}{\lambda}$$

or, equivalently,

$$0 < \lambda \leq \frac{1}{1+\gamma}.$$

For any $\varepsilon > 0$, $y_0 > 0$, $\kappa > 0$ and $\theta \in \mathbb{R}$, consider a "regularization" of the Eq. (4.1) of the form

$$Y_\varepsilon(t) = y_0 + \int_0^t \left(\frac{\kappa}{\left(Y_\varepsilon(s)\mathbb{1}_{\{Y_\varepsilon(s)>0\}} + \varepsilon\right)^\gamma} - \theta Y_\varepsilon(s) \right) ds + Z(t). \tag{4.10}$$

Note that, for any $\varepsilon > 0$, the SDE (4.10) has Lipschitz continuous drift and additive noise and hence, by the standard Picard iteration argument applied pathwisely, (4.10) has a unique and continuous solution $Y_\varepsilon = \{Y_\varepsilon(t),\ t \geq 0\}$ for each $\varepsilon > 0$. Moreover, by Lemma 3.2, for any $\varepsilon_1 < \varepsilon_2$,

$$Y_{\varepsilon_1}(t) \geq Y_{\varepsilon_2}(t), \quad t \in [0, T], \quad a.s., \tag{4.11}$$

and hence, for any $t \geq 0$, one can define the limit

$$Y(t) := \lim_{\varepsilon \to 0} Y_\varepsilon(t), \quad t \in [0, T]. \tag{4.12}$$

Notation 4.1 To emphasize the monotone nature of the convergence in (4.12), we will occasionally use the notation $Y_\varepsilon(t) \uparrow Y(t)$, $\varepsilon \to 0$.

At this point, it is not clear if the limit (4.12) is finite, so we need to check it separately. Before doing that, let us first give some auxiliary results.

Proposition 4.1 *Let $Y_1 = \{Y_1(t),\ t \ge 0\}$ be a stochastic process defined by*

$$Y_1(t) = y_0 + \int_0^t \left(\frac{\kappa}{(Y_1(s)\mathbb{1}_{\{Y_1(s)>0\}} + 1)^\gamma} - \theta Y_1(s) \right) ds + Z(t).$$

Let $\Lambda = \Lambda_\lambda$ be a random variable such that (3.2) holds. Then, with probability 1,

$$\max_{t\in[0,T]} |Y_1(t)| \le e^{|\theta|T} \left(y_0 + \kappa T + T^\lambda \Lambda \right).$$

Proof Fix $\omega \in \Omega$ such that the corresponding path $t \mapsto Z(\omega, t)$ is Hölder continuous; in particular,

$$\max_{t\in[0,T]} |Z(\omega, t)| \le \Lambda T^\lambda.$$

Taking into account that $\frac{\kappa}{(x\mathbb{1}_{\{x>0\}}+1)^\gamma} \le \kappa$ for any $x \in \mathbb{R}$, we can write

$$\begin{aligned} |Y_1(t)| &\le y_0 + \int_0^t \kappa ds + |\theta| \int_0^t |Y_1(s)| ds + T^\lambda \Lambda \\ &\le \left(y_0 + \kappa T + T^\lambda \Lambda \right) + |\theta| \int_0^t |Y_1(s)| ds. \end{aligned}$$

The result now follows from the Gronwall inequality. □

Proposition 4.2 *Let Y_ε be defined by (4.10) and let $\Lambda = \Lambda_\lambda$ be a random variable from (3.2). Then, with probability 1, for any $\varepsilon \in (0, 1]$ and $t \in [0, T]$,*

$$\max_{t\in[0,T]} |Y_\varepsilon(t)| \le e^{2|\theta|T} \left(y_0 + \left(\frac{2^\gamma}{y_0^\gamma} \vee 1 \right) \kappa T + T^\lambda \Lambda \right). \tag{4.13}$$

Proof Fix $\varepsilon \in (0, 1]$ and take $\omega \in \Omega$ such that the corresponding path $t \mapsto Z(\omega, t)$ is Hölder continuous so that

$$\max_{t\in[0,T]} |Z(\omega, t)| \le \Lambda T^\lambda.$$

Denote

$$\tau_{1,\varepsilon} := \sup \left\{ t \in [0, T] \;\middle|\; \forall s \in [0, t] :\ Y_\varepsilon(s) \ge \frac{y_0}{2} \right\}.$$

Our goal is to prove that

$$|Y_\varepsilon(t)| \le e^{|\theta|T}\left(y_0 + \left(\frac{2^\gamma}{y_0^\gamma} \vee 1\right)\kappa T + T^\lambda \Lambda\right) + |\theta| \int_0^t |Y_\varepsilon(s)| ds \tag{4.14}$$

for any $t \in [0, T]$ and then apply Gronwall inequality. We will separate two cases: $t \in [0, \tau_{1,\varepsilon}]$ and $t \in [\tau_{1,\varepsilon}, T]$.

Case 1: $t \in [0, \tau_{1,\varepsilon}]$. In this case, $Y_\varepsilon(s) \ge \frac{y_0}{2}$ for all $s \in [0, t]$ and hence we can write

$$\begin{aligned}
|Y_\varepsilon(t)| &= \left| y_0 + \int_0^t \left(\frac{\kappa}{(Y_\varepsilon(s)\mathbb{1}_{\{Y_\varepsilon(s)>0\}} + \varepsilon)^\gamma} - \theta Y_\varepsilon(s) \right) ds + Z(t) \right| \\
&\le y_0 + \int_0^t \frac{\kappa}{(Y_\varepsilon(s)\mathbb{1}_{\{Y_\varepsilon(s)>0\}} + \varepsilon)^\gamma} ds + |\theta| \int_0^t |Y_\varepsilon(s)| ds + T^\lambda \Lambda \\
&\le y_0 + \kappa \int_0^t \frac{2^\gamma}{y_0^\gamma} ds + |\theta| \int_0^t |Y_\varepsilon(s)| ds + T^\lambda \Lambda \\
&\le \left(y_0 + \frac{2^\gamma \kappa T}{y_0^\gamma} + T^\lambda \Lambda \right) + |\theta| \int_0^t |Y_\varepsilon(s)| ds \\
&\le e^{|\theta|T}\left(y_0 + \left(\frac{2^\gamma}{y_0^\gamma} \vee 1\right)\kappa T + T^\lambda \Lambda\right) + |\theta| \int_0^t |Y_\varepsilon(s)| ds
\end{aligned} \tag{4.15}$$

as required.

Case 2: $t \in [\tau_{1,\varepsilon}, T]$. Consider

$$\tau_{2,\varepsilon}(t) := \sup\left\{ s \in [\tau_{1,\varepsilon}, t] \;\middle|\; |Y_\varepsilon(s)| < \frac{y_0}{2} \right\}.$$

If $\tau_{2,\varepsilon}(t) = t$, then, by definition,

$$|Y_\varepsilon(t)| < \frac{y_0}{2}$$

and (4.14) holds automatically. If $\tau_{2,\varepsilon}(t) < t$, by continuity, $Y_\varepsilon(\tau_{2,\varepsilon}(t)) = \frac{y_0}{2}$ and, moreover, $|Y_\varepsilon(s)| \ge \frac{y_0}{2}$ for all $s \in [\tau_{2,\varepsilon}(t), t]$. In this case, since Y_ε is continuous, we have only two possibilities.

(1) If $Y_\varepsilon(s) \le -\frac{y_0}{2}$ for all $s \in [\tau_{2,\varepsilon}(t), t]$, then, for such s,

$$Y_1(s) \le Y_\varepsilon(s) < 0$$

and hence, by Proposition 4.1,

$$\begin{aligned}|Y_\varepsilon(s)| &\le |Y_1(s)| \le e^{|\theta|T}\left(y_0+\kappa T+T^\lambda\Lambda\right)\\ &\le e^{|\theta|T}\left(y_0+\left(\frac{2^\gamma}{y_0^\gamma}\vee 1\right)\kappa T+T^\lambda\Lambda\right)+|\theta|\int_0^s |Y_\varepsilon(u)|du.\end{aligned} \tag{4.16}$$

(2) If $Y_\varepsilon(s)\ge \frac{y_0}{2}$ for all $s\in[\tau_{2,\varepsilon}(t),t]$, then, just like in Case 1 above,

$$\begin{aligned}|Y_\varepsilon(t)| &\le \left|Y_\varepsilon(\tau_{2,\varepsilon}(t))\right| + \left|\int_{\tau_{2,\varepsilon}(t)}^t \frac{\kappa}{(Y_\varepsilon(s)\mathbb{1}_{\{Y_\varepsilon(s)>0\}}+\varepsilon)^\gamma}ds\right|\\ &\quad+|\theta|\int_{\tau_{2,\varepsilon}(t)}^t |Y_\varepsilon(s)|ds+\left|Z(t)-Z(\tau_{2,\varepsilon}(t))\right|\\ &\le \frac{y_0}{2}+\kappa\int_{\tau_{2,\varepsilon}(t)}^t \frac{2^\gamma}{y_0^\gamma}ds+T^\lambda\Lambda+|\theta|\int_{\tau_{2,\varepsilon}(t)}^t |Y_\varepsilon(s)|ds \qquad (4.17)\\ &\le \left(y_0+\frac{2^\gamma\kappa T}{y_0^\gamma}+T^\lambda\Lambda\right)+|\theta|\int_0^t |Y_\varepsilon(s)|ds\\ &\le e^{|\theta|T}\left(y_0+\left(\frac{2^\gamma}{y_0^\gamma}\vee 1\right)\kappa T+T^\lambda\Lambda\right)+|\theta|\int_0^t |Y_\varepsilon(s)|ds.\end{aligned}$$

Hence, summarizing (4.15), (4.16) and (4.17), we see that (4.14) holds for any $t\in[0,T]$ and, by Gronwall inequality,

$$|Y_\varepsilon(t)|\le e^{2|\theta|T}\left(y_0+\left(\frac{2^\gamma}{y_0^\gamma}\vee 1\right)\kappa T+T^\lambda\Lambda\right),$$

which immediately implies (4.13). □

Theorem 4.3 *Under the assumptions specified above, with probability 1,*

$$\sup_{t\in[0,T]}|Y(t)|\le e^{2|\theta|T}\left(y_0+\left(\frac{2^\gamma}{y_0^\gamma}\vee 1\right)\kappa T+T^\lambda\Lambda\right)<\infty. \tag{4.18}$$

Proof By definition, for any $t\ge 0$,

$$Y_\varepsilon(t)\uparrow Y(t)\quad a.s.,\quad \varepsilon\downarrow 0,$$

so (4.18) follows immediately from Proposition 4.2. □

The next step would be to show that the limit process Y from (4.12) is non-negative.

Theorem 4.4 *Under the assumptions specified above, with probability 1,*

$$Y(t) \geq 0, \quad t \in [0, T].$$

Proof Fix $\omega \in \Omega$ such that the corresponding path $t \mapsto Z(\omega, t)$ is Hölder continuous and observe that, on this ω, the Eq. (4.10) has a unique and continuous solution for all $\varepsilon > 0$.

Assume that there exists $\tau \in (0, T]$ such that $Y(\tau) < 0$. Then, for all $0 < \varepsilon \leq 1$,

$$Y_1(\tau) \leq Y_\varepsilon(\tau) \leq Y(\tau) < 0,$$

where Y_1 is just as in Proposition 4.1. For any $0 < \varepsilon \leq 1$, denote

$$\tau_-(\varepsilon) := \sup\{t \in (0, \tau) \mid Y_\varepsilon(t) = 0\},$$

and observe that, by the continuity of Y_ε, $0 < \tau_-(\varepsilon) < \tau \leq T$. It is clear that, again by the continuity of Y_ε, $Y_\varepsilon(\tau_-(\varepsilon)) = 0$ and, in addition, $Y_1(t) \leq Y_\varepsilon(t) < 0$ for all $t \in (\tau_-(\varepsilon), \tau]$. By Proposition 4.1, for all $t \in (\tau_-(\varepsilon), \tau]$,

$$\begin{aligned} -\theta \int_{\tau_-(\varepsilon)}^{t} Y_\varepsilon(s)ds &\geq -|\theta| \int_{\tau_-(\varepsilon)}^{t} |Y_\varepsilon(s)|ds \geq -|\theta| \int_{\tau_-(\varepsilon)}^{t} |Y_1(s)|ds \\ &\geq -|\theta| \int_{\tau_-(\varepsilon)}^{t} |Y_1(s)|ds \\ &\geq -|\theta| e^{|\theta| T} \left(y_0 + \kappa T + T^\lambda \Lambda\right) (t - \tau_-(\varepsilon)) \end{aligned}$$

and, in addition,

$$Z(t) - Z(\tau_-(\varepsilon)) \geq -\Lambda |t - \tau_-(\varepsilon)|^\lambda,$$

so, for all $\varepsilon \in (0, 1]$,

$$\begin{aligned} 0 > Y(\tau) &\geq Y_\varepsilon(\tau) = Y_\varepsilon(\tau) - Y_\varepsilon(\tau_-(\varepsilon)) \\ &= \int_{\tau_-(\varepsilon)}^{\tau} \left(\frac{\kappa}{(Y_\varepsilon(s)\mathbb{1}_{\{Y_\varepsilon(s)>0\}} + \varepsilon)^\gamma} - \theta Y_\varepsilon(s)\right) ds + Z(\tau) - Z(\tau_-(\varepsilon)) \\ &\geq \left(\frac{\kappa}{\varepsilon^\gamma} - |\theta| e^{|\theta| T} \left(y_0 + \kappa T + T^\lambda \Lambda\right)\right) (\tau - \tau_-(\varepsilon)) - \Lambda |\tau - \tau_-(\varepsilon)|^\lambda \\ &\geq F(\varepsilon), \end{aligned}$$

where

$$F(\varepsilon) := \min_{t\in[\tau_-(\varepsilon),\infty)} \left(\left(\frac{\kappa}{\varepsilon^\gamma} - |\theta| e^{|\theta| T} \left(y_0 + \kappa T + T^\lambda \Lambda \right) \right) (t - \tau_-(\varepsilon)) - \Lambda (t - \tau_-(\varepsilon))^\lambda \right).$$

To summarize, under our assumption that $Y(\tau) < 0$,

$$F(\varepsilon) \le Y(\tau) < 0, \quad \varepsilon \in (0, 1]. \tag{4.19}$$

Our goal is to prove that (4.19) cannot be true. In order to do that, let us verify that $F(\varepsilon) \to 0$, $\varepsilon \to 0$, i.e. that there exists $\varepsilon_* \in (0, 1]$ such that $Y(\tau) < F(\varepsilon_*) < 0$ which immediately contradicts (4.19).

Let $\varepsilon_0 \in (0, 1]$ be such that $\kappa - |\theta| e^{|\theta| T} \left(y_0 + \kappa T + T^\lambda \Lambda \right) \varepsilon_0^\gamma > 0$. Then it is easy to check that, for all $\varepsilon < \varepsilon_0$, the minimum of the function

$$t \mapsto \left(\frac{\kappa}{\varepsilon^\gamma} - |\theta| e^{|\theta| T} \left(y_0 + \kappa T + T^\lambda \Lambda \right) \right) (t - \tau_-(\varepsilon)) - \Lambda (t - \tau_-(\varepsilon))^\lambda$$

is attained at

$$t_* = \tau_-(\varepsilon) + \left(\frac{\lambda \Lambda \varepsilon^\gamma}{\kappa - |\theta| e^{|\theta| T} \left(y_0 + \kappa T + T^\lambda \Lambda \right) \varepsilon^\gamma} \right)^{\frac{1}{1-\lambda}}$$

and hence, for all $\varepsilon < \varepsilon_0$,

$$\begin{aligned} F(\varepsilon) &= \min_{t\in[\tau_-(\varepsilon),\infty)} \left(\left(\frac{\kappa}{\varepsilon^\gamma} - |\theta| e^{|\theta| T} \left(y_0 + \kappa T + T^\lambda \Lambda \right) \right) \right. \\ &\quad \left. (t - \tau_-(\varepsilon)) - \Lambda (t - \tau_-(\varepsilon))^\lambda \right) \\ &= \left(\frac{\kappa}{\varepsilon^\gamma} - |\theta| e^{|\theta| T} \left(y_0 + \kappa T + T^\lambda \Lambda \right) \right) (t_* - \tau_-(\varepsilon)) - \Lambda (t_* - \tau_-(\varepsilon))^\lambda \\ &= (\lambda^{\frac{1}{1-\lambda}} - \lambda^{\frac{\lambda}{1-\lambda}}) \left(\frac{\Lambda}{\left(\kappa - |\theta| e^{|\theta| T} \left(y_0 + \kappa T + T^\lambda \Lambda \right) \varepsilon^\gamma \right)^\lambda} \right)^{\frac{1}{1-\lambda}} \\ &\quad \varepsilon^{\frac{\gamma\lambda}{1-\lambda}} \to 0, \quad \varepsilon \to 0, \end{aligned}$$

as required. Therefore (4.19) cannot be true and, on the $\omega \in \Omega$ fixed at the beginning of the proof, $Y(t) \ge 0$ for all $t \in [0, T]$. □

Our next objective is to understand if the limit process Y from (4.12) satisfies an equation similar to (4.1). First of all, note that, with probability 1,

$$\lim_{\varepsilon \to 0} \int_0^t Y_\varepsilon(s) ds = \int_0^t Y(s) ds. \tag{4.20}$$

Indeed, by Theorem 4.3, $\int_0^T |Y(s)|ds < \infty$ a.s. and, in addition, with probability 1, $Y_\varepsilon(t) \uparrow Y(t)$ as $\varepsilon \to 0$, for all $t \in [0, T]$. Therefore, (4.20) holds by the Beppo Levi theorem. Next, since

$$Y_\varepsilon(t) - y_0 + \theta \int_0^t Y_\varepsilon(s)ds - Z(t) = \int_0^t \frac{\kappa}{\left(Y_\varepsilon(s)\mathbb{1}_{\{Y_\varepsilon(s)>0\}} + \varepsilon\right)^\gamma} ds,$$

with probability 1, there exists the limit

$$\begin{aligned} R(t) &:= \lim_{\varepsilon \to 0} \int_0^t \frac{\kappa}{\left(Y_\varepsilon(s)\mathbb{1}_{\{Y_\varepsilon(s)>0\}} + \varepsilon\right)^\gamma} ds \\ &= \lim_{\varepsilon \to 0} \left(Y_\varepsilon(t) - y_0 + \theta \int_0^t Y_\varepsilon(s)ds - Z(t) \right) \\ &= Y(t) - y_0 + \theta \int_0^t Y(s)ds - Z(t) < \infty. \end{aligned} \tag{4.21}$$

Note that, by the Fatou lemma, for any $t \in [0, T]$,

$$\int_0^t \frac{\kappa}{Y^\gamma(s)} ds \le \liminf_{\varepsilon \to 0} \int_0^t \frac{\kappa}{\left(Y_\varepsilon(s)\mathbb{1}_{\{Y_\varepsilon(s)>0\}} + \varepsilon\right)^\gamma} ds = R(t) < \infty \quad a.s.,$$

so we can define

$$L(t) := R(t) - \int_0^t \frac{\kappa}{Y^\gamma(s)} ds \ge 0 \tag{4.22}$$

and observe that the limit process Y from (4.12) satisfies the equation

$$Y(t) = y_0 + \int_0^t \left(\frac{\kappa}{Y^\gamma(s)} - \theta Y(s) \right) ds + Z(t) + L(t). \tag{4.23}$$

It is tempting to hope that the process L defined by (4.22) is equal to zero. Indeed, if this were the case, the "bad" scenario $\gamma \le \frac{1-\lambda}{\lambda}$ considered in this section would not be particularly problematic, as the limiting Eq. (4.23) would fully coincide with its counterpart (4.1). However, it remains an open question whether $L(t) = 0$ for all $t \ge 0$. For now, we can only report that L exhibits properties similar to those of the Skorokhod reflection functions discussed in Chap. 2: with probability 1, it is non-decreasing, can increase only at zeros of Y and is continuous. All these claims are proved in the theorem below.

Theorem 4.5 *Let $0 < \gamma \le \frac{1-\lambda}{\lambda}$ and Y, R and L be defined by (4.12), (4.21) and (4.22) respectively. Fix $\omega \in \Omega$ such that $t \mapsto Z(\omega, t)$ is continuous. Then the following claims are true:*

(1) For all $0 \le t_1 \le t_2 \le T$, $L(\omega, t_2) \ge L(\omega, t_1)$ and $R(\omega, t_2) \ge R(\omega, t_1)$. In other words, both L and R are non-decreasing a.s.
(2) If $Y(\tau) = Y(\omega, \tau(\omega)) > 0$ for some $\tau = \tau(\omega) > 0$, then there exists a neighbourhood $(\tau_1, \tau_2) = (\tau_1(\omega), \tau_2(\omega)) \ni \tau$ such that for all $t_1, t_2 \in (\tau_1, \tau_2)$, $L(\omega, t_1) - L(\omega, t_2) = 0$, i.e.

$$R(\omega, t_2) - R(\omega, t_1) = \int_{t_1}^{t_2} \frac{\kappa}{Y^\gamma(\omega, s)} ds. \tag{4.24}$$

In other words, with probability 1, L can increase only at zeros of Y.
(3) The trajectory $t \mapsto Y(\omega, t)$, $t \in [0, T]$, is continuous, i.e., with probability 1, the limit process Y (and hence both R and L) has continuous paths.

Proof In this proof, for the notational convenience, we will omit ω in the parentheses.

(1) It is sufficient to prove that L is non-decreasing, since

$$R(t) = \int_0^t \frac{\kappa}{Y^\gamma(s)} ds + L(t),$$

where Y is non-negative, which immediately implies that R is non-decreasing, if L is. Observe that, for any $0 \le t_1 < t_2 \le T$,

$$\begin{aligned} L(t_2) - L(t_1) &= R(t_2) - R(t_1) - \int_{t_1}^{t_2} \frac{\kappa}{Y^\gamma(s)} ds \\ &= \lim_{\varepsilon \to 0} \int_{t_1}^{t_2} \frac{\kappa}{\left(Y_\varepsilon(s) \mathbb{1}_{\{Y_\varepsilon(s) > 0\}} + \varepsilon\right)^\gamma} ds - \int_{t_1}^{t_2} \frac{\kappa}{Y^\gamma(s)} ds \ge 0 \end{aligned}$$

by the Fatou lemma, i.e. the process L is indeed non-decreasing a.s.
(2) Let $\tau > 0$ be such that $Y(\tau) > 0$. Since $Y_\varepsilon(\tau) \uparrow Y(\tau)$ as $\varepsilon \to 0$, there exists $\varepsilon_0 > 0$ such that for all $\varepsilon \le \varepsilon_0$, $Y_\varepsilon(\tau) > 0$. Moreover, since $t \mapsto Y_{\varepsilon_0}(t)$ is continuous w.r.t. t, there exists a neighbourhood $(\tau_1, \tau_2) \ni \tau$ and some positive value $x > 0$ such that for all $t \in (\tau_1, \tau_2)$

$$Y_{\varepsilon_0}(t) > x.$$

In addition, Eq. (4.11) implies that for all $\varepsilon \le \varepsilon_0$

$$Y_\varepsilon(t) > x, \quad t \in (\tau_1, \tau_2).$$

Therefore, for any $\varepsilon \le \varepsilon_0$ and $t \in (\tau_1, \tau_2)$,

$$\frac{\kappa}{(Y_\varepsilon(t)\mathbb{1}_{\{Y_\varepsilon(t)>0\}} + \varepsilon)^\gamma} = \frac{\kappa}{(Y_\varepsilon(t) + \varepsilon)^\gamma} < \frac{\kappa}{x^\gamma}$$

whence, by the dominated convergence theorem, for any $t_1, t_2 \in (\tau_1, \tau_2), t_1 < t_2$,

$$\begin{aligned} R(t_2) - R(t_1) &= \lim_{\varepsilon \to 0} \int_{t_1}^{t_2} \frac{\kappa}{(Y_\varepsilon(s) + \varepsilon)^\gamma} ds \\ &= \int_{t_1}^{t_2} \frac{\kappa}{Y^\gamma(s)} ds, \end{aligned}$$

which implies the required claim.

(3) Note that the fact that R is non-decreasing implies that its discontinuities can only take the form of positive jumps and, moreover, left and right limits

$$R(t-) := \lim_{\delta \to 0+} R(t - \delta), \quad R(t+) := \lim_{\delta \to 0+} R(t + \delta)$$

are well-defined at any point $t \in (0, T)$. Next, since the limit process Y defined by (4.12) satisfies the equation

$$Y(t) = y_0 - \theta \int_0^t Y(s)ds + R(t) + Z(t), \quad t \in [0, T],$$

the points of discontinuity of Y coincide with the ones of R, can occur only in the form of positive jumps of the same size as the corresponding jumps of R and, finally, one can define limits

$$Y(t-) := \lim_{\delta \to 0+} Y(t - \delta), \quad Y(t+) := \lim_{\delta \to 0+} Y(t + \delta).$$

Assume that $\tau \in [0, T)$ is a point of discontinuity of R (and hence of Y). Note that $Y(\tau) = 0$, otherwise, τ cannot be a point of discontinuity of Y and R by (4.24) proved above. Let $h > 0$ be such that

$$Y(\tau+) = R(\tau+) = h,$$

and $\delta > 0$ be such that for all $t \in (\tau, \tau + \delta)$

$$Y(t) > \frac{9h}{10}$$

(note that such δ exists because Y has the right-sided limit at τ). Choose $\delta_0 < \delta$ such that

$$\left(\frac{4^\gamma \kappa}{h^\gamma} + e^{2|\theta|T} \left(y_0 + \left(\frac{2^\gamma}{y_0^\gamma} \vee 1 \right) \kappa T + T^\lambda \Lambda \right) \right) \delta_0 + \Lambda \delta_0^\lambda < \frac{h}{4},$$

where $\lambda \in (0, 1)$ is the order of Hölder continuity of Z and $\Lambda > 0$ is from (3.2). Since $Y_\varepsilon(\tau + \delta_0) \uparrow Y(\tau + \delta_0)$ as $\varepsilon \to 0$, there exists $0 < \varepsilon_0 < 1$ such that for all $\varepsilon \in (0, \varepsilon_0]$

$$Y_\varepsilon(\tau + \delta_0) > \frac{4h}{5}.$$

Furthermore, $Y_\varepsilon(\tau) \uparrow Y(\tau) = 0$ as $\varepsilon \to 0$, hence $Y_\varepsilon(\tau) \le 0$ for all $\varepsilon > 0$ and therefore for any $\varepsilon \in (0, \varepsilon_0]$ one can define

$$\tau_\varepsilon^- := \sup \left\{ s \in (\tau, \tau + \delta_0) : \; Y_\varepsilon(s) = \frac{h}{4} \right\}$$

and

$$\tau_\varepsilon^+ := \inf \left\{ s \in (\tau_\varepsilon^-, \tau + \delta_0) : \; Y_\varepsilon(s) = \frac{3h}{4} \right\}.$$

By continuity, it is evident that $Y_\varepsilon(\tau_\varepsilon^-) = \frac{h}{4}$, $Y_\varepsilon(\tau_\varepsilon^+) = \frac{3h}{4}$ and $Y_\varepsilon(t) \ge \frac{h}{4}$ for any $t \in [\tau_\varepsilon^-, \tau_\varepsilon^+]$. Therefore, for any $\varepsilon \in (0, \varepsilon_0]$, by Proposition 4.2,

$$\begin{aligned}
\frac{h}{2} &= Y_\varepsilon(\tau_\varepsilon^+) - Y_\varepsilon(\tau_\varepsilon^-) \\
&= \int_{\tau_\varepsilon^-}^{\tau_\varepsilon^+} \frac{\kappa}{(Y_\varepsilon(s) \mathbb{1}_{\{Y_\varepsilon(s)>0\}} + \varepsilon)^\gamma} ds - \theta \int_{\tau_\varepsilon^-}^{\tau_\varepsilon^+} Y_\varepsilon(s) ds + \left(Z(\tau_\varepsilon^+) - Z(\tau_\varepsilon^-) \right) \\
&\le \int_{\tau_\varepsilon^-}^{\tau_\varepsilon^+} \frac{\kappa}{(Y_\varepsilon(s) \mathbb{1}_{\{Y_\varepsilon(s)>0\}} + \varepsilon)^\gamma} ds + |\theta| \int_{\tau_\varepsilon^-}^{\tau_\varepsilon^+} Y_\varepsilon(s) ds + \left(Z(\tau_\varepsilon^+) - Z(\tau_\varepsilon^-) \right) \\
&\le \left(\frac{4^\gamma \kappa}{h^\gamma} + e^{2|\theta|T} \left(y_0 + \left(\frac{2^\gamma}{y_0^\gamma} \vee 1 \right) \kappa T + T^\lambda \Lambda \right) \right) (\tau_\varepsilon^+ - \tau_\varepsilon^-) + \Lambda (\tau_\varepsilon^+ - \tau_\varepsilon^-)^\lambda \\
&\le \left(\frac{4^\gamma \kappa}{h^\gamma} + e^{2|\theta|T} \left(y_0 + \left(\frac{2^\gamma}{y_0^\gamma} \vee 1 \right) \kappa T + T^\lambda \Lambda \right) \right) \delta_0 + \Lambda \delta_0^\lambda \\
&< \frac{h}{4},
\end{aligned}$$

which gives a contradiction. Therefore, $h = 0$ and $Y(\tau) = Y(\tau+)$, i.e. Y (and hence R) is right-continuous.

It remains to notice that $Y(t-) = Y(t)$ for any $t \in (0, T]$. Indeed, as mentioned above, Y is continuous at t, if $Y(t) > 0$. If $Y(t) = 0$, since $Y(t) \geq 0$ can only have positive jumps,

$$Y(t) - Y(t-) = -Y(t-)$$

and, since Y can potentially have only positive jumps, $Y(t-) = 0$. □

Note that the pre-limit processes Y_ε given by (4.10) are monotonically non-decreasing w.r.t. ε. Moreover, with probability 1, their pointwise limit is continuous function by Theorem 4.5. Therefore, by Dini's theorem, we immediately obtain the uniform convergence which is summarized in the following corollary.

Corollary 4.1 *For any $T > 0$,*

$$\sup_{t\in[0,T]} |Y(t) - Y_\varepsilon(t)| \to 0 \quad a.s. \text{ as } \quad \varepsilon \downarrow 0.$$

References

1. Azmoodeh, E., Sottinen, T., Viitasaari, L., Yazigi, A.: Necessary and sufficient conditions for Hölder continuity of Gaussian processes. Statist. Probab. Lett. **94**, 230–235 (2014). https://doi.org/10.1016/j.spl.2014.07.030
2. Di Nunno, G., Mishura, Y., Yurchenko-Tytarenko, A.: Sandwiched SDEs with unbounded drift driven by Hölder noises. Adv. Appl. Probab. **55**(3) (2023)
3. Friz, P.K., Hairer, M.: A Course on Rough Paths. Springer International Publishing (2014). https://doi.org/10.1007/978-3-319-08332-2
4. Friz, P.K., Victoir, N.B.: Multidimensional stochastic processes as rough paths: theory and applications. In: Cambridge Studies in Advanced Mathematics. Cambridge University Press (2010). https://doi.org/10.1017/CBO9780511845079
5. Mishura, Y., Yurchenko-Tytarenko, A.: Fractional Cox–Ingersoll–Ross process with non-zero "mean". Mod. Stoch. Theory Appl. **5**(1), 99–111 (2018). https://doi.org/10.15559/18-VMSTA97
6. Mishura, Y., Yurchenko-Tytarenko, A.: Fractional Cox–Ingersoll–Ross process with small Hurst indices. Mod. Stoch. Theory Appl. **6**(1), 13–39 (2018). https://doi.org/10.15559/18-VMSTA126
7. Young, L.C.: An inequality of the Hölder type, connected with Stieltjes integration. Acta Math. **67**, 251–282 (1936). https://doi.org/10.1007/bf02401743
8. Zähle, M.: Integration with respect to fractal functions and stochastic calculus. I. Prob. Theory Relat. Fields **111**(3), 333–374 (1998)

Chapter 5
Drift-Implicit Euler Scheme

Abstract In this chapter, we turn to a more practical question: how to simulate sample paths of sandwiched processes introduced in Chapter 3. Namely, we suggest the so-called *drift-implicit* or *backward* Euler numerical scheme that has an advantage of preserving the bounds of the sandwiched processes, both in one-sided and in two-sided cases. We provide a detailed convergence analysis of the method and support our findings with simulations.

5.1 Introduction

Having the theoretical fundamentals established in the previous chapters, we are now ready to delve into a more application-oriented topic: simulation of paths of sandwiched processes (3.1). Clearly, there is little hope to find the explicit solution to the stochastic integrodifferential equation (3.1), so one has to use some numerical approximation techniques. However, there are two difficulties coming from the shape of the equation and the properties of the solution:

(1) we are interested in drift functions b with an explosive behaviour that is described in e.g. Assumption 3.2 (for the one-sided case) or Assumption 3.5 (for the two-sided case), and this creates a number of technical challenges: more specifically, one cannot use the classical proof for convergence of Euler approximations with Gronwall inequality;
(2) since the main distinctive feature of sandwiched processes is the existence of deterministic bounds for their trajectories, one naturally wants a numerical scheme that preserves the "sandwich" properties (3.29) or (3.51) – and, generally, the classical Euler schemes cannot do that.

Luckily, there is a modification of the Euler discretization called the *drift-implicit* or *backward* Euler scheme initially introduced for the classical Cox-Ingersoll-Ross process in [1, 2, 4] and extended to the case of the fractional Brownian motion in [6–8]: it allows to solve the two problems mentioned above, and, in this chapter, we will explore it in full detail.

G. Di Nunno et al., *Volterra Volatility Models*, Springer Finance,
https://doi.org/10.1007/978-3-032-26576-0_5

5.2 General Construction of Drift-Implicit Euler Approximations

First of all, let us discuss the general construction of the scheme and the intuition behind it, without delving into much technical detail. Consider a generic sandwiched process given by the stochastic integrodifferential equation

$$Y(t) = y_0 + \int_0^t b(s, Y(s))ds + \int_0^t \mathcal{G}(t, s)a(s, Y(s))ds + Z(t), \quad t \in [0, T], \tag{5.1}$$

where all the parameters satisfy one of the assumptions from Chapter 2 that ensure existence, uniqueness and boundedness of the solution (i.e. either of the properties (3.29) or (3.51)). For some $N \in \mathbb{N}$, take a uniform partition $\{0 = t_0 < t_1 < \cdots < t_N = T\}$ with the diameter $t_{n+1} - t_n = \frac{T}{N} =: \Delta_N$, $n = 0, ..., N-1$, and define

$$\widehat{Y}(0) = y_0, \tag{5.2}$$

$$\begin{aligned}\widehat{Y}(t_n) &= \widehat{Y}(t_{n-1}) + b(t_n, \widehat{Y}(t_n))\Delta_N + Z(t_n) - Z(t_{n-1}) \\ &\quad + \sum_{i=1}^{n} a(t_{i-1}, \widehat{Y}(t_{i-1})) \int_{t_{i-1}}^{t_i} (\mathcal{G}(t_n, s) - \mathcal{G}(t_{n-1}, s))\, ds\end{aligned} \tag{5.3}$$

for $n = 1, ..., N$. Note that $\widehat{Y}(t_n)$ is present on *both left- and right-hand sides* of (5.3), i.e. the expression (5.3) *should be interpreted as an equation with respect to* $\widehat{Y}(t_n)$:

$$\begin{aligned}\widehat{Y}(t_n) - b(t_n, \widehat{Y}(t_n))\Delta_N &= \widehat{Y}(t_{n-1}) + Z(t_n) - Z(t_{n-1}) \\ &\quad + \sum_{i=1}^{n} a(t_{i-1}, \widehat{Y}(t_{i-1})) \int_{t_{i-1}}^{t_i} (\mathcal{G}(t_n, s) - \mathcal{G}(t_{n-1}, s))\, ds.\end{aligned} \tag{5.4}$$

An immediate question arises: does this equation have a solution? It turns out that, with some extra conditions on b and large enough values of N, it does and, moreover, this solution lies within the same bounds as the original sandwiched process! To understand the intuition behind it, let us consider the two-sided case and assume that $\frac{\partial b}{\partial y}(t, y)$ exists and is bounded from above with some constant C_b for all admissible (t, y). Then, if $N > C_b T$,

$$\frac{\partial}{\partial y}(y - b(t, y)\Delta_N) = 1 - \frac{\partial b}{\partial y}(t, y)\Delta_N \geq 1 - \frac{C_b T}{N} > 0,$$

i.e., for each $t \in [0, T]$, the function $y \mapsto y - b(t, y)\Delta_N$ is strictly increasing. In addition, $b(t, y) \to +\infty$ as $y \to \varphi(t)+$ and $b(t, y) \to -\infty$ as $y \to \psi(t)-$, so

$$\begin{aligned} y - b(t, y)\Delta_N &\to -\infty, \quad y \to \varphi(t)+, \\ y - b(t, y)\Delta_N &\to +\infty, \quad y \to \psi(t)-. \end{aligned}$$

In other words, for each $t \in [0, T]$ and for all large enough N, the function $y \mapsto y - b(t, y)\Delta_N$ is a one-to-one correspondence between $(\varphi(t), \psi(t))$ and $\mathbb{R}$ and hence the equation

$$y - b(t, y)\Delta_N = z$$

has a unique solution $y = y(z) \in (\varphi(t), \psi(t))$ for all $z \in \mathbb{R}$. This immediately yields that the scheme (5.2)–(5.3) is correctly defined for the two-sided case and gives an approximation $\widehat{Y}(t_n)$ such that

$$\varphi(t_n) < \widehat{Y}(t_n) < \psi(t_n), \quad n = 0, 1, ..., N.$$

For the one-sided case, the idea is almost identical: the only difference is that one has to ensure that $y - b(t, y)\Delta_N \to +\infty$ as $y \to +\infty$, which can be done by exploiting the specific conditions from Assumption 3.2 (we will discuss this matter in full detail in the dedicated Sect. 5.4).

Our next goal is to prove the convergence of the drift-implicit Euler scheme (5.2)–(5.3) to the solution of the original stochastic integro-differential equation.

Notation 5.1 1. In what follows, by C we will denote any positive deterministic constant the exact value of which is not important in the context. Note that C may change from line to line (or even within one line).
2. Throughout proofs in this chapter, we fix $\omega \in \Omega$ such that $Z(\omega, t)$, $t \in [0, T]$, is λ-Hölder continuous (for simplicity of notation, we will omit ω in the brackets).

5.3 Two-Sided Case

Let us start with a simpler two-sided case. Just as in Section 3.5, fix two λ-Hölder-continuous functions φ, ψ: $[0, T] \to \mathbb{R}$ such that $\varphi(t) < \psi(t)$, $t \in [0, T]$,

$$|\varphi(t_2) - \varphi(t_1)| + |\psi(t_2) - \psi(t_1)| \le L_{\varphi,\psi}|t_2 - t_1|^{\lambda}, \quad t_1, t_2 \in [0, T], \tag{5.5}$$

for some constant $L_{\varphi,\psi} > 0$, take a starting point $y_0 \in (\varphi(0), \psi(0))$, and denote

$$\begin{aligned} \mathcal{D}_{x_1,x_2} &:= \{(t, y) \mid t \in [0, T], y \in (\varphi(t) + x_1, \psi(t) - x_2)\}, \\ \overline{\mathcal{D}_{x_1,x_2}} &:= \{(t, y) \mid t \in [0, T], y \in [\varphi(t) + x_1, \psi(t) - x_2]\}. \end{aligned}$$

Next, consider the two-sided sandwiched process (5.1) driven by a λ-Hölder continuous noise Z, $\lambda \in (0, 1)$. As explained in Theorem 3.7, the coefficients satisfying Assumptions 3.1, 3.4, 3.5 and 3.6 ensure that (5.1) has the unique continuous solution $Y = \{Y(t),\ t \in [0, T]\}$ which, in addition, has the property

$$\varphi(t) < Y(t) < \psi(t), \quad t \in [0, T].$$

In order to get our first convergence result, we will need some extra technical conditions on a, b and $\mathcal{G}$, as summarized in the following theorem.

Theorem 5.1 *Let Assumptions 3.1, 3.4, 3.5 and 3.6 hold and, in addition,*

(B5) *the partial derivative $\frac{\partial b}{\partial y}$ exists, is continuous on $\mathcal{D}_{0,0}$ and there is a constant $C_b > 0$ such that for all $t \in [0, T]$*

$$\frac{\partial b}{\partial y}(t, y) < C_b, \quad (t, y) \in \mathcal{D}_{0,0}.$$

Then the following claims are true.

1. *The drift-implicit Euler scheme* (5.2)–(5.3) *is well-defined for all uniform partitions $\{0 = t_0 < t_1 < \cdots < t_N = T\}$, $t_{n+1} - t_n = \frac{T}{N}$, such that $N > C_b T$ and, moreover,*
$$\varphi(t_n) < \widehat{Y}(t_n) < \psi(t_n), \quad n = 0, 1, ..., N.$$
2. *For all $n = 0, 1, ..., N$ and $\omega \in \Omega$ such that the corresponding path $\{Z(\omega, t),\ t \in [0, T]\}$ is λ-Hölder continuous, there exists a deterministic constant C_1 that does not depend on the partition and*

$$\varphi(t_n) + \frac{C_1}{(1 + \Lambda)^{\frac{1}{\gamma\lambda+\lambda-1}}} \le \widehat{Y}(t_n) \le \psi(t_n) - \frac{C_1}{(1 + \Lambda)^{\frac{1}{\gamma\lambda+\lambda-1}}}. \tag{5.6}$$

Remark 5.1 The conditions **(A2)** and **(B2')** from Theorem 5.2 above are refinements of Assumptions 3.6 and 3.5**(B2)** that impose some standard restrictions on the variability of a and b w.r.t. t. The regularity condition **(B5)** is needed to make sure that the drift-implicit scheme is well-defined.

Proof First of all, we note that the condition **(B5)** in combination with the arguments from Sect. 5.2 immediately yields the first claim. As for the second statement, let us prove that

$$\widehat{Y}(t_n) \ge \varphi(t_n) + \frac{C_1}{(1 + \Lambda)^{\frac{1}{\gamma\lambda+\lambda-1}}}, \quad n = 0, 1, 2, ..., N,$$

as the other side of the inequality (5.6) can be obtained in a similar manner.

Next, let us prove (5.6). Fix any $N > C_b T$, denote the diameter of partition $\Delta_N := \frac{T}{N}$ and put

$$\beta := \frac{\lambda^{\frac{\lambda}{1-\lambda}} - \lambda^{\frac{1}{1-\lambda}}}{c^{\frac{\lambda}{1-\lambda}}} > 0,$$

where c is from Assumption 3.5(**B3**),

$$L_2 := L_{\varphi,\psi} + \max_{(t,y)\in\overline{\mathcal{D}_{0,0}}} |a(t,y)|\sqrt{T}\ell_\lambda + (2\beta)^{\lambda-1}\left(\frac{(y_0-\varphi(0))\wedge y_* \wedge (\psi(0)-y_0)}{2}\right)^{1-\lambda-\gamma\lambda} > 0,$$

with the constants y_* and γ also from Assumption 3.5(**B3**), $L_{\varphi,\psi} > 0$ being from (5.5), and $\ell_\lambda > 0$ being from Assumption 3.4. Note also that $\max_{(t,y)\in\overline{\mathcal{D}_{0,0}}} |a(t,y)| < \infty$ since a is continuous on the compact set $\overline{\mathcal{D}_{0,0}}$ by Assumption 3.6. Finally, define

$$\varepsilon := \frac{1}{(2\beta)^{\frac{1-\lambda}{\gamma\lambda+\lambda-1}}(L_2+\Lambda)^{\frac{1}{\gamma\lambda+\lambda-1}}}.$$

Note that it is easy to check that $\varepsilon < y_0 - \varphi(0)$, $\varepsilon < \psi(0) - y_0$ and $\varepsilon < y_*$.

For any $n = 1, ..., N$, consider the last point of the partition before t_n, for which $\widehat{Y}(t_k)$ stays above ε, i.e.

$$\kappa(n) := \max\{k = 0, ..., n-1 \mid \widehat{Y}(t_k) \ge \varphi(t_k) + \varepsilon\}$$

(such point exists since $\widehat{Y}(t_0) - \varphi(0) = y_0 - \varphi(0) > \varepsilon$). Then, for all $k = \kappa(n) + 1, ..., n$ we have that $\widehat{Y}(t_k) < \varepsilon < y_*$ and therefore, using Assumption 3.5(**B3**), we obtain that, with probability 1,

$$\begin{aligned}
\widehat{Y}(t_n) - \varphi(t_n) &= \widehat{Y}(t_{\kappa(n)}) - \varphi(t_n) + \Delta_N \sum_{k=\kappa(n)+1}^{n} b(t_k, \widehat{Y}(t_k)) + Z(t_n) - Z(t_{\kappa(n)}) \\
&\quad + \sum_{i=1}^{n} a(t_{i-1}, \widehat{Y}(t_{i-1})) \int_{t_{i-1}}^{t_i} \big(\mathcal{G}(t_n, s) - \mathcal{G}(t_{\kappa(n)}, s)\big)\, ds \\
&\ge \varepsilon + \varphi(t_{\kappa(n)}) - \varphi(t_n) + \frac{c}{\varepsilon^\gamma}(t_n - t_{\kappa(n)}) + Z(t_n) - Z(t_{\kappa(n)}) \\
&\quad + \sum_{i=1}^{n} a(t_{i-1}, \widehat{Y}(t_{i-1})) \int_{t_{i-1}}^{t_i} \big(\mathcal{G}(t_n, s) - \mathcal{G}(t_{\kappa(n)}, s)\big)\, ds \\
&\ge \varepsilon + \frac{c}{\varepsilon^\gamma}(t_n - t_{\kappa(n)}) - (L_{\varphi,\psi} + \Lambda)(t_n - t_{\kappa(n)})^\lambda \\
&\quad + \sum_{i=1}^{n} a(t_{i-1}, \widehat{Y}(t_{i-1})) \int_{t_{i-1}}^{t_i} \big(\mathcal{G}(t_n, s) - \mathcal{G}(t_{\kappa(n)}, s)\big)\, ds.
\end{aligned}$$

By Assumption 3.4, we can also write

$$\begin{aligned}
&\left|\sum_{i=1}^{n} a(t_{i-1}, \widehat{Y}(t_{i-1})) \int_{t_{i-1}}^{t_i} \left(\mathcal{G}(t_n, s) - \mathcal{G}(t_{\kappa(n)}, s)\right) ds\right| \\
&\quad \le \sum_{i=1}^{n} |a(t_{i-1}, \widehat{Y}(t_{i-1}))| \int_{t_{i-1}}^{t_i} \left|\mathcal{G}(t_n, s) - \mathcal{G}(t_{\kappa(n)}, s)\right| ds \\
&\quad \le \max_{(t,y)\in\overline{\mathcal{D}_{0,0}}} |a(t, y)| \sum_{i=1}^{n} \int_{t_{i-1}}^{t_i} \left|\mathcal{G}(t_n, s) - \mathcal{G}(t_{\kappa(n)}, s)\right| ds \\
&\quad \le \max_{(t,y)\in\overline{\mathcal{D}_{0,0}}} |a(t, y)| \sqrt{T} \sqrt{\int_0^T \left|\mathcal{G}(t_n, s) - \mathcal{G}(t_{\kappa(n)}, s)\right|^2 ds} \\
&\quad \le \max_{(t,y)\in\overline{\mathcal{D}_{0,0}}} |a(t, y)| \sqrt{T} \ell_\lambda |t_n - t_{\kappa(n)}|^\lambda,
\end{aligned}$$

so, summarizing all of the above, we get

$$\begin{aligned}
\widehat{Y}(t_n) - \varphi(t_n) &\ge \varepsilon + \frac{c}{\varepsilon^\gamma}(t_n - t_{\kappa(n)}) - (L_{\varphi,\psi} + \Lambda + \max_{(t,y)\in\overline{\mathcal{D}_{0,0}}} |a(t, y)| \sqrt{T} \ell_\lambda)(t_n - t_{\kappa(n)})^\lambda \\
&\ge \varepsilon + \frac{c}{\varepsilon^\gamma}(t_n - t_{\kappa(n)}) - (L_2 + \Lambda)(t_n - t_{\kappa(n)})^\lambda.
\end{aligned}$$

Consider a function $F_\varepsilon : \mathbb{R}_+ \to \mathbb{R}$ such that

$$F_\varepsilon(t) = \varepsilon + \frac{c}{\varepsilon^\gamma} t - (L_2 + \Lambda) t^\lambda.$$

It is straightforward to verify that F_ε attains its minimum at

$$t_* := \left(\frac{\lambda}{c}\right)^{\frac{1}{1-\lambda}} \varepsilon^{\frac{\gamma}{1-\lambda}} (L_2 + \Lambda)^{\frac{1}{1-\lambda}}$$

and, taking into account the explicit form of ε,

$$\begin{aligned}
F_\varepsilon(t_*) &= \varepsilon + \frac{\lambda^{\frac{1}{1-\lambda}}}{c^{\frac{\lambda}{1-\lambda}}} \varepsilon^{\frac{\gamma\lambda}{1-\lambda}} (L_2 + \Lambda)^{\frac{1}{1-\lambda}} - \frac{\lambda^{\frac{\lambda}{1-\lambda}}}{c^{\frac{\lambda}{1-\lambda}}} \varepsilon^{\frac{\gamma\lambda}{1-\lambda}} (L_2 + \Lambda)^{\frac{1}{1-\lambda}} \\
&= \varepsilon - \beta \varepsilon^{\frac{\gamma\lambda}{1-\lambda}} (L_2 + \Lambda)^{\frac{1}{1-\lambda}} = \frac{1}{2^{\frac{\gamma\lambda}{\gamma\lambda+\lambda-1}} \beta^{\frac{1-\lambda}{\gamma\lambda+\lambda-1}} (L_2 + \Lambda)^{\frac{1}{\gamma\lambda+\lambda-1}}} \\
&= \frac{\varepsilon}{2}.
\end{aligned}$$

In other words, even if $\widehat{Y}(t_n) < \varphi(t_n) + \varepsilon$, we still have that, with probability 1,

$$\widehat{Y}(t_n) - \varphi(t_n) \geq F_\varepsilon(t_n - t_{\kappa(n)}) \geq F_\varepsilon(t_*) = \frac{\varepsilon}{2},$$

and thus, with probability 1, for any $n = 0, 1, ..., N$

$$\begin{aligned}\widehat{Y}(t_n) &\geq \varphi(t_n) + \frac{\varepsilon}{2} = \varphi(t_n) + \frac{1}{2^{\frac{\gamma\lambda}{\gamma\lambda+\lambda-1}} \beta^{\frac{1-\lambda}{\gamma\lambda+\lambda-1}} (L_2 + \Lambda)^{\frac{1}{\gamma\lambda+\lambda-1}}} \\ &=: \varphi(t_n) + \frac{L_1}{(L_2 + \Lambda)^{\frac{1}{\gamma\lambda+\lambda-1}}},\end{aligned}$$

where $L_1 := \frac{1}{2^{\frac{\gamma\lambda}{\gamma\lambda+\lambda-1}} \beta^{\frac{1-\lambda}{\gamma\lambda+\lambda-1}}}$, which yields (5.6) as required. □

Having Theorem 5.1, we are now ready to prove our first result for convergence of the drift-implicit Euler scheme (5.2)–(5.3).

Theorem 5.2 *Let Assumptions 3.1, 3.4, 3.5 and 3.6 hold and* $Y = \{Y(t),\ t \in [0, T]\}$ *be the unique continuous solution to* (5.1). *Assume also that*

(a1) *the partial derivative* $\frac{\partial a}{\partial y}$ *exists and is continuous on* $\overline{\mathcal{D}_{0,0}}$*;*
(a2) *there exists a constant* L_a *such that, for any* $(t_1, y_1), (t_2, y_2) \in \overline{\mathcal{D}_{0,0}}$,

$$|a(t_1, y_2) - a(t_2, y_1)| \leq L_a(|t_1 - t_2|^\lambda + |y_2 - y_1|);$$

(B2') *for any* $\varepsilon \in (0, \frac{1}{2}\min_{t\in[0,T]} |\psi(t) - \varphi(t)|)$, *there is a constant* $c_\varepsilon > 0$ *such that for any* $(t_1, y_1), (t_2, y_2) \in \overline{\mathcal{D}_{\varepsilon,\varepsilon}}$

$$|b(t_1, y_1) - b(t_2, y_2)| \leq c_\varepsilon \left(|t_1 - t_2|^\lambda + |y_1 - y_2|\right);$$

(B5) *the partial derivative* $\frac{\partial b}{\partial y}$ *exists, is continuous on* $\mathcal{D}_{0,0}$ *and there is a constant* $C_b > 0$ *such that for all* $t \in [0, T]$

$$\frac{\partial b}{\partial y}(t, y) < C_b, \quad (t, y) \in \mathcal{D}_{0,0}.$$

Then, with probability 1, the drift-implicit Euler approximation $\widehat{Y}$ *defined by* (5.2)–(5.3) *converges to the solution of* (5.1) *in the following sense:*

$$\sup_{n=0,1,\dots,N} |Y(t_n) - \widehat{Y}(t_n)| \to 0, \quad N \to \infty. \tag{5.7}$$

Proof Fix any $N > C_b T$, denote the diameter of partition $\Delta_N := \frac{T}{N}$ and put $e_n := Y(t_n) - \widehat{Y}(t_n)$, $\Delta Z_n := Z(t_n) - Z(t_{n-1})$. We can write

$$\begin{aligned}
e_n &= Y(t_{n-1}) + \int_{t_{n-1}}^{t_n} b(s, Y(s))ds + \Delta Z_n + \int_0^{t_n} (\mathcal{G}(t_n, s) - \mathcal{G}(t_{n-1}, s))\, a(s, Y(s))ds \\
&\quad - \widehat{Y}(t_{n-1}) - b(t_n, \widehat{Y}(t_n))\Delta_N - \Delta Z_n \\
&\quad - \sum_{i=1}^{n} a(t_{i-1}, \widehat{Y}(t_{i-1})) \int_{t_{i-1}}^{t_i} (\mathcal{G}(t_n, s) - \mathcal{G}(t_{n-1}, s))\, ds \\
&= e_{n-1} + \big(b(t_n, Y(t_n)) - b(t_n, \widehat{Y}(t_n))\big)\, \Delta_N \\
&\quad + \sum_{i=1}^{n} \int_{t_{i-1}}^{t_i} (\mathcal{G}(t_n, s) - \mathcal{G}(t_{n-1}, s)) \big(a(t_{i-1}, Y(t_{i-1})) - a(t_{i-1}, \widehat{Y}(t_{i-1}))\big)\, ds \\
&\quad + \int_{t_{n-1}}^{t_n} (b(s, Y(s)) - b(t_n, Y(t_n)))ds \\
&\quad + \sum_{i=1}^{n} \int_{t_{i-1}}^{t_i} (\mathcal{G}(t_n, s) - \mathcal{G}(t_{n-1}, s))\, (a(s, Y(s)) - a(t_{i-1}, Y(t_{i-1})))\, ds.
\end{aligned} \tag{5.8}$$

By the mean value theorem, for any $i = 1, \ldots, N$, we can write

$$\big(a(t_i, Y(t_i)) - a(t_i, \widehat{Y}(t_i))\big) = \frac{\partial a}{\partial y}(t_i, \Theta_{a,i})(Y(t_i) - \widehat{Y}(t_i)) = \frac{\partial a}{\partial y}(t_i, \Theta_{a,i})e_i,$$

and

$$\big(b(t_i, Y(t_i)) - b(t_i, \widehat{Y}(t_i))\big) = \frac{\partial b}{\partial y}(t_i, \Theta_{b,i})(Y(t_i) - \widehat{Y}(t_i)) = \frac{\partial b}{\partial y}(t_i, \Theta_{b,i})e_i,$$

with $\Theta_{a,i}, \Theta_{b,i} \in (Y(t_i) \wedge \widehat{Y}(t_i), Y(t_i) \vee \widehat{Y}(t_i)) \subset (\varphi(t_i), \psi(t_i))$. Using this, we can rewrite (5.8) as follows:

$$\begin{aligned}
&\left(1-\frac{\partial b}{\partial y}(t_n, \Theta_{b,n})\Delta_N\right)e_n \\
&\quad = e_{n-1} + \sum_{i=1}^{n}\left(\frac{\partial a}{\partial y}(t_{i-1}, \Theta_{a,i-1})\int_{t_{i-1}}^{t_i}(\mathcal{G}(t_n, s) - \mathcal{G}(t_{n-1}, s))\,ds\right)e_{i-1} \\
&\qquad + \int_{t_{n-1}}^{t_n}(b(s, Y(s)) - b(t_n, Y(t_n)))ds \\
&\qquad + \sum_{i=1}^{n}\int_{t_{i-1}}^{t_i}(\mathcal{G}(t_n, s) - \mathcal{G}(t_{n-1}, s))\,(a(s, Y(s)) - a(t_{i-1}, Y(t_{i-1})))\,ds,
\end{aligned} \tag{5.9}$$

where

$$1 - \frac{\partial b}{\partial y}(t_n, \Theta_{b,n})\Delta_N > 1 - C_b\Delta_N > 0$$

since $N > C_b T$.

Next, for any $k = 1, ..., n$, denote

$$\zeta_0 := 1, \quad \zeta_k := \prod_{i=1}^{k}\left(1 - \frac{\partial b}{\partial y}(t_i, \Theta_{b,i})\Delta_N\right)$$

and define $\tilde{e}_n := \zeta_n e_n$. By multiplying both sides of (5.9) by ζ_{n-1}, we obtain that

$$\begin{aligned}
\tilde{e}_n &= \tilde{e}_{n-1} + \zeta_{n-1}\sum_{i=1}^{n}\left(\frac{\partial a}{\partial y}(t_{i-1}, \Theta_{a,i-1})\int_{t_{i-1}}^{t_i}(\mathcal{G}(t_n, s) - \mathcal{G}(t_{n-1}, s))\,ds\right)e_{i-1} \\
&\quad + \zeta_{n-1}\int_{t_{n-1}}^{t_n}(b(s, Y(s)) - b(t_n, Y(t_n)))ds \\
&\quad + \zeta_{n-1}\sum_{i=1}^{n}\int_{t_{i-1}}^{t_i}(\mathcal{G}(t_n, s) - \mathcal{G}(t_{n-1}, s))\,(a(s, Y(s)) - a(t_{i-1}, Y(t_{i-1})))\,ds,
\end{aligned} \tag{5.10}$$

and, expanding the terms $\tilde{e}_{i-1}$ in (5.10) one by one for $i = n, n-1, ..., 1$ and taking into account that $\tilde{e}_0 = 0$, we obtain that

$$\tilde{e}_n = \sum_{k=1}^{n}\sum_{i=1}^{k}\zeta_{k-1}\int_{t_{i-1}}^{t_i}(\mathcal{G}(t_k,s)-\mathcal{G}(t_{k-1},s))\,ds\frac{\partial a}{\partial y}(t_{i-1},\Theta_{a,i-1})e_{i-1}$$
$$+\sum_{k=1}^{n}\zeta_{k-1}\int_{t_{k-1}}^{t_k}(b(s,Y(s))-b(t_k,Y(t_k)))ds$$
$$+\sum_{k=1}^{n}\sum_{i=1}^{k}\zeta_{k-1}\int_{t_{i-1}}^{t_i}(\mathcal{G}(t_k,s)-\mathcal{G}(t_{k-1},s))\,(a(s,Y(s))-a(t_{i-1},Y(t_{i-1})))\,ds.$$

Therefore

$$e_n = \sum_{k=1}^{n}\sum_{i=1}^{k}\frac{\zeta_{k-1}}{\zeta_n}\int_{t_{i-1}}^{t_i}(\mathcal{G}(t_k,s)-\mathcal{G}(t_{k-1},s))\,ds\frac{\partial a}{\partial y}(t_{i-1},\Theta_{a,i-1})e_{i-1}$$
$$+\sum_{k=1}^{n}\frac{\zeta_{k-1}}{\zeta_n}\int_{t_{k-1}}^{t_k}(b(s,Y(s))-b(t_k,Y(t_k)))ds$$
$$+\sum_{k=1}^{n}\sum_{i=1}^{k}\frac{\zeta_{k-1}}{\zeta_n}\int_{t_{i-1}}^{t_i}(\mathcal{G}(t_k,s)-\mathcal{G}(t_{k-1},s))\,(a(s,Y(s))-a(t_{i-1},Y(t_{i-1})))\,ds. \tag{5.11}$$

It follows from **(A1)** that

$$\left|\sum_{k=1}^{n}\sum_{i=1}^{k}\frac{\zeta_{k-1}}{\zeta_n}\int_{t_{i-1}}^{t_i}(\mathcal{G}(t_k,s)-\mathcal{G}(t_{k-1},s))\,ds\frac{\partial a}{\partial y}(t_{i-1},\Theta_{a,i-1})e_{i-1}\right|$$
$$=\left|\sum_{i=1}^{n}\left(\int_{t_{i-1}}^{t_i}\sum_{k=i}^{n}\frac{\zeta_{k-1}}{\zeta_n}(\mathcal{G}(t_k,s)-\mathcal{G}(t_{k-1},s))\,ds\right)\frac{\partial a}{\partial y}(t_{i-1},\Theta_{a,i-1})e_{i-1}\right|$$
$$\le C\sum_{i=1}^{n}\left|\int_{t_{i-1}}^{t_i}\sum_{k=i}^{n}\frac{\zeta_{k-1}}{\zeta_n}(\mathcal{G}(t_k,s)-\mathcal{G}(t_{k-1},s))\,ds\right||e_{i-1}|. \tag{5.12}$$

Note that

$$\frac{\zeta_k}{\zeta_n}-\frac{\zeta_{k-1}}{\zeta_n}=\frac{\zeta_{k-1}\left(1-\frac{\partial b}{\partial y}(t_k,\Theta_{b,k})\Delta_N\right)-\zeta_{k-1}}{\zeta_n}$$
$$=-\frac{\partial b}{\partial y}(t_k,\Theta_{b,k})\frac{\zeta_{k-1}}{\zeta_n}\Delta_N.$$

Therefore, by Abel's summation-by-parts formula, for any $s\in(t_{i-1},t_i)$,

$$\begin{aligned}&\sum_{k=i}^{n}\frac{\zeta_{k-1}}{\zeta_n}\left(\mathcal{G}(t_k,s)-\mathcal{G}(t_{k-1},s)\right)\\&\quad=\frac{\zeta_{n-1}}{\zeta_n}\mathcal{G}(t_n,s)-\frac{\zeta_{i-1}}{\zeta_n}\mathcal{G}(t_{i-1},s)-\sum_{k=i+1}^{n}\mathcal{G}(t_{k-1},s)\left(\frac{\zeta_{k-1}}{\zeta_n}-\frac{\zeta_{k-2}}{\zeta_n}\right)\\&\quad=\frac{\zeta_{n-1}}{\zeta_n}\mathcal{G}(t_n,s)+\sum_{k=i+1}^{n}\mathcal{G}(t_{k-1},s)\frac{\partial b}{\partial y}(t_{k-1},\Theta_{b,k-1})\frac{\zeta_{k-2}}{\zeta_n}\Delta_N,\end{aligned}\tag{5.13}$$

and hence we can re-write (5.12) as follows:

$$\begin{aligned}&\left|\sum_{k=1}^{n}\sum_{i=1}^{k}\frac{\zeta_{k-1}}{\zeta_n}\int_{t_{i-1}}^{t_i}\left(\mathcal{G}(t_k,s)-\mathcal{G}(t_{k-1},s)\right)ds\frac{\partial a}{\partial y}(t_{i-1},\Theta_{a,i-1})e_{i-1}\right|\\&\qquad\le C\sum_{i=1}^{n}\left(\int_{t_{i-1}}^{t_i}\left|\frac{\zeta_{n-1}}{\zeta_n}\right|\left|\mathcal{G}(t_n,s)\right|ds\right)|e_{i-1}|\\&\qquad\quad+C\Delta_N\sum_{i=1}^{n}\left(\sum_{k=i+1}^{n}\left|\frac{\partial b}{\partial y}(t_{k-1},\Theta_{b,k-1})\right|\left|\frac{\zeta_{k-2}}{\zeta_n}\right|\int_{t_{i-1}}^{t_i}|\mathcal{G}(t_{k-1},s)|ds\right)|e_{i-1}|.\end{aligned}\tag{5.14}$$

Note that, by Theorem 3.8 and Theorem 5.1, there exists a constant $C_1 > 0$ that does not depend on the partition such that, for any $n = 0, 1, 2, ..., N$,

$$\begin{aligned}\varphi(t_n)+\frac{C_1}{(1+\Lambda)^{\frac{1}{\gamma\lambda+\lambda-1}}}&\le Y(t_n)\le\psi(t_n)-\frac{C_1}{(1+\Lambda)^{\frac{1}{\gamma\lambda+\lambda-1}}},\\\varphi(t_n)+\frac{C_1}{(1+\Lambda)^{\frac{1}{\gamma\lambda+\lambda-1}}}&\le\widehat{Y}(t_n)\le\psi(t_n)-\frac{C_1}{(1+\Lambda)^{\frac{1}{\gamma\lambda+\lambda-1}}},\end{aligned}$$

which, in turn, implies that

$$\varphi(t_n)+\frac{C_1}{(1+\Lambda)^{\frac{1}{\gamma\lambda+\lambda-1}}}\le\Theta_{b,n}\le\psi(t_n)-\frac{C_1}{(1+\Lambda)^{\frac{1}{\gamma\lambda+\lambda-1}}}.$$

In other words, for any $n = 0, 1, 2, ..., N$, $(t_n, \Theta_{b,n}) \in \overline{\mathcal{D}_{\xi,\xi}}$ with

$$\xi:=\frac{C_1}{(1+\Lambda)^{\frac{1}{\gamma\lambda+\lambda-1}}},\tag{5.15}$$

and, by the continuity of $\frac{\partial b}{\partial y}$ on the compact set $\overline{\mathcal{D}_{\xi,\xi}}$ thanks to **(B5)**,

$$\sup_{N>C_bT}\max_{n=0,1,\dots,N}\left|\frac{\partial b}{\partial y}(t_{k-1},\Theta_{b,k-1})\right| \le \max_{(t,y)\in\overline{\mathcal{D}_{\xi,\xi}}}\left|\frac{\partial b}{\partial y}(t,y)\right| =: \eta_b < \infty.$$

In addition, for any $k, n = 0, \dots, N$, $k < n$,

$$\begin{aligned}\frac{\zeta_k}{\zeta_n} &= \prod_{i=k+1}^{n}\left(1-\frac{\partial b}{\partial y}(t_i,\Theta_{b,i})\Delta_N\right)^{-1} \le \prod_{i=k+1}^{N}(1-C_b\Delta_N)^{-1}\\ &\le (1-C_b\Delta_N)^{-N} = \left(1-\frac{C_bT}{N}\right)^{-N} \to e^{C_bT}, \quad N\to\infty,\end{aligned}$$

whence there exists a constant C that does not depend on k, n or N such that

$$\frac{\zeta_k}{\zeta_n} \le C. \tag{5.16}$$

Summarizing all of the above, we can continue (5.14) by writing

$$\begin{aligned}&\left|\sum_{k=1}^{n}\sum_{i=1}^{k}\frac{\zeta_{k-1}}{\zeta_n}\int_{t_{i-1}}^{t_i}(\mathcal{G}(t_k,s)-\mathcal{G}(t_{k-1},s))\,ds\frac{\partial a}{\partial y}(t_{i-1},\Theta_{a,i-1})e_{i-1}\right|\\ &\quad\le C\sum_{i=1}^{n}\left(\int_{t_{i-1}}^{t_i}|\mathcal{G}(t_n,s)|\,ds+\Delta_N\eta_b\int_{t_{i-1}}^{t_i}\sum_{k=i+1}^{n}|\mathcal{G}(t_{k-1},s)|ds\right)|e_{i-1}|,\end{aligned} \tag{5.17}$$

where the constant $C > 0$ is deterministic and does not depend on the partition whereas η_b is a random variable.

Next, since $(t, Y(t)) \in \overline{\mathcal{D}_{\xi,\xi}}$ for all $t \in [0, T]$, by **(B2')** as well as (5.16) and Theorem 3.9, we can deduce that

$$\begin{aligned}&\left|\sum_{k=1}^{n}\frac{\zeta_{k-1}}{\zeta_n}\int_{t_{k-1}}^{t_k}(b(s,Y(s))-b(t_k,Y(t_k)))ds\right|\\ &\quad\le Cc_\xi\sum_{k=1}^{n}\int_{t_{k-1}}^{t_k}|s-t_k|^\lambda\,ds + Cc_\xi\sum_{k=1}^{n}\int_{t_{k-1}}^{t_k}|Y(s)-Y(t_i)|\,ds\\ &\quad\le Cc_\xi\sum_{k=1}^{n}\int_{t_{k-1}}^{t_k}|s-t_k|^\lambda\,ds + Cc_\xi\Upsilon\sum_{i=1}^{n}\int_{t_{i-1}}^{t_i}|s-t_i|^\lambda\,ds\\ &\quad= Cc_\xi(1+\Upsilon)\sum_{i=1}^{n}\frac{1}{(1+\lambda)}\Delta_N^{1+\lambda}\\ &\quad\le Cc_\xi(1+\Upsilon)\Delta_N^{\lambda},\end{aligned} \tag{5.18}$$

where c_ξ is from **(B2')** that corresponds to $\varepsilon = \xi$ and Υ is from Theorem 3.9 and has the form $\Upsilon = C(1 + \Lambda + c_\xi + c_\xi \Lambda)$.

Finally, using (5.13) and **(A2)** again,

$$\begin{aligned}
&\left|\sum_{k=1}^{n}\sum_{i=1}^{k}\frac{\zeta_{k-1}}{\zeta_n}\int_{t_{i-1}}^{t_i}\left(\mathcal{G}(t_k,s)-\mathcal{G}(t_{k-1},s)\right)\left(a(s,Y(s))-a(t_{i-1},Y(t_{i-1}))\right)ds\right|\\
&=\left|\sum_{i=1}^{n}\int_{t_{i-1}}^{t_i}\left(\sum_{k=i}^{n}\frac{\zeta_{k-1}}{\zeta_n}\left(\mathcal{G}(t_k,s)-\mathcal{G}(t_{k-1},s)\right)\right)\left(a(s,Y(s))-a(t_{i-1},Y(t_{i-1}))\right)ds\right|\\
&\le\left|\sum_{i=1}^{n}\int_{t_{i-1}}^{t_i}\frac{\zeta_{n-1}}{\zeta_n}\mathcal{G}(t_n,s)\left(a(s,Y(s))-a(t_{i-1},Y(t_{i-1}))\right)ds\right|\\
&\quad+\left|\sum_{i=1}^{n}\int_{t_{i-1}}^{t_i}\left(\sum_{k=i+1}^{n}\mathcal{G}(t_{k-1},s)\frac{\partial b}{\partial y}(t_{k-1},\Theta_{b,k-1})\frac{\zeta_{k-2}}{\zeta_n}\Delta_N\right)\left(a(s,Y(s))-a(t_{i-1},Y(t_{i-1}))\right)ds\right|\\
&\le C\sum_{i=1}^{n}\int_{t_{i-1}}^{t_i}|\mathcal{G}(t_n,s)|\left|a(s,Y(s))-a(t_{i-1},Y(t_{i-1}))\right|ds\\
&\quad+C\eta_b\Delta_N\sum_{i=1}^{n}\int_{t_{i-1}}^{t_i}\sum_{k=i+1}^{n}|\mathcal{G}(t_{k-1},s)|\left|a(s,Y(s))-a(t_{i-1},Y(t_{i-1}))\right|ds\\
&\le C\sum_{i=1}^{n}\int_{t_{i-1}}^{t_i}|\mathcal{G}(t_n,s)|\left|Y(s)-Y(t_{i-1})\right|ds+C\sum_{i=1}^{n}\int_{t_{i-1}}^{t_i}|\mathcal{G}(t_n,s)|\left|s-t_{i-1}\right|^{\lambda}ds\\
&\quad+C\eta_b\Delta_N\sum_{i=1}^{n}\int_{t_{i-1}}^{t_i}\sum_{k=i+1}^{n}|\mathcal{G}(t_{k-1},s)|\left|Y(s)-Y(t_{i-1})\right|ds\\
&\quad+C\eta_b\Delta_N\sum_{i=1}^{n}\int_{t_{i-1}}^{t_i}\sum_{k=i+1}^{n}|\mathcal{G}(t_{k-1},s)|\left|s-t_{i-1}\right|^{\lambda}ds,
\end{aligned}$$

so, exploiting the Hölder continuity of Y guaranteed by Theorem 3.9, we can finally write

$$
\begin{aligned}
&\left|\sum_{k=1}^{n}\sum_{i=1}^{k}\frac{\zeta_{k-1}}{\zeta_n}\int_{t_{i-1}}^{t_i}\big(\mathcal{G}(t_k,s)-\mathcal{G}(t_{k-1},s)\big)\big(a(s,Y(s))-a(t_{i-1},Y(t_{i-1}))\big)\,ds\right| \\
&\le C(1+\Upsilon)\sum_{i=1}^{n}\int_{t_{i-1}}^{t_i}|\mathcal{G}(t_n,s)|\,|s-t_{i-1}|^{\lambda}\,ds \\
&\quad + C\Delta_N\eta_b(1+\Upsilon)\sum_{i=1}^{n}\int_{t_{i-1}}^{t_i}\sum_{k=i+1}^{n}|\mathcal{G}(t_{k-1},s)|\,|s-t_{i-1}|^{\lambda}\,ds \\
&\le C(1+\Upsilon)\Delta_N^{\lambda}\int_0^{t_n}|\mathcal{G}(t_n,s)|ds + C\Delta_N^{1+\lambda}\eta_b(1+\Upsilon)\sum_{k=2}^{n}\sum_{i=1}^{k-1}\int_{t_{i-1}}^{t_i}|\mathcal{G}(t_{k-1},s)|ds \\
&\le C(1+\Upsilon)\Delta_N^{\lambda} + C\Delta_N^{1+\lambda}\eta_b(1+\Upsilon)\sum_{k=2}^{n}\int_0^{t_{k-1}}|\mathcal{G}(t_{k-1},s)|ds \\
&\le C(1+\eta_b+\Upsilon+\eta_b\Upsilon)\Delta_N^{\lambda}.
\end{aligned}
\tag{5.19}
$$

Therefore, by (5.11), (5.17), (5.18) and (5.19),

$$
\begin{aligned}
|e_n| &\le Cc_{\xi}(1+\Upsilon)\Delta_N^{\lambda} + C(1+\eta_b+\Upsilon+\eta_b\Upsilon)\Delta_N^{\lambda} \\
&\quad + C\sum_{i=1}^{n}\left(\int_{t_{i-1}}^{t_i}|\mathcal{G}(t_n,s)|\,ds + \Delta_N\eta_b\int_{t_{i-1}}^{t_i}\sum_{k=i+1}^{n}|\mathcal{G}(t_{k-1},s)|ds\right)|e_{i-1}| \\
&\le C(1+c_{\xi}+\eta_b+\Upsilon+c_{\xi}\Upsilon+\eta_b\Upsilon)\Delta_N^{\lambda} \\
&\quad + C\sum_{i=1}^{n}\left(\int_{t_{i-1}}^{t_i}|\mathcal{G}(t_n,s)|\,ds + \Delta_N\eta_b\int_{t_{i-1}}^{t_i}\sum_{k=i+1}^{n}|\mathcal{G}(t_{k-1},s)|ds\right)|e_{i-1}|.
\end{aligned}
\tag{5.20}
$$

Also note that according to Assumption 3.4,

$$
\begin{aligned}
&\sum_{i=1}^{n}\left(\int_{t_{i-1}}^{t_i}|\mathcal{G}(t_n,s)|\,ds + \Delta_N\eta_b\int_{t_{i-1}}^{t_i}\sum_{k=i+1}^{n}|\mathcal{G}(t_{k-1},s)|ds\right) \\
&\quad = \int_0^{t_n}|\mathcal{G}(t_n,s)|\,ds + \Delta_N\eta_b\sum_{k=i+1}^{n}\int_0^{t_n}|\mathcal{G}(t_{k-1},s)|ds \\
&\quad \le \sqrt{t_n}\sqrt{\int_0^{t_n}|\mathcal{G}(t_n,s)|^2\,ds} + \Delta_N\eta_b\sum_{k=i+1}^{n}\sqrt{t_{k-1}}\sqrt{\int_0^{t_{k-1}}|\mathcal{G}(t_{k-1},s)|^2ds} \\
&\quad \le \ell_{\lambda}T^{\frac{1}{2}+\lambda} + \ell_{\lambda}T^{\frac{1}{2}+\lambda}\eta_b\sum_{k=1}^{N}\Delta_N = (1+T\eta_b)\ell_{\lambda}T^{\frac{1}{2}+\lambda},
\end{aligned}
$$

and hence, by the discrete version of the Gronwall inequality, for any $n = 0, 1, ..., N$,

$$|e_n| \le C(1 + c_\xi + \eta_b + \Upsilon + c_\xi\Upsilon + \eta_b\Upsilon)e^{(1+T\eta_b)\ell_\lambda T^{\frac{1}{2}+\lambda}}\Delta_N^\lambda, \tag{5.21}$$

where the constant C does not depend on the partition. Hence, with probability 1, for any $N > C_b T$,

$$\sup_{n=0,1,...,N} |Y(t_n) - \widehat{Y}(t_n)| \le C(1 + c_\xi + \eta_b + \Upsilon + c_\xi\Upsilon + \eta_b\Upsilon)e^{(1+T\eta_b)\ell_\lambda T^{\frac{1}{2}+\lambda}}\Delta_N^\lambda, \tag{5.22}$$

which immediately yields the proof. □

An attentive reader would notice that the proof of Theorem 5.2 actually gives something more than the convergence (5.7): it actually says that, with probability 1,

$$\sup_{n=0,1,...,N} |Y(t_n) - \widehat{Y}(t_n)| \le C(\omega)\Delta_N^\lambda,$$

where $C(\omega)$ does not depend on the partition, but depends on ω. It is thus tempting to try choosing $C(\omega)$ in such a way that $\omega \mapsto C(\omega)$ is a proper random variable (i.e. is measurable) and find out the conditions supplying the upper bound

$$\mathbb{E}[C^r] < \infty$$

for some $r > 0$. In this case, we would immediately get L^r-convergence for (5.2)–(5.3) and obtain an estimate of its rate. However, there is one important complication: by (5.22), our current expression for $C(\omega)$ contains the multiplier

$$\exp\left\{\max_{(t,y)\in\overline{\mathcal{D}_{\xi,\xi}}} \left|\frac{\partial b}{\partial y}(t, y)\right|\right\},$$

which happens to be quite problematic in terms of integrability. However, with some extra conditions on $\mathcal{G}$, it is possible to completely avoid this value, as explained in the theorem below.

Theorem 5.3 *Let Assumptions 3.1, 3.4, 3.5 and 3.6 hold and $Y = \{Y(t),\ t \in [0, T]\}$ be the unique continuous solution to* (5.1). *Assume also that*

- ***(a1)*** *the partial derivative $\frac{\partial a}{\partial y}$ exists and is continuous on $\overline{\mathcal{D}_{0,0}}$;*
- ***(a2)*** *there exists a constant L_a such that, for any $(t_1, y_1), (t_2, y_2) \in \overline{\mathcal{D}_{0,0}}$,*

$$|a(t_1, y_2) - a(t_2, y_1)| \le L_a(|t_1 - t_2|^\lambda + |y_2 - y_1|);$$

(B2") *there are constants* $c > 0$ *and* $p > 1$ *such that for any* $\varepsilon \in (0, \frac{1}{2}\min_{t\in[0,T]}|\psi(t) - \varphi(t)|)$ *and any* $(t_1, y_1), (t_2, y_2) \in \overline{\mathcal{D}_{\varepsilon,\varepsilon}}$

$$|b(t_1, y_1) - b(t_2, y_2)| \le \frac{c}{\varepsilon^p}\left(|t_1 - t_2|^\lambda + |y_1 - y_2|\right);$$

(B5) *the partial derivative* $\frac{\partial b}{\partial y}$ *exists, is continuous on* $\mathcal{D}_{0,0}$ *and there is a constant* $C_b > 0$ *such that for all* $t \in [0, T]$

$$\frac{\partial b}{\partial y}(t, y) < C_b, \quad (t, y) \in \mathcal{D}_{0,0}.$$

(G3) *the mapping* $t \mapsto \mathcal{G}(t, s)$ *is non-decreasing (or, alternatively, non-increasing) on* $[0, T]$ *for all* $s \in [0, T]$.

Then, if the random variable Λ *from Assumption 3.1(Z2) has all moments, for all* $r > 0$*, the drift-implicit Euler approximation* $\widehat{Y}$ *defined by* (5.2)–(5.3) *satisfies*

$$\mathbb{E}\left[\sup_{n=0,1,\dots,N}|Y(t_n) - \widehat{Y}(t_n)|^r\right] \le \frac{C}{N^{\lambda r}}, \tag{5.23}$$

where $C > 0$ *is a constant that does not depend on the partition.*

Proof Assume that the mapping $t \mapsto \mathcal{G}(t, s)$ is non-decreasing on $[0, T]$ for all $s \in [0, T]$ (the non-increasing case can be treated identically). Proceeding exactly as in the proof of Theorem 5.3, we can write

$$\begin{aligned} e_n = &\sum_{k=1}^{n}\sum_{i=1}^{k}\frac{\zeta_{k-1}}{\zeta_n}\int_{t_{i-1}}^{t_i}(\mathcal{G}(t_k, s) - \mathcal{G}(t_{k-1}, s))\,ds\frac{\partial a}{\partial y}(t_{i-1}, \Theta_{a,i-1})e_{i-1} \\ &+ \sum_{k=1}^{n}\frac{\zeta_{k-1}}{\zeta_n}\int_{t_{k-1}}^{t_k}(b(s, Y(s)) - b(t_k, Y(t_k)))ds \\ &+ \sum_{k=1}^{n}\sum_{i=1}^{k}\frac{\zeta_{k-1}}{\zeta_n}\int_{t_{i-1}}^{t_i}(\mathcal{G}(t_k, s) - \mathcal{G}(t_{k-1}, s))\,(a(s, Y(s)) - a(t_{i-1}, Y(t_{i-1})))\,ds. \end{aligned} \tag{5.24}$$

Just like in (5.12), we can write

$$\left|\sum_{k=1}^{n}\sum_{i=1}^{k}\frac{\zeta_{k-1}}{\zeta_n}\int_{t_{i-1}}^{t_i}(\mathcal{G}(t_k,s)-\mathcal{G}(t_{k-1},s))\,ds\frac{\partial a}{\partial y}(t_{i-1},\Theta_{a,i-1})e_{i-1}\right|$$
$$\le C\sum_{i=1}^{n}\left|\int_{t_{i-1}}^{t_i}\sum_{k=i}^{n}\frac{\zeta_{k-1}}{\zeta_n}(\mathcal{G}(t_k,s)-\mathcal{G}(t_{k-1},s))\,ds\right||e_{i-1}|$$

and, since all $\frac{\zeta_{k-1}}{\zeta_n}$ are positive and $t\mapsto\mathcal{G}(t,s)$ is non-decreasing, we can omit the absolute value above and write

$$\left|\sum_{k=1}^{n}\sum_{i=1}^{k}\frac{\zeta_{k-1}}{\zeta_n}\int_{t_{i-1}}^{t_i}(\mathcal{G}(t_k,s)-\mathcal{G}(t_{k-1},s))\,ds\frac{\partial a}{\partial y}(t_{i-1},\Theta_{a,i-1})e_{i-1}\right|$$
$$\le C\sum_{i=1}^{n}\int_{t_{i-1}}^{t_i}\sum_{k=i}^{n}\frac{\zeta_{k-1}}{\zeta_n}(\mathcal{G}(t_k,s)-\mathcal{G}(t_{k-1},s))\,ds|e_{i-1}|.$$

Furthermore, since (5.16) holds, one can write

$$\begin{aligned}\left|\sum_{k=1}^{n}\sum_{i=1}^{k}\frac{\zeta_{k-1}}{\zeta_n}\int_{t_{i-1}}^{t_i}(\mathcal{G}(t_k,s)-\mathcal{G}(t_{k-1},s))\,ds\frac{\partial a}{\partial y}(t_{i-1},\Theta_{a,i-1})e_{i-1}\right|&\\
\le C\sum_{i=1}^{n}\int_{t_{i-1}}^{t_i}\sum_{k=i}^{n}(\mathcal{G}(t_k,s)-\mathcal{G}(t_{k-1},s))\,ds|e_{i-1}|&\\
= C\sum_{i=1}^{n}\int_{t_{i-1}}^{t_i}(\mathcal{G}(t_n,s)-\mathcal{G}(t_{i-1},s))\,ds|e_{i-1}|&\\
= C\sum_{i=1}^{n}\int_{t_{i-1}}^{t_i}\mathcal{G}(t_n,s)ds|e_{i-1}|.&\end{aligned}\tag{5.25}$$

Next, proceeding just like in (5.18) and using the explicit expressions for ξ from (5.15) and c_ξ given by **(B2")**, we can write

$$
\begin{aligned}
&\left|\sum_{k=1}^{n} \frac{\zeta_{k-1}}{\zeta_n} \int_{t_{k-1}}^{t_k} (b(s, Y(s)) - b(t_k, Y(t_k)))ds\right| \\
&\qquad \le C(1+\Lambda)^{\frac{p}{\gamma\lambda+\lambda-1}} \sum_{k=1}^{n} \int_{t_{k-1}}^{t_k} |s - t_k|^{\lambda}\, ds \\
&\qquad\quad + C(1+\Lambda)^{\frac{p}{\gamma\lambda+\lambda-1}} \sum_{k=1}^{n} \int_{t_{k-1}}^{t_k} |Y(s) - Y(t_i)|\, ds \\
&\qquad \le C(1+\Lambda)^{\frac{p}{\gamma\lambda+\lambda-1}} \sum_{k=1}^{n} \int_{t_{k-1}}^{t_k} |s - t_k|^{\lambda}\, ds \\
&\qquad\quad + C(1+\Lambda)^{\frac{p}{\gamma\lambda+\lambda-1}} \Upsilon \sum_{i=1}^{n} \int_{t_{i-1}}^{t_i} |s - t_i|^{\lambda}\, ds \\
&\qquad = C(1+\Lambda)^{\frac{p}{\gamma\lambda+\lambda-1}} (1+\Upsilon) \sum_{i=1}^{n} \frac{1}{(1+\lambda)} \Delta_N^{1+\lambda} \\
&\qquad \le C(1+\Lambda)^{\frac{p}{\gamma\lambda+\lambda-1}} (1+\Upsilon) \Delta_N^{\lambda},
\end{aligned}
\tag{5.26}
$$

where Υ is from Theorem 3.9 and has the form

$$
\Upsilon = C\left(1 + \Lambda + (1+\Lambda)^{\frac{p}{\gamma\lambda+\lambda-1}} + \Lambda(1+\Lambda)^{\frac{p}{\gamma\lambda+\lambda-1}}\right).
$$

Finally, using again the monotonicity of $t \mapsto \mathcal{G}(t, s)$, (5.16), **(A2)** and Theorem 3.9,

$$
\begin{aligned}
&\left|\sum_{k=1}^{n}\sum_{i=1}^{k}\frac{\zeta_{k-1}}{\zeta_n}\int_{t_{i-1}}^{t_i}(\mathcal{G}(t_k,s)-\mathcal{G}(t_{k-1},s))\,(a(s,Y(s))-a(t_{i-1},Y(t_{i-1})))\,ds\right|\\
&\le \sum_{i=1}^{n}\int_{t_{i-1}}^{t_i}\left(\sum_{k=i}^{n}\frac{\zeta_{k-1}}{\zeta_n}(\mathcal{G}(t_k,s)-\mathcal{G}(t_{k-1},s))\right)|a(s,Y(s))-a(t_{i-1},Y(t_{i-1}))|ds\\
&\le C\sum_{i=1}^{n}\int_{t_{i-1}}^{t_i}\left(\sum_{k=i}^{n}(\mathcal{G}(t_k,s)-\mathcal{G}(t_{k-1},s))\right)|s-t_{i-1}|^{\lambda}ds\\
&\quad + C\sum_{i=1}^{n}\int_{t_{i-1}}^{t_i}\left(\sum_{k=i}^{n}(\mathcal{G}(t_k,s)-\mathcal{G}(t_{k-1},s))\right)|Y(s)-Y(t_{i-1})|ds\\
&\le C\Delta_N^{\lambda}\sum_{i=1}^{n}\int_{t_{i-1}}^{t_i}\mathcal{G}(t_n,s)ds + C\Upsilon\Delta_N^{\lambda}\sum_{i=1}^{n}\int_{t_{i-1}}^{t_i}\mathcal{G}(t_n,s)ds\\
&\le C(1+\Upsilon)\int_0^{t_n}\mathcal{G}(t_n,s)ds\,\Delta_N^{\lambda}\le C(1+\Upsilon)\Delta_N^{\lambda}.
\end{aligned}
$$

Summarizing all of the above, we obtain a refinement of (5.20) of the form

$$
\begin{aligned}
|e_n| &\le C(1+\Lambda)^{\frac{p}{\gamma\lambda+\lambda-1}}(1+\Upsilon)\Delta_N^{\lambda}+C(1+\Upsilon)\Delta_N^{\lambda}\\
&\quad + C\sum_{i=1}^{n}\int_{t_{i-1}}^{t_i}\mathcal{G}(t_n,s)ds|e_{i-1}|,
\end{aligned}
$$

which, by the discrete Gronwall inequality, yields

$$
|e_n|\le C\left((1+\Lambda)^{\frac{p}{\gamma\lambda+\lambda-1}}+1\right)(1+\Upsilon)\exp\left\{C\int_0^T\mathcal{G}(T,s)ds\right\}\Delta_N^{\lambda}.
$$

It remains to notice that, if Λ has all the moments, the random variable

$$
C\left((1+\Lambda)^{\frac{p}{\gamma\lambda+\lambda-1}}+1\right)(1+\Upsilon)\exp\left\{C\int_0^T\mathcal{G}(T,s)ds\right\}
$$

also does so, which yields that, for any $r > 0$,

$$\mathbb{E}\left[\sup_{n=0,1,\dots,N} |Y(t_n) - \widehat{Y}(t_n)|^r\right] \le \frac{C}{N^{\lambda r}}.$$

□

The drift-implicit Euler approximation has another advantage: since it preserves the "sandwich" property and (5.6) holds, the values $\frac{1}{\widehat{Y}-\varphi}$ and $\frac{1}{\psi-\widehat{Y}}$ approximate $\frac{1}{Y-\varphi}$ and $\frac{1}{\psi-Y}$ respectively. This claim is summarized in the following theorem.

Theorem 5.4 *1. Under assumptions of Theorem 5.2, with probability 1,*

$$\sup_{n=0,1,\dots,N}\left|\frac{1}{Y(t_n)-\varphi(t_n)} - \frac{1}{\widehat{Y}(t_n)-\varphi(t_n)}\right| \to 0, \quad N \to \infty,$$

and

$$\sup_{n=0,1,\dots,N}\left|\frac{1}{\psi(t_n)-Y(t_n)} - \frac{1}{\psi(t_n)-\widehat{Y}(t_n)}\right| \to 0, \quad N \to \infty.$$

2. *Under assumptions of Theorem 5.3, if Λ from Assumption 3.1(**Z2**) has moments of all orders, for any $r > 0$, there exists a constant $C > 0$ that does not depend on the partition such that*

$$\mathbb{E}\left[\sup_{n=0,1,\dots,N}\left|\frac{1}{Y(t_n)-\varphi(t_n)} - \frac{1}{\widehat{Y}(t_n)-\varphi(t_n)}\right|^r\right] \le \frac{C}{N^{\lambda r}}$$

and

$$\mathbb{E}\left[\sup_{n=0,1,\dots,N}\left|\frac{1}{\psi(t_n)-Y(t_n)} - \frac{1}{\psi(t_n)-\widehat{Y}(t_n)}\right|^r\right] \le \frac{C}{N^{\lambda r}}.$$

Proof Let us prove the convergences for $\frac{1}{Y-\varphi}$, and the ones for $\frac{1}{\psi-Y}$ can be obtained in an identical manner.

By Theorem 3.57 and Theorem 5.1, with probability 1, for all $n = 0, ..., N$,

$$\begin{aligned}\left|\frac{1}{Y(t_n)-\varphi(t_n)} - \frac{1}{\widehat{Y}(t_n)-\varphi(t_n)}\right|^r &= \frac{|Y(t_n)-\widehat{Y}(t_n)|^r}{(Y_{t_n}-\varphi(t_n))^r(\widehat{Y}_{t_n}-\varphi(t_n))^r} \\ &\le C(L_2+\Lambda)^{\frac{2r}{\gamma\lambda+\lambda-1}} \sup_{n=0,1,\dots,N}|Y(t_n)-\widehat{Y}(t_n)|^r.\end{aligned}$$

Then the first claim follows from Theorem 5.2 whereas Theorem 5.3 yields the second claim. □

5.4 One-Sided Case

The drift-implicit Euler approximation scheme described in Sect. 5.3 for the two-sided case can also be adapted for the one-sided setting described in Section 3.3. However, the two-sided sandwich case turns out to be slightly simpler: in this situation, the process Y is bounded (which helps in the definition of the drift-implicit scheme) and, moreover, the behaviour of Y is similar near both φ and ψ so that it was sufficient to analyze only one of the bounds. In the one-sided case, we do not have the property

$$Y(t) < \psi(t), \quad t \in [0, T],$$

so the approach from Sect. 5.3 has to be adjusted.

First of all, let us discuss the definition of the drift-implicit Euler scheme (5.2)–(5.3). Fix a λ-Hölder continuous function φ: $[0, T] \to \mathbb{R}$ and recall the notation

$$\mathcal{D}_x := \{(t, y) \mid t \in [0, T], y \in (\varphi(t) + x, \infty)\},$$
$$\overline{\mathcal{D}_x} := \{(t, y) \mid t \in [0, T], y \in [\varphi(t) + x, \infty)\}.$$

As outlined in Sect. 5.2, we have to make sure that the equation (5.4) has a unique solution $\widehat{Y}(t_n) > \varphi(t_n)$ for all $n = 0, 1, ..., N$.

Theorem 5.5 *Let Assumptions 3.1, 3.4, 3.2 and 3.3 hold and, in addition,*

(A5) *the partial derivative $\frac{\partial b}{\partial y}$ exists, is continuous on $\mathcal{D}_0$ and there is a constant $C_b > 0$ such that for all $t \in [0, T]$*

$$\frac{\partial b}{\partial y}(t, y) < C_b, \quad (t, y) \in \mathcal{D}_0.$$

Then the following claims are true:

1. *The drift-implicit Euler scheme* (5.2)–(5.3) *is well-defined for all uniform partitions $\{0 = t_0 < t_1 < \cdots < t_N = T\}$, $t_{n+1} - t_n = \frac{T}{N} =: \Delta_N$, such that $N > C_b T$ and, moreover,*

$$\widehat{Y}(t_n) > \varphi(t_n), \quad n = 0, 1, ..., N.$$

2. *For all $N > C_b T$ and $n = 0, 1, ..., N$ and $\omega \in \Omega$ such that the corresponding path $\{Z(\omega, t),\ t \in [0, T]\}$ is λ-Hölder continuous, there exists a deterministic constant C_1 that does not depend on the partition and such that*

$$\widehat{Y}(t_n) \ge \varphi(t_n) + \frac{C_1}{(1 + \Lambda)^{\frac{1}{\gamma\lambda+\lambda-1}}}; \tag{5.27}$$

3. *If, in addition, $N > c_{y_0-\varphi(0)} T$, where $c_{y_0-\varphi(0)}$ is from Assumption 3.2**(A2)**, i.e.*

$$|b(t, y_1) - b(t, y_2)| \le c_{y_0-\varphi(0)} |y_1 - y_2|, \quad (t, y_1), (t, y_2) \in \mathcal{D}_{y_0-\varphi(0)},$$

then the following estimate holds with probability 1:

$$\widehat{Y}(t_n) \le C_2(1 + \Lambda), \quad n = 0, 1, ..., N, \tag{5.28}$$

where the constant C_2 does not depend on the partition. Moreover, (5.28) *is true on all $\omega \in \Omega$ such that the corresponding path $\{Z(\omega, t),\ t \in [0, T]\}$ is λ-Hölder continuous.*

Proof Fix a uniform partition with $N > C_b T$. In order to prove the first claim, we need to check that the equation (5.4) has a unique solution $\widehat{Y}(t_n) > \varphi(t_n)$ for all $n = 1, ..., N$, and, in order to do that, it is sufficient to check that:

(i) the function $y \mapsto y - b(t, y)\Delta_N$ is strictly increasing;
(ii) $y - b(t, y)\Delta_N \to -\infty$ as $y \to \varphi(t)+$;
(iii) $y - b(t, y)\Delta_N \to +\infty$, $y \to \infty$.

Condition (i) follows directly from the Assumption **(A5)** above and the fact that $N > C_b T$ whereas condition (ii) is also satisfied by Assumption 3.2**(A3)**. Let us check (iii). By Assumption 3.2**(A2)**, for any $(s, y_1), (s, y_2) \in \overline{\mathcal{D}_1} := \{(u, y) \in [0, T] \times \mathbb{R}_+,\ y \in [\varphi(u) + 1, \infty)\}$ we have that

$$|b(s, y_1) - b(s, y_2)| \le c_1 |y_1 - y_2|,$$

i.e.

$$\sup_{(s,y)\in\overline{\mathcal{D}_1}} \left| \frac{\partial b}{\partial y}(s, y) \right| < \infty.$$

Using this and the mean value theorem, we can easily deduce that, for any $t \in [0, T]$ and any positive $y \ge \varphi(t) + 1$,

$$\begin{aligned} b(t, y) &= b(t, \varphi(t) + 1) + \frac{\partial b}{\partial y}(t, \theta_y)(y - 1 - \varphi(t)) \\ &\le \max_{s\in[0,T]} b(s, \varphi(s) + 1) + \max_{s\in[0,T]} |1 + \varphi(s)| \sup_{(s,y)\in\overline{\mathcal{D}_1}} \left| \frac{\partial b}{\partial y}(s, y) \right| + \frac{\partial b}{\partial y}(t, \theta_y) y \\ &\le C + C_b y \end{aligned}$$

for some constant $C > 0$. Therefore, if $N > C_b T$, for all positive $y \ge \varphi(t) + 1$,

$$y - b(t, y)\Delta_N \ge -C\Delta_N + (1 - C_b\Delta_N)y$$

and hence

$$y - b(t, y)\Delta_N \to \infty, \quad y \to \infty,$$

as required.

Next, the inequality (5.27) can be proven with a method identical to Theorem 5.5, so we leave the proof as an exercise.

Finally, let us prove (5.28). Fix a partition with the size N satisfying both $N > C_b T$ and $N > c_{y_0-\varphi(0)} T$ and consider $n = 0, 1, \ldots, N-1$ such that $\widehat{Y}(t_{n+1}) > \varphi(t_{n+1}) + (y_0 - \varphi(0))$ (if such n does not exist, the claim of our theorem holds automatically and there is nothing to prove). Put

$$\kappa(n) := \max\{k = 0, 1, \ldots, n \mid \widehat{Y}(t_k) \le \varphi(t_k) + (y_0 - \varphi(0))\},$$

and note that, since $\widehat{Y}(t_0) = y_0$, such $\kappa(n)$ exists and is well-defined. Observe that $(t_k, \widehat{Y}(t_k)) \in \mathcal{D}_{Y(0)-\varphi(0)}$ for any $k = \kappa(n) + 1, \ldots, n+1$. By Assumption 3.2(**A2**), for all $(t, y) \in \mathcal{D}_{Y(0)-\varphi(0)}$,

$$\begin{aligned}
|b(t,y)| &\le |b(t,y) - b(t, \varphi(t) + (y_0 - \varphi(0)))| + |b(t, \varphi(t) + (y_0 - \varphi(0)))| \\
&\le c_{y_0-\varphi(0)} |y - \varphi(t) - (y_0 - \varphi(0))| + \max_{s\in[0,T]} |b(s, \varphi(s) + (y_0 - \varphi(0)))| \\
&\le c_{y_0-\varphi(0)} |y| + c_{y_0-\varphi(0)} \max_{s\in[0,T]} |\varphi(s) + (y_0 - \varphi(0))| \\
&\quad + \max_{s\in[0,T]} |b(s, \varphi(s) + (y_0 - \varphi(0)))|,
\end{aligned}$$

i.e. there exists a constant $C > 0$ that does not depend on the partition such that, for all $t \in [0, T]$ and $y > \varphi(t) + (y_0 - \varphi(0))$,

$$|b(t,y)| \le C + c_{y_0-\varphi(0)} |y|. \tag{5.29}$$

In addition, by Assumption 3.3, for any $(t, y) \in \overline{\mathcal{D}_0}$,

$$\begin{aligned}
|a(t,y)| &\le |a(t,y) - a(t, \varphi(t))| + |a(t, \varphi(t))| \\
&\le L_a |y - \varphi(t)| + \max_{s\in[0,T]} |a(s, \varphi(s))| \\
&\le C(1 + |y|).
\end{aligned}$$

Next, observe that, for any $k = \kappa(n), \ldots, n$, we have

$$\begin{aligned}
\widehat{Y}(t_{k+1}) &= \widehat{Y}(t_{\kappa(n)}) + \sum_{i=\kappa(n)+1}^{k+1} b(t_i, \widehat{Y}(t_i))\Delta_N + Z(t_{k+1}) - Z(t_{\kappa(n)}) \\
&\quad + \sum_{i=1}^{k+1} a(t_{i-1}, \widehat{Y}(t_{i-1})) \int_{t_{i-1}}^{t_i} \big(\mathcal{G}(t_{k+1}, s) - \mathcal{G}(t_{\kappa(n)}, s)\big)\, ds \\
&\le \varphi(t_{\kappa(n)}) + (y_0 - \varphi(0)) + \sum_{i=\kappa(n)+1}^{k+1} b(t_i, \widehat{Y}(t_i))\Delta_N + \Lambda(t_k - t_{\kappa(n)})^{\lambda} \\
&\quad + \sum_{i=1}^{k+1} a(t_{i-1}, \widehat{Y}(t_{i-1})) \int_{t_{i-1}}^{t_i} \big(\mathcal{G}(t_{k+1}, s) - \mathcal{G}(t_{\kappa(n)}, s)\big)\, ds
\end{aligned}$$

$$\leq \left| \max_{s\in[0,T]} \varphi(s) + (y_0 - \varphi(0)) \right| + T^\lambda \Lambda + \sum_{i=\kappa(n)+1}^{k+1} b(t_i, \widehat{Y}(t_i))\Delta_N$$

$$+ \sum_{i=1}^{k+1} a(t_{i-1}, \widehat{Y}(t_{i-1})) \int_{t_{i-1}}^{t_i} \left(\mathcal{G}(t_{k+1}, s) - \mathcal{G}(t_{\kappa(n)}, s)\right) ds,$$

therefore

$$|\widehat{Y}(t_{k+1})| \leq \left| \max_{s\in[0,T]} \varphi(s) + (y_0 - \varphi(0)) \right| + T^\lambda \Lambda + \sum_{i=\kappa(n)+1}^{k+1} |b(t_i, \widehat{Y}(t_i))|\Delta_N$$

$$+ \sum_{i=1}^{k+1} |a(t_{i-1}, \widehat{Y}(t_{i-1}))| \int_{t_{i-1}}^{t_i} \left|\mathcal{G}(t_{k+1}, s) - \mathcal{G}(t_{\kappa(n)}, s)\right| ds$$

$$\leq \left| \max_{s\in[0,T]} \varphi(s) + (y_0 - \varphi(0)) \right| + T^\lambda \Lambda + \sum_{i=\kappa(n)+1}^{k+1} \left(C + c_{y_0-\varphi(0)}|\widehat{Y}(t_i)|\right) \Delta_N$$

$$+ \sum_{i=1}^{k+1} C(1 + |\widehat{Y}(t_{i-1})|) \int_{t_{i-1}}^{t_i} \left|\mathcal{G}(t_{k+1}, s) - \mathcal{G}(t_{\kappa(n)}, s)\right| ds$$

$$\leq C + T^\lambda \Lambda + c_{y_0-\varphi(0)}|\widehat{Y}(t_{k+1})|\Delta_N + C \int_0^{t_{k+1}} \left|\mathcal{G}(t_{k+1}, s) - \mathcal{G}(t_{\kappa(n)}, s)\right| ds$$

$$+ c_{y_0-\varphi(0)} \sum_{i=0}^{k} |\widehat{Y}(t_i)|\Delta_N + \sum_{i=0}^{k} |\widehat{Y}(t_i)| \int_{t_i}^{t_{i+1}} \left|\mathcal{G}(t_{k+1}, s) - \mathcal{G}(t_{\kappa(n)}, s)\right| ds$$

$$\leq C + T^\lambda \Lambda + c_{y_0-\varphi(0)}|\widehat{Y}(t_{k+1})|\Delta_N + C \int_0^{t_{k+1}} \left|\mathcal{G}(t_{k+1}, s) - \mathcal{G}(t_{\kappa(n)}, s)\right| ds$$

$$+ \sum_{i=0}^{k} |\widehat{Y}(t_i)| \left(c_{y_0-\varphi(0)}\Delta_N + C \int_{t_i}^{t_{i+1}} \left|\mathcal{G}(t_{k+1}, s) - \mathcal{G}(t_{\kappa(n)}, s)\right| ds \right).$$

By Assumption 3.4,

$$\int_0^{t_{k+1}} \left|\mathcal{G}(t_{k+1}, s) - \mathcal{G}(t_{\kappa(n)}, s)\right| ds \leq \sqrt{T} \sqrt{\int_0^{t_{k+1}} \left|\mathcal{G}(t_{k+1}, s) - \mathcal{G}(t_{\kappa(n)}, s)\right|^2 ds}$$

$$\leq \sqrt{T} \ell_\lambda T^\lambda,$$

so, overall, there exists a deterministic constant $C > 0$ such that

$$
\begin{aligned}
&(1 - c_{y_0-\varphi(0)}\Delta_N)|\widehat{Y}(t_{k+1})| \\
&\quad \le C + T^\lambda \Lambda + \sum_{i=0}^{k} |\widehat{Y}(t_i)| \left(c_{y_0-\varphi(0)}\Delta_N + C \int_{t_i}^{t_{i+1}} \left|\mathcal{G}(t_{k+1}, s) - \mathcal{G}(t_{\kappa(n)}, s)\right| ds \right).
\end{aligned}
$$

Note that $N > c_{y_0-\varphi(0)}T$,

$$
1 - c_{y_0-\varphi(0)}\Delta_N \ge 1 - \frac{c_{y_0-\varphi(0)}T}{[c_{y_0-\varphi(0)}T] + 1} = \frac{1 - \{c_{y_0-\varphi(0)}T\}}{[c_{y_0-\varphi(0)}T] + 1} > 0,
$$

where $[x]$ denotes the greatest integer less than or equal to x and $\{x\} := x - [x]$. Therefore

$$
\begin{aligned}
|\widehat{Y}(t_{k+1})| &\le \frac{[c_{y_0-\varphi(0)}T] + 1}{1 - \{c_{y_0-\varphi(0)}T\}} \left(C + T^\lambda \Lambda\right) \\
&\quad + \frac{[c_{y_0-\varphi(0)}T] + 1}{1 - \{c_{y_0-\varphi(0)}T\}} \sum_{i=0}^{k} c_{y_0-\varphi(0)} |\widehat{Y}(t_i)| \Delta_N \\
&\quad + C \frac{[c_{y_0-\varphi(0)}T] + 1}{1 - \{c_{y_0-\varphi(0)}T\}} \sum_{i=0}^{k} |\widehat{Y}(t_i)| \int_{t_i}^{t_{i+1}} \left|\mathcal{G}(t_{k+1}, s) - \mathcal{G}(t_{\kappa(n)}, s)\right| ds \\
&\le C(1 + \Lambda) + C \sum_{i=0}^{k} |\widehat{Y}(t_i)| \left(c_{y_0-\varphi(0)}\Delta_N + \int_{t_i}^{t_{i+1}} \left|\mathcal{G}(t_{k+1}, s) - \mathcal{G}(t_{\kappa(n)}, s)\right| ds \right).
\end{aligned}
$$

Using the discrete version of the Gronwall inequality, we now obtain that for all $k = \kappa(n), \ldots, n$

$$
\begin{aligned}
&|\widehat{Y}(t_{k+1})| \\
&\quad \le C(1 + \Lambda) \exp\left\{ \sum_{i=0}^{k} \left(c_{y_0-\varphi(0)}\Delta_N + \int_{t_i}^{t_{i+1}} \left|\mathcal{G}(t_{k+1}, s) - \mathcal{G}(t_{\kappa(n)}, s)\right| ds \right) \right\} \\
&\quad \le C_2(1 + \Lambda)
\end{aligned}
$$

for some constant $C_2 > 0$, which ends the proof. □

Having estimates (5.27) and (5.28) in place, we are ready to formulate and prove the main convergence results for the one-sided case.

Theorem 5.6 *Let Assumptions 3.1, 3.4, 3.2 and 3.3 hold and $Y = \{Y(t),\ t \in [0, T]\}$ be the unique continuous solution to* (5.1)*. Assume also that*

(a1) *the partial derivative $\frac{\partial a}{\partial y}$ exists and is continuous on $\overline{\mathcal{D}_0}$;*
(a2) *there exists a constant L_a such that, for any $(t_1, y_1), (t_2, y_2) \in \overline{\mathcal{D}_0}$,*

$$|a(t_1, y_2) - a(t_2, y_1)| \le L_a(|t_1 - t_2|^\lambda + |y_2 - y_1|);$$

(b2') *for any $\varepsilon > 0$, there is a constant $c_\varepsilon > 0$ such that for any $(t_1, y_1), (t_2, y_2) \in \overline{\mathcal{D}_\varepsilon}$*

$$|b(t_1, y_1) - b(t_2, y_2)| \le c_\varepsilon \left(|t_1 - t_2|^\lambda + |y_1 - y_2|\right);$$

(b5) *the partial derivative $\frac{\partial b}{\partial y}$ exists, is continuous on $\mathcal{D}_0$ and there is a constant $C_b > 0$ such that for all $t \in [0, T]$*

$$\frac{\partial b}{\partial y}(t, y) < C_b, \quad (t, y) \in \mathcal{D}_0.$$

Then, with probability 1, the drift-implicit Euler approximation $\widehat{Y}$ defined by (5.2)–(5.3) *converges to the solution of* (5.1) *in the following sense:*

$$\sup_{n=0,1,\dots,N} |Y(t_n) - \widehat{Y}(t_n)| \to 0, \quad N \to \infty. \tag{5.30}$$

Proof Fix any $N > \max\left\{c_{y_0-\varphi(0)}, C_b\right\} T$, where $c_{y_0-\varphi(0)}$ is from Assumption 3.2(**A2**), denote the diameter of the partition $\Delta_N := \frac{T}{N}$ and put $e_n := Y(t_n) - \widehat{Y}(t_n)$, $\Delta Z_n := Z(t_n) - Z(t_{n-1})$. Proceeding just like in the proof of Theorem 5.2, we can arrive to

$$\begin{aligned} e_n &= \sum_{k=1}^{n}\sum_{i=1}^{k} \frac{\zeta_{k-1}}{\zeta_n} \int_{t_{i-1}}^{t_i} (\mathcal{G}(t_k, s) - \mathcal{G}(t_{k-1}, s))\, ds \frac{\partial a}{\partial y}(t_{i-1}, \Theta_{a,i-1}) e_{i-1} \\ &+ \sum_{k=1}^{n} \frac{\zeta_{k-1}}{\zeta_n} \int_{t_{k-1}}^{t_k} (b(s, Y(s)) - b(t_k, Y(t_k))) ds \\ &+ \sum_{k=1}^{n}\sum_{i=1}^{k} \frac{\zeta_{k-1}}{\zeta_n} \int_{t_{i-1}}^{t_i} (\mathcal{G}(t_k, s) - \mathcal{G}(t_{k-1}, s)) \left(a(s, Y(s)) - a(t_{i-1}, Y(t_{i-1}))\right) ds, \end{aligned} \tag{5.31}$$

where

$$\zeta_0 := 1, \quad \zeta_k := \prod_{i=1}^{k}\left(1 - \frac{\partial b}{\partial y}(t_i, \Theta_{b,i})\Delta_N\right),$$

and $\Theta_{b,i} \in (Y(t_i) \wedge \widehat{Y}(t_i), Y(t_i) \vee \widehat{Y}(t_i)) \subset (\varphi(t_i), \infty)$. Note that, just like in Theorem 5.2, with probability 1,

$$\max_{k=1,\dots,n} \left| \frac{\zeta_{k-1}}{\zeta_n} \right| < C$$

for some constant C that does not depend on the partition. Next, similarly to (5.12)–(5.14), there is a constant $C > 0$ that does not depend on the partition such that

$$\begin{aligned} &\left| \sum_{k=1}^{n} \sum_{i=1}^{k} \frac{\zeta_{k-1}}{\zeta_n} \int_{t_{i-1}}^{t_i} (\mathcal{G}(t_k, s) - \mathcal{G}(t_{k-1}, s))\, ds \frac{\partial a}{\partial y}(t_{i-1}, \Theta_{a,i-1}) e_{i-1} \right| \\ &\quad \le C \sum_{i=1}^{n} \left| \int_{t_{i-1}}^{t_i} \sum_{k=i}^{n} \frac{\zeta_{k-1}}{\zeta_n} (\mathcal{G}(t_k, s) - \mathcal{G}(t_{k-1}, s))\, ds \right| |e_{i-1}| \\ &\quad \le C \sum_{i=1}^{n} \left(\int_{t_{i-1}}^{t_i} \left| \frac{\zeta_{n-1}}{\zeta_n} \right| |\mathcal{G}(t_n, s)|\, ds \right) |e_{i-1}| \\ &\quad + C\Delta_N \sum_{i=1}^{n} \left(\sum_{k=i+1}^{n} \left| \frac{\partial b}{\partial y}(t_{k-1}, \Theta_{b,k-1}) \right| \left| \frac{\zeta_{k-2}}{\zeta_n} \right| \int_{t_{i-1}}^{t_i} |\mathcal{G}(t_{k-1}, s)| ds \right) |e_{i-1}|, \end{aligned} \tag{5.32}$$

where, as always, $C > 0$ denotes some deterministic constant that does not depend on N and may vary from line to line.

By Theorem 3.3 and Theorem 5.5, there exist constants $C_1, C_2 > 0$ that does not depend on the partition and such that

$$\begin{aligned} \varphi(t_n) + \frac{C_1}{(1+\Lambda)^{\frac{1}{\gamma\lambda+\lambda-1}}} &\le Y(t_n) \le C_2(1+\Lambda), \\ \varphi(t_n) + \frac{C_1}{(1+\Lambda)^{\frac{1}{\gamma\lambda+\lambda-1}}} &\le \widehat{Y}(t_n) \le C_2(1+\Lambda), \end{aligned}$$

which, in turn, implies that

$$\varphi(t_n) + \frac{C_1}{(1+\Lambda)^{\frac{1}{\gamma\lambda+\lambda-1}}} \le \Theta_{b,n} \le C_2(1+\Lambda).$$

In other words, for any $n = 0, 1, 2, \dots, N$, $(t_n, \Theta_{b,n}) \in \{(t, y) \mid t \in [0, T],\ y \in [\varphi(t) + \xi, C_2(1+\Lambda)]\}$ with

$$\xi := \frac{C_1}{(1+\Lambda)^{\frac{1}{\gamma\lambda+\lambda-1}}} \tag{5.33}$$

and, by the continuity of $\frac{\partial b}{\partial y}$ on the compact set $\{(t, y) \mid t \in [0, T],\ y \in [\varphi(t) + \xi, C_2(1 + \Lambda)]\}$ thanks to **(A5)**,

$$\sup_{N > \max\left\{c_{y_0-\varphi(0)}, C_b\right\}T} \max_{n=0,1,\ldots,N} \left|\frac{\partial b}{\partial y}(t_{k-1}, \Theta_{b,k-1})\right| \le \eta_b < \infty$$

with $\eta_b := \max_{\{(t,y) \mid t\in[0,T],\ y\in[\varphi(t)+\xi, C_2(1+\Lambda)]\}} \left|\frac{\partial b}{\partial y}(t, y)\right|$. Therefore, proceeding just like in Theorem 5.2,

$$\begin{aligned} &\left|\sum_{k=1}^{n}\sum_{i=1}^{k}\frac{\zeta_{k-1}}{\zeta_n}\int_{t_{i-1}}^{t_i}(\mathcal{G}(t_k, s) - \mathcal{G}(t_{k-1}, s))\,ds\frac{\partial a}{\partial y}(t_{i-1}, \Theta_{a,i-1})e_{i-1}\right| \\ &\qquad \le C\sum_{i=1}^{n}\left(\int_{t_{i-1}}^{t_i}|\mathcal{G}(t_n, s)|\,ds + \Delta_N\eta_b\int_{t_{i-1}}^{t_i}\sum_{k=i+1}^{n}|\mathcal{G}(t_{k-1}, s)|ds\right)|e_{i-1}|. \end{aligned} \tag{5.34}$$

Next, following the same idea as in (5.18), it is straightforward to obtain that

$$\begin{aligned} &\left|\sum_{k=1}^{n}\frac{\zeta_{k-1}}{\zeta_n}\int_{t_{k-1}}^{t_k}(b(s, Y(s)) - b(t_k, Y(t_k)))ds\right| \\ &\le Cc_\xi(1 + \Upsilon)\Delta_N^\lambda, \end{aligned} \tag{5.35}$$

where c_ξ is from **(A2')** that corresponds to $\varepsilon = \xi$ and Υ is from Theorem 3.4 and has the form $\Upsilon = C(1 + \Lambda + c_\xi + c_\xi\Lambda)$. Finally, just like in (5.19),

$$\begin{aligned} &\left|\sum_{k=1}^{n}\sum_{i=1}^{k}\frac{\zeta_{k-1}}{\zeta_n}\int_{t_{i-1}}^{t_i}(\mathcal{G}(t_k, s) - \mathcal{G}(t_{k-1}, s))\,(a(s, Y(s)) - a(t_{i-1}, Y(t_{i-1})))\,ds\right| \\ &\le C(1 + \eta_b + \Upsilon + \eta_b\Upsilon)\Delta_N^\lambda. \end{aligned} \tag{5.36}$$

and, summarizing the estimates above,

$$\begin{aligned} |e_n| &\le C(1 + c_\xi + \eta_b + \Upsilon + c_\xi\Upsilon + \eta_b\Upsilon)\Delta_N^\lambda \\ &\quad + C\sum_{i=1}^{n}\left(\int_{t_{i-1}}^{t_i}|\mathcal{G}(t_n, s)|\,ds + \Delta_N\eta_b\int_{t_{i-1}}^{t_i}\sum_{k=i+1}^{n}|\mathcal{G}(t_{k-1}, s)|ds\right)|e_{i-1}|. \end{aligned} \tag{5.37}$$

It remains to use the discrete Gronwall inequality and obtain that, for all $N > \max\left\{c_{y_0-\varphi(0)}, C_b\right\}T$ and $n = 0, 1, \ldots, N$, with probability 1,

$$\sup_{n=0,1,\dots,N} |Y(t_n) - \widehat{Y}(t_n)| \le C(1 + c_\xi + \eta_b + \Upsilon + c_\xi \Upsilon + \eta_b \Upsilon) e^{(1+T\eta_b)\ell_\lambda T^{\frac{1}{2}+\lambda}} \Delta_N^\lambda, \tag{5.38}$$

which immediately yields the proof. □

Finally, let us also prove the one-sided counterpart of Theorem 5.3.

Theorem 5.7 *Let Assumptions 3.1, 3.4, 3.2 and 3.3 hold and* $Y = \{Y(t),\ t \in [0, T]\}$ *be the unique continuous solution to* (5.1). *Assume also that*

(a1) *the partial derivative* $\frac{\partial a}{\partial y}$ *exists and is continuous on* $\overline{\mathcal{D}_0}$*;*
(a2) *there exists a constant* L_a *such that, for any* $(t_1, y_1), (t_2, y_2) \in \overline{\mathcal{D}_0}$,

$$|a(t_1, y_2) - a(t_2, y_1)| \le L_a(|t_1 - t_2|^\lambda + |y_2 - y_1|);$$

(b2") *there are constants* $c > 0$ *and* $p > 1$ *such that for any* $\varepsilon > 0$ *and any* $(t_1, y_1), (t_2, y_2) \in \overline{\mathcal{D}_\varepsilon}$

$$|b(t_1, y_1) - b(t_2, y_2)| \le \frac{c}{\varepsilon^p} \left(|t_1 - t_2|^\lambda + |y_1 - y_2|\right);$$

(b5) *the partial derivative* $\frac{\partial b}{\partial y}$ *exists, is continuous on* $\mathcal{D}_0$ *and there is a constant* $C_b > 0$ *such that for all* $t \in [0, T]$

$$\frac{\partial b}{\partial y}(t, y) < C_b, \quad (t, y) \in \mathcal{D}_0.$$

(G3) *the mapping* $t \mapsto \mathcal{G}(t, s)$ *is non-decreasing (or, alternatively, non-increasing) on* $[0, T]$ *for all* $s \in [0, T]$.

Then, if the random variable Λ *from Assumption 3.1(Z2) has all moments, for all* $r > 0$, *the drift-implicit Euler approximation* $\widehat{Y}$ *defined by* (5.2)–(5.3) *satisfies*

$$\mathbb{E}\left[\sup_{n=0,1,\dots,N} |Y(t_n) - \widehat{Y}(t_n)|^r\right] \le \frac{C}{N^{\lambda r}}, \tag{5.39}$$

where $C > 0$ *is a constant that does not depend on the partition.*

Proof The result can be obtained by adjusting the proof of Theorem 5.6 in a way similar to Theorem 5.3 and hence is left as an exercise. □

5.5 Examples and Simulations

The drift-implicit Euler scheme (5.2)–(5.3) requires to solve N equations of the form (5.4) which, in general, can be challenging from the computational point of view. However, in some cases that are relevant for applications these equations have a simple explicit solution.

Remark 5.2 Regarding the numerical examples that follow, we remark that:

(1) all the simulations are performed in the R programming language on the system with Intel Core i9-9900K CPU and 64 Gb RAM;
(2) in order to simulate paths of fractional Brownian motion, R package somebm is used;
(3) in Example 5.3, discrete samples of the multifractional Brownian motion (mBm) values are simulated using the Cholesky decomposition of the corresponding covariance matrix (for covariance structure of the mBm, see e.g. [3, Proposition 4]).

Example 5.1 (*Generalized CIR processes*) Consider a one-sided case with $\varphi \equiv 0$. Assume that Z satisfies Assumption 3.1 with $\lambda \in \left(\frac{1}{2}, 1\right)$ and consider a sandwiched SDE of the form,

$$Y(t) = Y(0) + \int_0^t \left(\frac{\kappa_1}{Y(s)} - \kappa_2 Y(s)\right) ds + Z(t), \quad t \in [0, T], \tag{5.40}$$

where $y_0, \kappa_1, \kappa_2 > 0$ and $\gamma > \frac{1}{\lambda} - 1$. This process fits into the framework of Sect. 5.4 and the equation for $\widehat{Y}(t_{k+1})$ from (5.4) reads as follows:

$$\widehat{Y}(t_{k+1}) = \widehat{Y}(t_k) + \left(\frac{\kappa_1}{\widehat{Y}(t_{k+1})} - \kappa_2 \widehat{Y}(t_{k+1})\right) \Delta_N + Z(t_{k+1}) - Z(t_k).$$

It is easy to see that it has a unique positive solution

$$\widehat{Y}(t_{k+1}) = \frac{\widehat{Y}(t_k) + (Z(t_{k+1}) - Z(t_k)) + \sqrt{\left(\widehat{Y}(t_k) + (Z(t_{k+1}) - Z(t_k))\right)^2 + 4\kappa_1 \Delta_N (1 + \kappa_2 \Delta_N)}}{2(1 + \kappa_2 \Delta_N)}.$$

Fig. 5.1 contains 10 sample paths of the process (5.40) driven by a fractional Brownian motion with $H = 0.7$ (see Example 3.5). Note that in this case, as explained in Example 3.2, the Hölder constant Λ has moments of all positive orders. Therefore, Theorem 5.7 holds and, for any $r > 0$,

$$\mathbb{E}\left[\sup_{n=0,1,\ldots,N} |Y(t_n) - \widehat{Y}(t_n)|^r\right] \le \frac{C}{N^{\lambda r}}.$$

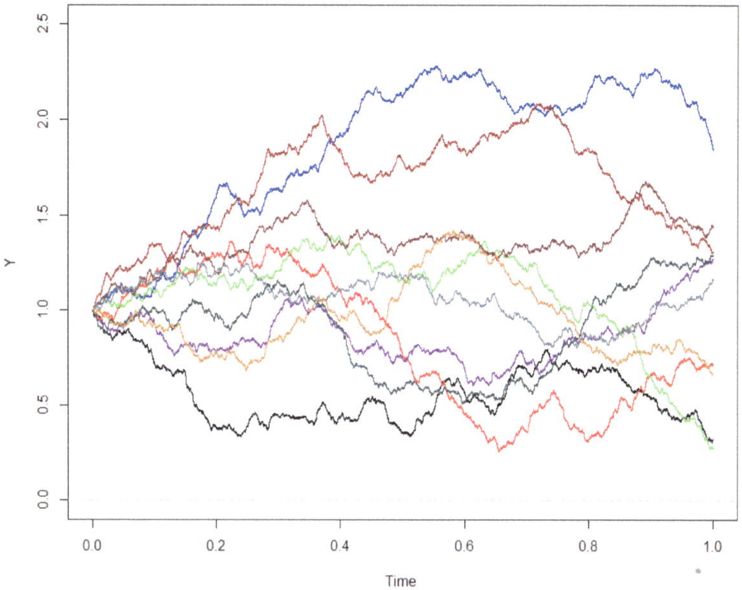

Fig. 5.1 Ten sample paths of (5.40) generated using the drift-implicit Euler approximation scheme; $N = 10000$, $T = 1$, $y_0 = \kappa_1 = \kappa_2 = 1$, Z is a fractional Brownian motion with $H = 0.7$

In all simulations, we take $N = 10000$, $T = 1$ and $y_0 = \kappa_1 = \kappa_2 = 1$. Based on 10000 simulations, the average time for simulating one path is 0.005388308 s.

Example 5.2 *(Sandwiched process of the TSB type)* As an example of the two-sided case, consider a sandwiched SDE of the form

$$Y(t) = y_0 + \int_0^t \left(\frac{\kappa_1}{Y(s) - \varphi(s)} - \frac{\kappa_2}{\psi(s) - Y(s)} - \kappa_3 Y(s) \right) ds + Z(t), \quad t \in [0, T], \tag{5.41}$$

where Z satisfies Assumption 3.1 with $\lambda \in \left(\frac{1}{2}, 1\right)$, φ and ψ are λ-Hölder continuous functions on $[0, T]$, $\varphi < \psi$, $\varphi(0) < y_0 < \psi(0)$ and $\kappa_1, \kappa_2 > 0$, $\kappa_3 \in \mathbb{R}$. This equation fits into the framework of Sect. 5.3 and the scheme (5.2)–(5.3) leads to N cubic equations of the form

$$\widehat{Y}^3(t_{n+1}) + B_{2,n}\widehat{Y}^2(t_{n+1}) + B_{1,n}\widehat{Y}(t_{n+1}) + B_{0,n} = 0, \quad n = 0, \ldots, N-1, \tag{5.42}$$

where

$$B_{0,n} := \frac{-\varphi(t_{n+1})\psi(t_{n+1})\left(\widehat{Y}(t_n) + \Delta Z_n\right) + \Delta_N\left(\kappa_1\psi(t_{n+1}) + \kappa_2\varphi(t_{n+1})\right)}{1 + \Delta_N\kappa_3},$$

$$B_{1,n} := \varphi(t_{n+1})\psi(t_{n+1}) + \frac{(\varphi(t_{n+1}) + \psi(t_{n+1}))(\widehat{Y}(t_n) + \Delta Z_n) - \Delta_N(\kappa_1 + \kappa_2)}{1 + \Delta_N\kappa_3},$$

$$B_{2,n} := -\varphi(t_{n+1}) - \psi(t_{n+1}) - \frac{\widehat{Y}(t_n) + \Delta Z_n}{1 + \Delta_N\kappa_3},$$

Note this equation can be solved explicitly using, e.g., the celebrated Cardano method. Namely, define

$$Q_{1,n} := B_{1,n} - \frac{B_{2,n}^2}{3}, \quad Q_{2,n} := \frac{2B_{2,n}^3}{27} - \frac{B_{2,n}B_{1,n}}{3} + B_{0,n}$$

and put

$$Q_n := \left(\frac{Q_{1,n}}{3}\right)^3 + \left(\frac{Q_{2,n}}{2}\right)^2,$$

$$\alpha_n := \sqrt[3]{-\frac{Q_{2,n}}{2} + \sqrt{Q_n}}, \quad \beta_n := \sqrt[3]{-\frac{Q_{2,n}}{2} - \sqrt{Q_n}},$$

where among possible complex values of α_n and β_n one should take those for which $\alpha_n\beta_n = -\frac{Q_{1,n}}{3}$. Then the three roots of the cubic equation (5.42) are

$$y_{1,n} = \alpha_n + \beta_n, \qquad y_{2,n} = -\frac{\alpha_n + \beta_n}{2} + i\frac{\alpha_n - \beta_n}{2}\sqrt{3},$$
$$y_{3,n} = -\frac{\alpha_n + \beta_n}{2} - i\frac{\alpha_n - \beta_n}{2}\sqrt{3},$$

and $\widehat{Y}(t_{n+1})$ is equal to the root which belongs to $(\varphi(t_{n+1}), \psi(t_{n+1}))$ (note that there is exactly one root in that interval).

Figure 5.2 contains 10 sample paths of the process (5.41) driven by a fractional Brownian motion with $H = 0.7$. In all simulations, we take $\varphi \equiv -1$, $\psi \equiv 1$, $N = 10000$, $T = 1$ and $y_0 = 0$, $\kappa_1 = \kappa_2 = \frac{1}{2}$, $\kappa_3 = 0$. Simulation is performed by direct implementation of the Cardano's method in R. Based on 10000 simulations, the average time for simulating one path is 0.03700142 s.

In both Examples 5.1 and 5.2, equations for computing $\widehat{Y}$ could be explicitly solved but the Hölder continuity of the noise could not be less then $1/2$. The next example shows that the drift-implicit Euler scheme can be applied in the rough case as well.

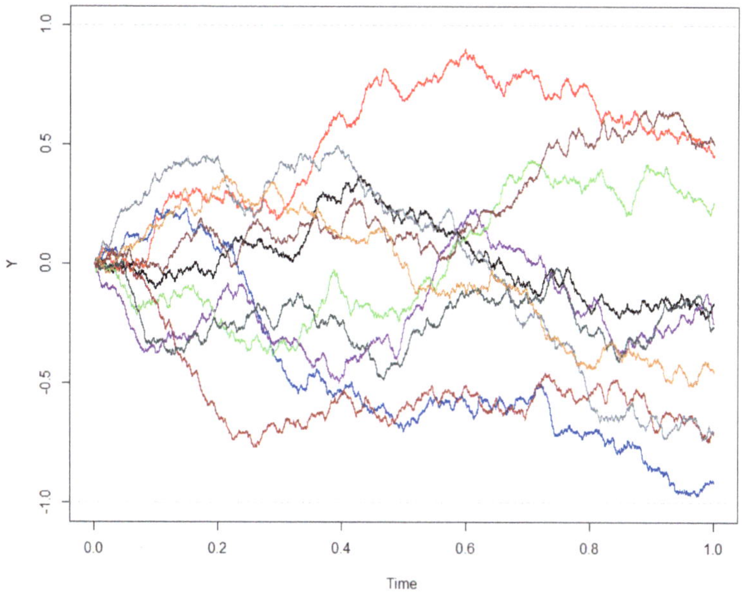

Fig. 5.2 Ten sample paths of (5.41) generated using the drift-implicit Euler approximation scheme; $\varphi \equiv -1$, $\psi \equiv 1$, $N = 10000$, $T = 1$, $y_0 = 0$, $\kappa_1 = \kappa_2 = \frac{1}{2}$, $\kappa_3 = 0$, Z is a fractional Brownian motion with $H = 0.7$

Example 5.3 *(Sandwiched process driven by multifractional Brownian motion)* Let $Z = \{Z(t),\ t \in [0, T]\}$ be a multifractional Brownian motion as defined in Example 3.7 with functional Hurst parameter $H(t) = \frac{1}{5}\sin(2\pi t) + \frac{1}{2}$ (note that the lowest value of the functional Hurst parameter is $H\left(\frac{3}{4}\right) = 0.3 > \frac{1}{5}$). Consider the sandwiched SDE of the form

$$Y(t) = Y(0) + \int_0^t \left(\frac{\kappa_1}{(Y(s) - \varphi(s))^4} - \frac{\kappa_2}{(\psi(s) - Y(s))^4} \right) ds + Z(t), \quad t \in [0, T]. \tag{5.43}$$

In this case, Theorem 3.7 guarantees existence and uniqueness of the solution and, moreover, by Theorem 5.3, for any $r > 0$,

$$\mathbb{E}\left[\sup_{n=0,1,\ldots,N} |Y(t_n) - \widehat{Y}(t_n)|^r \right] \le \frac{C}{N^{\lambda r}}.$$

On Fig. 5.3, one can see paths of the process (5.43) with $\kappa_1 = \kappa_2 = 1$, $\varphi(t) = \sin(10t)$, $\psi(t) = \sin(10t) + 2$. Unlike Example 5.1 or Example 5.2, we cannot solve the equation (5.4) explicitly and have to use approximate solutions. For that matter, we use the uniroot function of the stats R package. Based on 10000 simulations, the average time for simulating one path is 0.4968714 s.

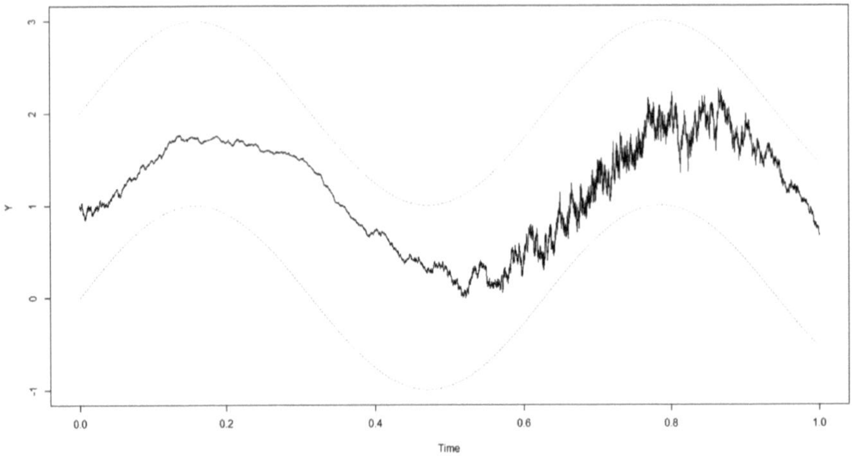

Fig. 5.3 A sample path of (5.43) generated using the backward Euler approximation scheme; $N = 10000, T = 1, Y(0) = 1, \kappa_1 = \kappa_2 = 1, \varphi(t) = \sin(10t), \psi(t) = \sin(10t) + 2$, Z is a multifractional Brownian motion with functional Hurst parameter $H(t) = \frac{1}{5}\sin(2\pi t) + \frac{1}{2}$

5.6 Notes

1. Note that the drift-implicit Euler scheme for (5.40) driven by the fractional Brownian motion was the main subject of [6, 8], but in both cases the convergence of $\widehat{Y}$ to Y is established only on $[0, T]$ with T being small (see e.g. [6, Eq. (8) and Remark 3.1]). Our results fill this gap and prove that convergence holds on arbitrary $[0, T]$ for any model parameters.
2. It should be noted though that the convergence rate in Theorem 5.3 and Theorem 5.7 is not optimal and can be improved for the fractional Brownian driver. It is well-known that paths of a fractional Brownian motion are Hölder continuous up to (but not including) its Hurst index H and whence Theorem 5.3 and Theorem 5.7 indicate that the exact convergence speed of the drift-implicit Euler scheme is better than $O\left(\Delta_N^{\lambda}\right)$ for any $\lambda \in (0, H)$. However, if one has more detailed information about the noise Z beyond Hölder continuity, it is possible to obtain a better rate. For example, [6] uses the results on the modulus of the continuity of the fractional Brownian motion and establishes that the exact speed of convergence is $O\left(\Delta_N^H \sqrt{|\log(\Delta_N)|}\right)$ (provided that T is small enough).

References

1. Alfonsi, A.: On the discretization schemes for the CIR (and Bessel squared) processes. Monte Carlo Methods Appl. **11**(4), 355–384 (2005). https://doi.org/10.1163/156939605777438569, https://doi.org/10.1163/156939605777438569

2. Alfonsi, A.: Strong order one convergence of a drift implicit Euler scheme: application to the CIR process. Statist. Probab. Lett. **83**(2), 602–607 (2013). https://doi.org/10.1016/j.spl.2012.10.034, https://doi.org/10.1016/j.spl.2012.10.034
3. Ayache, A., Cohen, S., Vehel, J.L.: The covariance structure of multifractional Brownian motion, with application to long range dependence. In: 2000 IEEE International Conference on Acoustics, Speech, and Signal Processing. Proceedings (Cat. No.00CH37100). IEEE (2002)
4. Dereich, S., Neuenkirch, A., Szpruch, L.: An Euler-type method for the strong approximation of the Cox–Ingersoll–Ross process. Proc. R. Soc. A. Math. Phys. Eng. Sci. **468**(2140), 1105–1115 (2012)
5. Di Nunno, G., Mishura, Y., Yurchenko-Tytarenko, A.: Drift-implicit Euler scheme for sandwiched processes driven by Hölder noises. Numer. Algorithms **93**(2), 459–491 (2023). https://doi.org/10.1007/s11075-022-01424-6. http://dx.doi.org/10.1007/s11075-022-01424-6
6. Hong, J., Huang, C., Kamrani, M., Wang, X.: Optimal strong convergence rate of a backward Euler type scheme for the Cox–Ingersoll–Ross model driven by fractional Brownian motion. Stoch. Process. Their Appl. **130**(5), 2675–2692 (2020). https://doi.org/10.1016/j.spa.2019.07.014, http://www.sciencedirect.com/science/article/pii/S0304414918305283
7. Kubilius, K., Medžiūnas, A.: Positive solutions of the fractional SDEs with non-Lipschitz diffusion coefficient. Mathematics **9**(1) (2020)
8. Zhang, S.Q., Yuan, C.: Stochastic differential equations driven by fractional Brownian motion with locally Lipschitz drift and their implicit Euler approximation. iN: Proceedings of the Royal Society of Edinburgh: Section A Mathematics, pp. 1–27 (2020). https://doi.org/10.1017/prm.2020.60

Chapter 6
Gaussian Volterra Drivers

Abstract In this chapter, we study sandwiched stochastic equations driven by Hölder-continuous Gaussian Volterra processes. We establish the Malliavin differentiability of their solutions as well as construct Markovian approximations of their non-Markovian dynamics. These results serve as the foundation for the numerical methods developed in Chaps. 8 and 9.

6.1 Introduction

Most of the results presented in Chaps. 2–5 concern sandwiched stochastic equations of the form

$$Y(t) = y_0 + \int_0^t b(s, Y(s))ds + \int_0^t \mathcal{G}(t, s)a(s, Y(s))ds + Z(t), \quad t \in [0, T], \tag{6.1}$$

driven by generic Hölder continuous noises $Z = \{Z(t),\ t \in [0, T]\}$. In particular, we have not imposed any specific assumptions on the distribution of Z. So far, this level of generality has been sufficient for our purposes; for example, our existence and uniqueness results rely only on the condition that the trajectories of Z do not "vary too much", as ensured by the Hölder continuity assumed in Assumption 3.1.

In this chapter, however, we aim to explore more intricate properties of the solution, and, naturally, this requires narrowing the class of potential driving noises. One class of stochastic processes particularly suited for our purposes is *Gaussian Volterra processes* of the form

$$Z(t) = \int_0^t \mathcal{K}(t, s)dB(s), \quad t \in [0, T], \tag{6.2}$$

G. Di Nunno et al., *Volterra Volatility Models*, Springer Finance,
https://doi.org/10.1007/978-3-032-26576-0_6

where B is a Brownian motion and $\mathcal{K}$ is a suitable Volterra kernel. This choice of driver is not random: Gaussian Volterra processes are prominent in the volatility modeling literature because of their mathematical convenience and exceptional versatility in replicating various phenomena observed in real markets.

6.2 Gaussian Volterra Drivers and Sandwiched Equations

First, let us specify our setup more precisely. Throughout this chapter, we consider a classical Wiener space $(\Omega, \mathcal{F}, \mathbb{P})$, where $\Omega := C_0([0,T]) = \{f \in C([0,T]) \mid f(0) = 0\}$, $\mathbb{P}$ is the classical Wiener measure, and $\mathcal{F}$ is the $\mathbb{P}$-augmented Borel sigma-algebra. Let B denote the corresponding Brownian motion, i.e., $B\colon [0,T] \times \Omega \to \mathbb{R}$, $B(t,\omega) = \omega(t)$. Additionally, let $\mathbb{F} = \{\mathcal{F}_t,\ t \in [0,T]\}$ be the $\mathbb{P}$-augmented filtration generated by B.

Since we intend to work with processes of the form (6.2), we must impose certain assumptions on the kernel $\mathcal{K}$ to ensure that the integral in (6.2) is well-defined. Specifically, the measurable function $\mathcal{K}\colon [0,T]^2 \to \mathbb{R}$ must satisfy the following assumption.

Assumption 6.1 $\mathcal{K}\colon [0,T]^2 \to \mathbb{R}$ is a Volterra kernel, i.e. $\mathcal{K}(t,s) = 0$ if $s \ge t$, and

(K1) $\int_0^T \int_0^T \mathcal{K}^2(t,s)dsdt < \infty$ and

$$\int_0^T \mathcal{K}^2(t,s)ds < \infty, \quad \int_0^T \mathcal{K}^2(t,s)dt < \infty, \quad \forall s,t \in [0,T];$$

(K2) there exists $H \in (0,1)$ such that for any $0 \le t_1 \le t_2 \le T$ and any $\lambda \in (0,H)$

$$\int_0^T (\mathcal{K}(t_2,s) - \mathcal{K}(t_1,s))^2 ds \le \ell_\lambda^2 |t_2 - t_1|^{2\lambda},$$

for some constant and $\ell_\lambda > 0$ depending on λ.

Remark 6.1 Note that items **(K1)** and **(K2)** of Assumption 6.1 jointly imply that

$$\sup_{t\in[0,T]} \int_0^T \mathcal{K}^2(t,s)ds < \infty. \tag{6.3}$$

Next, let $z : [0,T] \to \mathbb{R}$ be a deterministic H-Hölder continuous function, where the value $H \in (0,1)$ is the same as in Assumption 6.1**(K2)** and $z(0) = 0$. Thanks to

Assumption 6.1(**K1**), we can define a stochastic process $\widetilde{Z} = \{\widetilde{Z}(t),\ t \in [0, T]\}$ of the form

$$\widetilde{Z}(t) = z(t) + \int_0^t \mathcal{K}(t, s) dB(s), \quad t \in [0, T], \tag{6.4}$$

and, moreover, for any $\lambda \in (0, H)$ and $0 \le t_1 \le t_2 \le T$,

$$\begin{aligned}
\mathbb{E}\left[|\widetilde{Z}(t_2) - \widetilde{Z}(t_1)|^2\right] &= |z(t_2) - z(t_1)|^2 + \mathbb{E}\left[\left|\int_0^{t_2} (\mathcal{K}(t_2, s) - \mathcal{K}(t_1, s))\, dB(s)\right|^2\right] \\
&\quad + 2(z(t_2) - z(t_1))\mathbb{E}\left[\int_0^{t_2} (\mathcal{K}(t_2, s) - \mathcal{K}(t_1, s))\, dB(s)\right] \\
&= |z(t_2) - z(t_1)|^2 + \mathbb{E}\left[\int_0^{t_2} (\mathcal{K}(t_2, s) - \mathcal{K}(t_1, s))^2\, ds\right] \\
&\le C|t_2 - t_1|^{2\lambda} + \ell_\lambda^2 |t_2 - t_1|^{2\lambda}.
\end{aligned}$$

By the Kolmogorov-Chentsov test from Theorem A.2, this calculation means that $\widetilde{Z}$ has a modification with λ-Hölder continuous paths for all $\lambda \in (0, H)$, i.e. there is a positive random variable $\widetilde{\Lambda} = \widetilde{\Lambda}_\lambda$ such that for all $0 \le t_1 \le t_2 \le T$

$$|\widetilde{Z}(t_2) - \widetilde{Z}(t_1)| \le \widetilde{\Lambda} |t_2 - t_1|^\lambda. \tag{6.5}$$

In addition, as noted in Theorem A.5, this $\widetilde{\Lambda}$ can be chosen to have moments of all orders. In other words, the noise $\widetilde{Z}$ defined by (6.4) fits the scope of Chaps. 2–5 including the more restrictive Theorems 5.3 and 5.7.

Keeping these observations in mind, we can now use $\widetilde{Z}$ as an additive driving noise in (3.1), i.e. consider the integrodifferential equation of the form

$$Y(t) = y_0 + \int_0^t b(s, Y(s))ds + \int_0^t \mathcal{G}(t, s)a(s, Y(s))ds + z(t) + \int_0^t \mathcal{K}(t, s)dB(s), \tag{6.6}$$

where b, a and $\mathcal{G}$ are chosen as described in Assumptions 3.2, 3.3 and 3.4 respectively for the one-sided case or Assumptions 3.5, 3.6 and 3.4 for the two-sided case. Then, as we saw in Chap. 3, (6.6) has a unique continuous solution that is either

$$Y(t) > \varphi(t), t \in [0, T], \quad a.s.$$

or

$$\varphi(t) < Y(t) < \psi(t), t \in [0, T], \quad a.s.$$

for the chosen deterministic Hölder continuous functions φ, ψ.

Remark 6.2 Naturally, we require that the initial value satisfies $y_0 > \varphi(0)$ in the one-sided case or $\varphi(0) < y_0 < \varphi(0)$ in the two-sided setting. Throughout this chapter, we always assume that one of these conditions holds.

For the notational convenience, we can put

$$y_0(t) := y_0 + z(t), \quad Z(t) := \widetilde{Z}(t) - z(t) = \int_0^t \mathcal{K}(t,s)dB(s),$$

and rewrite (6.6) as

$$Y(t) = y_0(t) + \int_0^t b(s, Y(s))ds + \int_0^t \mathcal{G}(t,s)a(s, Y(s))ds + Z(t). \tag{6.7}$$

Notation 6.1 Since both $\widetilde{Z}$ and z are λ-Hölder continuous for all $\lambda \in (0, H)$, so is Z. In accordance with the notation from the previous chapters, we denote $\Lambda = \Lambda_\lambda$ a random variable such that for all $0 \le t_1 \le t_2 \le T$

$$|Z(t_2) - Z(t_1)| \le \Lambda |t_2 - t_1|^\lambda. \tag{6.8}$$

In what follows, we also take Λ that has moments of all orders. In addition, for a given $\lambda \in (0, H)$, we denote by $L_0 = L_0(\lambda) > 0$ the deterministic constant such that for all $0 \le t_1 \le t_2 \le T$

$$|y_0(t_2) - y_0(t_1)| \le L_0 |t_2 - t_1|^\lambda. \tag{6.9}$$

In fact, we can put

$$\widetilde{\Lambda} := \Lambda + L_0 \tag{6.10}$$

because, for all $0 \le t_1 \le t_2 \le T$,

$$|\widetilde{Z}(t_2) - \widetilde{Z}(t_1)| = |(Z(t_2) + y_0(t_2)) - (Z(t_1) - y_0(t_1))| \le \widetilde{\Lambda}|t_2 - t_1|^\lambda.$$

Remark 6.3 As explained above, Kolmogorov-Chentsov theorem guarantees that $\widetilde{Z}$ and Z are λ-Hölder continuous for all $\lambda \in (0, H)$. Notice that the value H is not included in this interval: generally speaking, Assumption 6.1 *does not* imply H-Hölder continuity of Gaussian Volterra processes (6.2).

In Chap. 3, we have already established Assumptions 3.2, 3.3, 3.4 and Assumptions 3.5, 3.6, 3.4 that ensure the bounds for the process (6.7). However, since the noise term Z is now expressed in terms of the kernel $\mathcal{K}$, let us slightly reformulate these conditions and simultaneously take a chance to remind the reader about the desired properties of b, a, and $\mathcal{G}$.

As before, let φ, ψ be H-Hölder continuous functions on $[0, T]$, $\varphi < \psi$, and

$$\begin{aligned}\mathcal{D}_x &:= \{(t, y) \mid t \in [0, T], y \in (\varphi(t) + x, \infty)\},\\ \mathcal{D}_{x_1,x_2} &:= \{(t, y) \mid t \in [0, T], y \in (\varphi(t) + x_1, \psi(t) - x_2)\},\\ \overline{\mathcal{D}_x} &:= \{(t, y) \mid t \in [0, T], y \in [\varphi(t) + x, \infty)\},\\ \overline{\mathcal{D}_{x_1,x_2}} &:= \{(t, y) \mid t \in [0, T], y \in [\varphi(t) + x_1, \psi(t) - x_2]\}\end{aligned}$$

for any $x, x_1, x_2 \geq 0$.

Assumption 6.2 *(One-sided case)*

1. The function b: $\mathcal{D}_0 \to \mathbb{R}$ is such that

   (b1) $b \in C(\mathcal{D}_0)$;
   (b2) there are constants $c > 0$ and $p > 1$ such that for any $\varepsilon > 0$ and any $(t_1, y_1), (t_2, y_2) \in \overline{\mathcal{D}_\varepsilon}$

$$|b(t_1, y_1) - b(t_2, y_2)| \leq \frac{c}{\varepsilon^p}\left(|t_1 - t_2|^H + |y_1 - y_2|\right)$$

   with H being the same as in Assumption 6.1;
   (b3) there are such positive constants y_*, c and γ that for all $(t, y) \in \mathcal{D}_0 \setminus \mathcal{D}_{y_*}$:

$$b(t, y) \geq \frac{c}{(y - \varphi(t))^\gamma};$$

   (b4) the constant γ from assumption **(b3)** satisfies the condition

$$\gamma > \frac{1 - H}{H}$$

   with H being the same as in Assumption 6.1;
   (b5) the partial derivative $\frac{\partial b}{\partial y}$ exists, is continuous on $\mathcal{D}_0$ and there are constants $C_b, q > 0$ such that for all $t \in [0, T]$

$$-C_b\left(1 + \frac{1}{(y - \varphi(t))^q}\right) < \frac{\partial b}{\partial y}(t, y) < C_b, \quad (t, y) \in \mathcal{D}_0.$$

2. $\mathcal{G}$: $[0, T]^2 \to \mathbb{R}$ is a Volterra kernel, i.e. $\mathcal{G}(t, s) = 0$ if $s \geq t$, and

   (G1) $\int_0^T \int_0^T \mathcal{G}^2(t, s) ds dt < \infty$;
   (G2) for any $0 \leq t_1 \leq t_2 \leq T$ and any $\lambda \in (0, H)$

$$\int_0^T (\mathcal{G}(t_2, s) - \mathcal{G}(t_1, s))^2 ds \leq \ell_\lambda^2 |t_2 - t_1|^{2\lambda},$$

where $H \in (0, 1)$ and and $\ell_\lambda > 0$ are the same as in Assumption 6.1.

3. The function a is continuous on $\overline{\mathcal{D}_0}$ and

(a1) the partial derivative $\frac{\partial a}{\partial y}$ exists and is continuous on $\overline{\mathcal{D}_0}$;
(a2) there exists a constant L_a such that, for any $(t_1, y_1), (t_2, y_2) \in \overline{\mathcal{D}_0}$,

$$|a(t_1, y_2) - a(t_2, y_1)| \le L_a(|t_1 - t_2|^H + |y_2 - y_1|)$$

with H being the same as in Assumption 6.1.

Assumption 6.3 *(Two-sided case)*

1. The function $b: \mathcal{D}_{0,0} \to \mathbb{R}$ is such that

(b1) $b \in C(\mathcal{D}_{0,0})$;
(b2) there are constants $c > 0$ and $p > 1$ such that for any $\varepsilon \in (0, \frac{1}{2} \min_{t\in[0,T]} |\psi(t) - \varphi(t)|)$ and any $(t_1, y_1), (t_2, y_2) \in \overline{\mathcal{D}_{\varepsilon,\varepsilon}}$

$$|b(t_1, y_1) - b(t_2, y_2)| \le \frac{c}{\varepsilon^p} \left(|t_1 - t_2|^H + |y_1 - y_2|\right)$$

with H being the same as in Assumption 6.1;
(b3) there are constants $y_* \in (0, \frac{1}{2} \min_{t\in[0,T]} |\psi(t) - \varphi(t)|)$, $c > 0$ and $\gamma > 0$ that for all $(t, y) \in \mathcal{D}_{0,0} \setminus \mathcal{D}_{y_*,0}$:

$$b(t, y) \ge \frac{c}{(y - \varphi(t))^\gamma},$$

and for all $(t, y) \in \mathcal{D}_{0,0} \setminus \mathcal{D}_{0,y_*}$:

$$b(t, y) \le -\frac{c}{(\psi(t) - y)^\gamma}.$$

(b4) the constant γ from assumption **(b3)** satisfies the condition

$$\gamma > \frac{1 - H}{H}$$

with H being the same as in Assumption 6.1;
(b5) the partial derivative $\frac{\partial b}{\partial y}$ exists, is continuous on $\mathcal{D}_0$ and there are constants $C_b, q > 0$ such that for all $t \in [0, T]$

$$-C_b\left(1 + \frac{1}{(y - \varphi(t))^q} + \frac{1}{(\psi(t) - y)^q}\right) < \frac{\partial b}{\partial y}(t, y) < C_b, \quad (t, y) \in \mathcal{D}_{0,0}.$$

2. $\mathcal{G}$: $[0, T]^2 \to \mathbb{R}$ is a Volterra kernel, i.e. $\mathcal{G}(t, s) = 0$ if $s \geq t$, and

    (G1) $\int_0^T \int_0^T \mathcal{G}^2(t, s) ds dt < \infty$;
    (G2) for any $0 \leq t_1 \leq t_2 \leq T$ and any $\lambda \in (0, H)$

$$\int_0^T (\mathcal{G}(t_2, s) - \mathcal{G}(t_1, s))^2 ds \leq \ell_\lambda^2 |t_2 - t_1|^{2\lambda},$$

    where $H \in (0, 1)$ and and $\ell_\lambda > 0$ are the same as in Assumption 6.1;
    (G3) the mapping $t \mapsto \mathcal{G}(t, s)$ is non-decreasing (or, alternatively, non-increasing) on $[0, T]$ for all $s \in [0, T]$.

3. The function a is continuous on $\overline{\mathcal{D}_{0,0}}$ and

    (a1) the partial derivative $\frac{\partial a}{\partial y}$ exists and is continuous on $\overline{\mathcal{D}_{0,0}}$;
    (a2) there exists a constant L_a such that, for any $(t_1, y_1), (t_2, y_2) \in \overline{\mathcal{D}_{0,0}}$,

$$|a(t_1, y_2) - a(t_2, y_1)| \leq L_a(|t_1 - t_2|^H + |y_2 - y_1|)$$

    with H being the same as in Assumption 6.1.

Remark 6.4 Note that Assumption 6.2 implies Assumptions 3.2, 3.3, 3.4 whereas Assumption 6.3 implies Assumptions 3.5, 3.6, 3.4. In other words, we indeed stay within the framework established in Chap. 3 and the equation (6.7) has a unique continuous and sandwiched solution.

Example 6.1 Let $H_1 \in (0, 1)$, $H_2 \in (1/2, 1)$, $H := \min\{H_1, H_2\}$ and

$$\mathcal{K}(t, s) = \frac{1}{\Gamma\left(H_1 + \frac{1}{2}\right)}(t - s)^{H_1 - \frac{1}{2}} \mathbb{1}_{s<t}, \quad \mathcal{G}(t, s) = \frac{1}{\Gamma\left(H_2 + \frac{1}{2}\right)}(t - s)^{H_2 - \frac{1}{2}} \mathbb{1}_{s<t},$$

$$b(t, y) = \frac{\kappa_1}{(y - \varphi(s))^\gamma} - \frac{\kappa_2}{(\psi(s) - y)^\gamma}$$

with $\gamma > \frac{1-H}{H}$. Let also y_0 be an arbitrary H-Hölder continuous function such that $\varphi(s) < y_0(0) < \psi(0)$ and $a(t, y) = \sum_{n=0}^N a_n y^n$, $a_n \in \mathbb{R}$. Then, with such choice of the coefficients, Assumptions 6.1 and 6.3 hold.

The next proposition highlights the Hölder continuity of the solution to (6.7).

Proposition 6.1 *Let* $\lambda \in (0, H)$ *be such that* $\lambda > \frac{1}{1+\gamma}$ *and* $\widetilde{\Lambda}$ *be defined by* (6.10)*.Then, if the kernel* $\mathcal{K}$ *satisfies Assumption 6.1 and the coefficients of the sandwiched stochastic equation* (6.7) *satisfy either Assumption 6.2 or Assumption 6.3, the unique continuous solution* Y *of* (6.7) *is* λ*-Hölder continuous with*

$$|Y(t_1) - Y(t_2)| \le \Upsilon |t_1 - t_2|^\lambda, \quad t_1, t_2 \in [0, T], \tag{6.11}$$

where $\Upsilon := C(1 + \widetilde{\Lambda} + (1 + \widetilde{\Lambda})^{\frac{p}{\gamma\lambda+\lambda-1}} + \widetilde{\Lambda}(1 + \widetilde{\Lambda})^{\frac{p}{\gamma\lambda+\lambda-1}})$ *for some deterministic constant* $C > 0$ *and* p *is from Assumption 6.2 or Assumption 6.3. In particular, for any* $r \in \mathbb{R}$,

$$\mathbb{E}[\Upsilon^r] < \infty.$$

Proof For the one-sided case, the claim follows from Theorem 3.4 in combination with the explicit shape of the constant c_ξ given by Assumption 6.2(b2).

In the two-sided situation, the claim follows directly from Theorem 3.9 and Assumption 6.3(b2). □

In this Chapter, we aim to exploit the Gaussian Volterra structure of the driving noise in two ways:

- Gaussianity allows to use the *Malliavin calculus* (see Appendix A.3) – a powerful tool that will be used in the next chapters to study the regularity of probability laws associated to the sandwiched processes (6.7) as well as for numerical algorithms;
- Volterra kernel can be replaced by its approximation of a special shape that will allow us to recover some Markovian properties – those will turn out extremely useful for numerical computations of conditional expectations, especially for hedging purposes.

As we will see later, both of the aims specified above require to evaluate the error between the original sandwiched process (6.7) and the one obtained by perturbing the driving noise in the equation (6.7). This technical result is provided below.

Theorem 6.1 (Perturbation error in two-sided case) *Let Assumptions 6.1 and 6.3 hold and* Y *be the unique continuous solution to* (6.7). *Let* $\delta = \{\delta(t),\ t \in [0, T]\}$ *be a stochastic process starting at* 0 *with Hölder continuous paths up to the order* H, *i.e. for any* $\lambda \in (0, H)$ *there exists random variable* $\Lambda_\delta = \Lambda_{\delta,\lambda}$ *such that*

$$|\delta(t_2) - \delta(t_1)| \le \Lambda_\delta |t_2 - t_1|^\lambda, \quad t_1, t_2 \in [0, T].$$

Consider also a perturbed sandwiched process given by

$$Y_\delta(t) = y_0(t) + \int_0^t b(s, Y_\delta(s))ds + \int_0^t \mathcal{G}(t, s)a(s, Y_\delta(s))ds + Z(t) + \delta(t). \tag{6.12}$$

Then there exists a deterministic constant $C > 0$ *such that for all* $t \in [0, T]$

$$|Y(t) - Y_\delta(t)| \le C\left(|\delta(t)| + \xi \int_0^t |\delta(u)|du\right),$$

where $\xi := (1 + \Lambda + \Lambda_\delta)^{\frac{q}{\gamma\lambda+\lambda-1}}$ *with* $\lambda \in (0, H)$ *being such that* $\lambda > \frac{1}{1+\gamma}$ *and* q *being from Assumption 6.3(b5).*

Proof Without loss of generality, we assume that the mapping $t \mapsto \mathcal{G}(t, s)$ is non-decreasing, and the non-increasing case can be proved in a similar manner. Fix $t \in [0, T]$ and consider a uniform partition $0 = t_0 < t_1 < ... < t_N = t$, $t_n = \frac{n}{N}t$, of the interval $[0, t]$ such that the mesh $\Delta_N := \frac{t}{N}$ satisfies the condition

$$C_b \Delta_N < 1, \tag{6.13}$$

for C_b from Assumption 6.3(b5). Denote $e_{n,\delta} := Y(t_n) - Y_\delta(t_n)$ and observe that

$$\begin{aligned}
e_{N,\delta} = {} & Y(t_{N-1}) + \int_{t_{N-1}}^{t_N} b(s, Y(s))ds + Z(t_N) - Z(t_{N-1}) \\
& + \int_0^{t_N} (\mathcal{G}(t_N, s) - \mathcal{G}(t_{N-1}, s))\, a(s, Y(s))ds \\
& - Y_\delta(t_{N-1}) - \int_{t_{N-1}}^{t_N} b(s, Y_\delta(s))ds - (Z(t_N) - Z(t_{N-1})) - (\delta(t_N) - \delta(t_{N-1})) \\
& - \int_0^{t_N} (\mathcal{G}(t_N, s) - \mathcal{G}(t_{N-1}, s))\, a(s, Y_\delta(s))ds \\
= {} & e_{N-1,\delta} + \int_{t_{N-1}}^{t_N} (b(s, Y(s)) - b(t_N, Y(t_N)))\, ds \\
& - \int_{t_{N-1}}^{t_N} (b(s, Y_\delta(s)) - b(t_N, Y_\delta(t_N)))\, ds \\
& + \sum_{k=1}^N \int_{t_{k-1}}^{t_k} (\mathcal{G}(t_N, s) - \mathcal{G}(t_{N-1}, s))\, (a(s, Y(s)) - a(t_{k-1}, Y(t_{k-1})))ds \\
& - \sum_{k=1}^N \int_{t_{k-1}}^{t_k} (\mathcal{G}(t_N, s) - \mathcal{G}(t_{N-1}, s))\, (a(s, Y_\delta(s)) - a(t_{k-1}, Y_\delta(t_{k-1})))ds \\
& + \sum_{k=1}^N \int_{t_{k-1}}^{t_k} (\mathcal{G}(t_N, s) - \mathcal{G}(t_{N-1}, s))\, (a(t_{k-1}, Y(t_{k-1})) - a(t_{k-1}, Y_\delta(t_{k-1})))ds \\
& + (b(t_N, Y(t_N)) - b(t_N, Y_\delta(t_N)))\, \Delta_N - (\delta(t_N) - \delta(t_{N-1})).
\end{aligned}$$

Note that, for each $n = 1, ..., N$ and $s \in [0, T]$, there exist random values $\Theta^a_{n,\delta}(s)$ and $\Theta^b_{n,\delta}$ between $Y(t_n)$ and $Y_\delta(t_n)$ such that

$$a(s, Y(t_n)) - a(s, Y_\delta(t_n)) = \frac{\partial a}{\partial y}(s, \Theta^a_{n,\delta}(s))(Y(t_n) - Y_\delta(t_n)) = \frac{\partial a}{\partial y}(s, \Theta^a_{n,\delta}(s))e_{n,\delta},$$

and

$$b(t_n, Y(t_n)) - b(t_n, Y_\delta(t_n)) = \frac{\partial b}{\partial y}(t_n, \Theta^b_{n,\delta})(Y(t_n) - Y_\delta(t_n)) = \frac{\partial b}{\partial y}(t_n, \Theta^b_{n,\delta})e_{n,\delta},$$

thus

$$\begin{aligned}
\left(1 - \frac{\partial b}{\partial y}(t_N, \Theta^b_{N,\delta})\Delta_N\right) e_{N,\delta} &= e_{N-1,\delta} + \int_{t_{N-1}}^{t_N} (b(s, Y(s)) - b(t_N, Y(t_N)))\, ds \\
&\quad - \int_{t_{N-1}}^{t_N} (b(s, Y_\delta(s)) - b(t_N, Y_\delta(t_N)))\, ds \\
&\quad + \sum_{k=1}^{N} \int_{t_{k-1}}^{t_k} (\mathcal{G}(t_N, s) - \mathcal{G}(t_{N-1}, s))\,(a(s, Y(s)) - a(t_{k-1}, Y(t_{k-1})))ds \\
&\quad - \sum_{k=1}^{N} \int_{t_{k-1}}^{t_k} (\mathcal{G}(t_N, s) - \mathcal{G}(t_{N-1}, s))\,(a(s, Y_\delta(s)) - a(t_{k-1}, Y_\delta(t_{k-1})))ds \\
&\quad + \sum_{k=1}^{N} \left(\int_{t_{k-1}}^{t_k} (\mathcal{G}(t_N, s) - \mathcal{G}(t_{N-1}, s)) \frac{\partial a}{\partial y}(s, \Theta^a_{k-1,\delta}(s))ds \right) e_{k-1,\delta} \\
&\quad - (\delta(t_N) - \delta(t_{N-1})).
\end{aligned} \tag{6.14}$$

Observe that each $1 - \frac{\partial b}{\partial y}(t_N, \Theta^b_{N,\delta})\Delta_N > 0$ by (6.13) and denote

$$\zeta_{0,\delta} := 1, \quad \zeta_{n,\delta} := \prod_{k=1}^{n} \left(1 - \frac{\partial b}{\partial y}(t_k, \Theta^b_{k,\delta})\Delta_N\right),$$

$\widetilde{e}_{n,\delta} := \zeta_{n,\delta} e_{n,\delta}$. Multiplying the left- and right-hand sides of (6.14) by $\zeta_{N-1,\delta}$, we obtain the equalities

$$\widetilde{e}_{N,\delta} = \widetilde{e}_{N-1,\delta} + \zeta_{N-1,\delta} \int_{t_{N-1}}^{t_N} (b(s, Y(s)) - b(t_N, Y(t_N)))\, ds$$

$$
\begin{aligned}
&- \zeta_{N-1,\delta} \int_{t_{N-1}}^{t_N} (b(s, Y_\delta(s)) - b(t_N, Y_\delta(t_N)))\, ds \\
&+ \zeta_{N-1,\delta} \sum_{k=1}^{N} \int_{t_{k-1}}^{t_k} \big(\mathcal{G}(t_N, s) - \mathcal{G}(t_{N-1}, s)\big) (a(s, Y(s)) - a(t_{k-1}, Y(t_{k-1})))ds \\
&- \zeta_{N-1,\delta} \sum_{k=1}^{N} \int_{t_{k-1}}^{t_k} \big(\mathcal{G}(t_N, s) - \mathcal{G}(t_{N-1}, s)\big) (a(s, Y_\delta(s)) - a(t_{k-1}, Y_\delta(t_{k-1})))ds \\
&+ \zeta_{N-1,\delta} \sum_{k=1}^{N} \left(\int_{t_{k-1}}^{t_k} \big(\mathcal{G}(t_N, s) - \mathcal{G}(t_{N-1}, s)\big) \frac{\partial a}{\partial y}(s, \Theta^a_{k-1,\delta}(s))ds \right) e_{k-1,\delta} \\
&- \zeta_{N-1,\delta}(\delta(t_N) - \delta(t_{N-1})) \\
&= \widetilde{e}_{N-2,\delta} + \sum_{n=N-1}^{N} \zeta_{n-1,\delta} \int_{t_{n-1}}^{t_n} (b(s, Y(s)) - b(t_n, Y(t_n)))\, ds \\
&- \sum_{n=N-1}^{N} \zeta_{n-1,\delta} \int_{t_{n-1}}^{t_n} (b(s, Y_\delta(s)) - b(t_n, Y_\delta(t_n)))\, ds \\
&+ \sum_{n=N-1}^{N} \zeta_{n-1,\delta} \sum_{k=1}^{n} \int_{t_{k-1}}^{t_k} \big(\mathcal{G}(t_n, s) - \mathcal{G}(t_{n-1}, s)\big) (a(s, Y(s)) - a(t_{k-1}, Y(t_{k-1})))ds \\
&- \sum_{n=N-1}^{N} \zeta_{n-1,\delta} \sum_{k=1}^{n} \int_{t_{k-1}}^{t_k} \big(\mathcal{G}(t_n, s) - \mathcal{G}(t_{n-1}, s)\big) (a(s, Y_\delta(s)) - a(t_{k-1}, Y_\delta(t_{k-1})))ds \\
&+ \sum_{n=N-1}^{N} \zeta_{n-1,\delta} \sum_{k=1}^{n} \left(\int_{t_{k-1}}^{t_k} \big(\mathcal{G}(t_n, s) - \mathcal{G}(t_{n-1}, s)\big) \frac{\partial a}{\partial y}(s, \Theta^a_{k-1,\delta}(s))ds \right) e_{k-1,\delta} \\
&- \sum_{n=N-1}^{N} \zeta_{n-1,\delta}(\delta(t_n) - \delta(t_{n-1})) \\
&= \sum_{n=1}^{N} \zeta_{n-1,\delta} \int_{t_{n-1}}^{t_n} (b(s, Y(s)) - b(t_n, Y(t_n)))\, ds \\
&- \sum_{n=1}^{N} \zeta_{n-1,\delta} \int_{t_{n-1}}^{t_n} (b(s, Y_\delta(s)) - b(t_n, Y_\delta(t_n)))\, ds \\
&+ \sum_{n=1}^{N} \zeta_{n-1,\delta} \sum_{k=1}^{n} \int_{t_{k-1}}^{t_k} \big(\mathcal{G}(t_n, s) - \mathcal{G}(t_{n-1}, s)\big) (a(s, Y(s)) - a(t_{k-1}, Y(t_{k-1})))ds
\end{aligned}
$$

$$
\begin{aligned}
&- \sum_{n=1}^{N} \zeta_{n-1,\delta} \sum_{k=1}^{n} \int_{t_{k-1}}^{t_k} \big(\mathcal{G}(t_n, s) - \mathcal{G}(t_{n-1}, s)\big) (a(s, Y_\delta(s)) - a(t_{k-1}, Y_\delta(t_{k-1})))ds \\
&+ \sum_{n=1}^{N} \zeta_{n-1,\delta} \sum_{k=1}^{n} \left(\int_{t_{k-1}}^{t_k} \big(\mathcal{G}(t_n, s) - \mathcal{G}(t_{n-1}, s)\big) \frac{\partial a}{\partial y}(s, \Theta^a_{k-1,\delta}(s))ds \right) e_{k-1,\delta} \\
&- \sum_{n=1}^{N} \zeta_{n-1,\delta}(\delta(t_n) - \delta(t_{n-1})).
\end{aligned}
$$

It means that

$$
\begin{aligned}
e_{N,\delta} = &\sum_{n=1}^{N} \frac{\zeta_{n-1,\delta}}{\zeta_{N,\delta}} \int_{t_{n-1}}^{t_n} (b(s, Y(s)) - b(t_n, Y(t_n)))\, ds \\
&- \sum_{n=1}^{N} \frac{\zeta_{n-1,\delta}}{\zeta_{N,\delta}} \int_{t_{n-1}}^{t_n} (b(s, Y_\delta(s)) - b(t_n, Y_\delta(t_n)))\, ds \\
&+ \sum_{n=1}^{N} \frac{\zeta_{n-1,\delta}}{\zeta_{N,\delta}} \sum_{k=1}^{n} \int_{t_{k-1}}^{t_k} \big(\mathcal{G}(t_n, s) - \mathcal{G}(t_{n-1}, s)\big) (a(s, Y(s)) - a(t_{k-1}, Y(t_{k-1})))ds \\
&- \sum_{n=1}^{N} \frac{\zeta_{n-1,\delta}}{\zeta_{N,\delta}} \sum_{k=1}^{n} \int_{t_{k-1}}^{t_k} \big(\mathcal{G}(t_n, s) - \mathcal{G}(t_{n-1}, s)\big) (a(s, Y_\delta(s)) - a(t_{k-1}, Y_\delta(t_{k-1})))ds \\
&+ \sum_{n=1}^{N} \frac{\zeta_{n-1,\delta}}{\zeta_{N,\delta}} \sum_{k=1}^{n} \left(\int_{t_{k-1}}^{t_k} \big(\mathcal{G}(t_n, s) - \mathcal{G}(t_{n-1}, s)\big) \frac{\partial a}{\partial y}(s, \Theta^a_{k-1,\delta}(s))ds \right) e_{k-1,\delta} \\
&- \sum_{n=1}^{N} \frac{\zeta_{n-1,\delta}}{\zeta_{N,\delta}} (\delta(t_n) - \delta(t_{n-1})).
\end{aligned}
\tag{6.15}
$$

Let us analyze each term in the right-hand side of (6.15). First of all, fix $\lambda \in (0, H)$ such that $\lambda > \frac{1}{1+\gamma}$ and note that the perturbed equation (6.12) can be viewed as a basic sandwiched equation (6.1) driven by a λ-Hölder continuous noise $y_0(t) - y_0(0) + Z(t) + \delta(t)$, $t \in [0, T]$, for which

$$
|y_0(t_2) - y_0(t_1) + Z(t_2) - Z(t_1) + \delta(t_2) - \delta(t_1)| \le C(1 + \Lambda + \Lambda_\delta)|t_2 - t_1|^\lambda, \quad t_1, t_2 \in [0, T].
$$

Therefore, by Theorem 3.8, there exists a deterministic constant $C > 0$ such that, for all $t \in [0, T]$,

$$
\varphi(t) + \frac{1}{C(1 + \Lambda)^{\frac{1}{\gamma\lambda + \lambda - 1}}} \le Y(t) \le \psi(t) - \frac{1}{C(1 + \Lambda)^{\frac{1}{\gamma\lambda + \lambda - 1}}}, \tag{6.16}
$$

$$\varphi(t) + \frac{1}{C(1+\Lambda+\Lambda_\delta)^{\frac{1}{\gamma\lambda+\lambda-1}}} \le Y_\delta(t) \le \psi(t) - \frac{1}{C(1+\Lambda+\Lambda_\delta)^{\frac{1}{\gamma\lambda+\lambda-1}}}. \quad (6.17)$$

Since each $\Theta^b_{n,\delta}$ is between $Y(t_n)$ and $Y_\delta(t_n)$, it is true that

$$\varphi(t_n) + \frac{1}{C(1+\Lambda+\Lambda_\delta)^{\frac{1}{\gamma\lambda+\lambda-1}}} \le \Theta^b_{n,\delta} \le \psi(t_n) - \frac{1}{C(1+\Lambda+\Lambda_\delta)^{\frac{1}{\gamma\lambda+\lambda-1}}}.$$

Together with Assumption 6.3(b5), this implies that

$$-C_b\left(1+2C(1+\Lambda+\Lambda_\delta)^{\frac{q}{\gamma\lambda+\lambda-1}}\right) < \frac{\partial b}{\partial y}(t_n, \Theta^b_{n,\delta}) < C_b,$$

i.e., for some constant $C > 0$ that does not depend on the partition or δ,

$$\left|\frac{\partial b}{\partial y}(t_n, \Theta^b_{n,\delta})\right| \le C(1+\Lambda+\Lambda_\delta)^{\frac{q}{\gamma\lambda+\lambda-1}}.$$

Now, by the Abel's summation-by-parts formula,

$$\begin{aligned}
&\left|\sum_{n=1}^{N} \frac{\zeta_{n-1,\delta}}{\zeta_{N,\delta}}\Big(\delta(t_n) - \delta(t_{n-1})\Big)\right| \\
&= \left|\frac{\zeta_{N-1,\delta}}{\zeta_{N,\delta}}\delta(t_N) - \sum_{n=1}^{N-1}\delta(t_n)\left(\frac{\zeta_{n,\delta}}{\zeta_{N,\delta}} - \frac{\zeta_{n-1,\delta}}{\zeta_{N,\delta}}\right)\right| \\
&= \left|\frac{\zeta_{N-1,\delta}}{\zeta_{N,\delta}}\delta(t_N) + \sum_{n=1}^{N-1}\frac{\zeta_{n-1,\delta}}{\zeta_{N,\delta}}\frac{\partial b}{\partial y}(t_n, \Theta^b_{n,\delta})\delta(t_n)\Delta_N\right| \qquad (6.18) \\
&\le C|\delta(t_N)| + C\sum_{n=1}^{N-1}\left|\frac{\partial b}{\partial y}(t_n, \Theta^b_{n,\delta})\right| |\delta(t_n)|\Delta_N \\
&\le C|\delta(t_N)| + C(1+\Lambda+\Lambda_\delta)^{\frac{q}{\gamma\lambda+\lambda-1}}\sum_{n=1}^{N-1}|\delta(t_n)|\Delta_N.
\end{aligned}$$

Next, observe that, by Assumption 6.3(b5) and (6.13), for any $n = 1, \dots, N$

$$\frac{\zeta_{n-1,\delta}}{\zeta_{N,\delta}} = \prod_{k=n}^{N}\left(1 - \frac{\partial b}{\partial y}(t_k, \Theta_{k,\delta})\Delta_N\right)^{-1} \le \prod_{k=n}^{N}(1 - C_b\Delta_N)^{-1} \le (1 - C_b\Delta_N)^{-N}.$$

Next, note that $(1 - C_b\Delta_N)^{-N}$ converges to e^{C_bT} as $N \to \infty$ and hence is bounded w.r.t. N. Therefore, there exists a (non-random) constant $C > 0$ that does not depend on n, N or δ such that

$$0 < \frac{\zeta_{n-1,\delta}}{\zeta_{N,\delta}} \le C.$$

In addition, by (6.16), Assumption 6.3(**b2**) and Proposition 6.1, for any $s \in [0, T]$,

$$\begin{aligned}|b(s, Y(s)) - b(t_n, Y(t_n))| &\le C(1+\Lambda)^{\frac{p}{\gamma\lambda+\lambda-1}} \left(|s-t_n|^H + |Y(s) - Y(t_n)|\right)\\ &\le C(1+\Lambda)^{\frac{p}{\gamma\lambda+\lambda-1}} \left(|s-t_n|^H + \Upsilon|s-t_n|^\lambda\right)\\ &\le C(1+\Lambda)^{\frac{p}{\gamma\lambda+\lambda-1}} (1+\Upsilon)|s-t_n|^\lambda,\end{aligned}$$

which implies that

$$\begin{aligned}&\left|\sum_{n=1}^N \frac{\zeta_{n-1,\delta}}{\zeta_{N,\delta}} \int_{t_{n-1}}^{t_n} (b(s, Y(s)) - b(t_n, Y(t_n)))\, ds\right|\\ &\qquad \le \sum_{n=1}^N \left|\frac{\zeta_{n-1,\delta}}{\zeta_{N,\delta}}\right| \int_{t_{n-1}}^{t_n} |b(s, Y(s)) - b(t_n, Y(t_n))|\, ds\\ &\qquad \le C(1+\Lambda)^{\frac{p}{\gamma\lambda+\lambda-1}} (1+\Upsilon) \sum_{n=1}^N \int_{t_{n-1}}^{t_n} |s-t_n|^\lambda ds\\ &\qquad \le C(1+\Lambda)^{\frac{p}{\gamma\lambda+\lambda-1}} (1+\Upsilon)\Delta_N^\lambda.\end{aligned}$$

Similarly,

$$\begin{aligned}&\left|\sum_{n=1}^N \frac{\zeta_{n-1,\delta}}{\zeta_{N,\delta}} \int_{t_{n-1}}^{t_n} (b(s, Y_\delta(s)) - b(t_n, Y_\delta(t_n)))\, ds\right|\\ &\qquad \le C(1+\Lambda+\Lambda_\delta)^{\frac{p}{\gamma\lambda+\lambda-1}} (1+\Upsilon_\delta)\Delta_N^\lambda,\end{aligned}$$

where Υ_δ is such that

$$|Y_\delta(t_1) - Y_\delta(t_2)| \le \Upsilon_\delta |t_1 - t_2|^\lambda, \quad t_1, t_2 \in [0, T],$$

which is guaranteed by Proposition 6.1.

Next, recall that, by Assumption 6.3(G3) and a remark in the beginning of the proof, the mapping $t \mapsto \mathcal{G}(t, s)$ is non-decreasing.In addition to Assumption 6.3(a2) and Proposition 6.1, we can produce the relations:

$$\left|\sum_{n=1}^N \frac{\zeta_{n-1,\delta}}{\zeta_{N,\delta}} \sum_{k=1}^n \int_{t_{k-1}}^{t_k} (\mathcal{G}(t_n, s) - \mathcal{G}(t_{n-1}, s))\,(a(s, Y(s)) - a(t_{k-1}, Y(t_{k-1})))ds\right|$$

$$
\begin{aligned}
&= \left| \sum_{k=1}^{N} \int_{t_{k-1}}^{t_k} \left(\sum_{n=k}^{N} \frac{\zeta_{n-1,\delta}}{\zeta_{N,\delta}} \left(\mathcal{G}(t_n, s) - \mathcal{G}(t_{n-1}, s)\right) \right) (a(s, Y(s)) - a(t_{k-1}, Y(t_{k-1}))) ds \right| \\
&\leq \sum_{k=1}^{N} \int_{t_{k-1}}^{t_k} \left(\sum_{n=k}^{N} \frac{\zeta_{n-1,\delta}}{\zeta_{N,\delta}} \left(\mathcal{G}(t_n, s) - \mathcal{G}(t_{n-1}, s)\right) \right) |a(s, Y(s)) - a(t_{k-1}, Y(t_{k-1}))| \, ds \\
&\leq C \sum_{k=1}^{N} \int_{t_{k-1}}^{t_k} \left(\sum_{n=k}^{N} (\mathcal{G}(t_n, s) - \mathcal{G}(t_{n-1}, s)) \right) |s - t_{k-1}|^H \, ds \\
&\quad + C \sum_{k=1}^{N} \int_{t_{k-1}}^{t_k} \left(\sum_{n=k}^{N} (\mathcal{G}(t_n, s) - \mathcal{G}(t_{n-1}, s)) \right) |Y(s) - Y(t_{k-1})| \, ds \\
&\leq C \Delta_N^{\lambda} \sum_{i=1}^{n} \int_{t_{i-1}}^{t_i} \mathcal{G}(t_n, s) ds + C \Upsilon \Delta_N^{\lambda} \sum_{i=1}^{n} \int_{t_{i-1}}^{t_i} \mathcal{G}(t_n, s) ds \\
&\leq C(1 + \Upsilon) \int_{0}^{t_n} \mathcal{G}(t_n, s) ds \, \Delta_N^{\lambda} \leq C(1 + \Upsilon) \Delta_N^{\lambda}.
\end{aligned}
$$

Similarly,

$$
\begin{aligned}
&\left| \sum_{n=1}^{N} \frac{\zeta_{n-1,\delta}}{\zeta_{N,\delta}} \sum_{k=1}^{n} \int_{t_{k-1}}^{t_k} (\mathcal{G}(t_n, s) - \mathcal{G}(t_{n-1}, s)) \, (a(s, Y_\delta(s)) - a(t_{k-1}, Y_\delta(t_{k-1}))) ds \right| \\
&\quad \leq C(1 + \Upsilon_\delta) \Delta_N^{\lambda}.
\end{aligned}
$$

Finally,

$$
\begin{aligned}
&\left| \sum_{n=1}^{N} \frac{\zeta_{n-1,\delta}}{\zeta_{N,\delta}} \sum_{k=1}^{n} \left(\int_{t_{k-1}}^{t_k} \left(\mathcal{G}(t_n, s) - \mathcal{G}(t_{n-1}, s)\right) \frac{\partial a}{\partial y}(s, \Theta^a_{k-1,\delta}(s)) ds \right) e_{k-1,\delta} \right| \\
&\leq \sum_{k=1}^{N} \left(\int_{t_{k-1}}^{t_k} \left(\sum_{n=k}^{N} \frac{\zeta_{n-1,\delta}}{\zeta_{N,\delta}} \left(\mathcal{G}(t_n, s) - \mathcal{G}(t_{n-1}, s)\right) \right) \left| \frac{\partial a}{\partial y}(s, \Theta^a_{k-1,\delta}(s)) \right| ds \right) |e_{k-1,\delta}| \\
&\leq C \sum_{k=1}^{N} \left(\int_{t_{k-1}}^{t_k} \left(\sum_{n=k}^{N} \left(\mathcal{G}(t_n, s) - \mathcal{G}(t_{n-1}, s)\right) \right) ds \right) |e_{k-1,\delta}| \\
&\leq C \sum_{k=1}^{N} \int_{t_{k-1}}^{t_k} \left(\mathcal{G}(t_N, s) - \mathcal{G}(t_{k-1}, s)\right) ds |e_{k-1,\delta}|
\end{aligned}
$$

$$= C \sum_{k=1}^{N} \int_{t_{k-1}}^{t_k} \mathcal{G}(t_N, s) ds |e_{k-1,\delta}|.$$

Now, using (6.15) as well as the estimates of its right-hand side derived above, we can write

$$\begin{aligned} |e_{N,\delta}| \le C|\delta(t_N)| &+ C(1 + \Lambda + \Lambda_\delta)^{\frac{q}{\gamma\lambda+\lambda-1}} \sum_{n=1}^{N-1} |\delta(t_n)|\Delta_N \\ &+ C(1+\Lambda)^{\frac{p}{\gamma\lambda+\lambda-1}}(1+\Upsilon)\Delta_N^\lambda \\ &+ C(1+\Lambda+\Lambda_\delta)^{\frac{p}{\gamma\lambda+\lambda-1}}(1+\Upsilon_\delta)\Delta_N^\lambda \\ &+ C(1+\Upsilon)\Delta_N^\lambda + C(1+\Upsilon_\delta)\Delta_N^\lambda \\ &+ C\sum_{k=1}^{N} \int_{t_{k-1}}^{t_k} \mathcal{G}(t_N, s) ds |e_{k-1,\delta}|. \end{aligned}$$

This upper bound implies, by the discrete Gronwall inequality, that for all large enough N,

$$\begin{aligned} |e_{N,\delta}| &= |Y(t) - Y_\delta(t)| \\ &\le \exp\left\{\int_0^t \mathcal{G}(t,s)ds\right\} \left(C|\delta(t_N)| + C(1+\Lambda+\Lambda_\delta)^{\frac{q}{\gamma\lambda+\lambda-1}} \sum_{n=1}^{N-1} |\delta(t_n)|\Delta_N \right. \\ &\quad + C(1+\Lambda)^{\frac{p}{\gamma\lambda+\lambda-1}}(1+\Upsilon)\Delta_N^\lambda \\ &\quad + C(1+\Lambda+\Lambda_\delta)^{\frac{p}{\gamma\lambda+\lambda-1}}(1+\Upsilon_\delta)\Delta_N^\lambda \\ &\quad \left. + C(1+\Upsilon)\Delta_N^\lambda + C(1+\Upsilon_\delta)\Delta_N^\lambda \right). \end{aligned}$$

Moving $N \to \infty$, we obtain that

$$\begin{aligned} |Y(t) - Y_\delta(t)| &\le C \exp\left\{\int_0^t \mathcal{G}(t,s)ds\right\} \left(|\delta(t)| + (1+\Lambda+\Lambda_\delta)^{\frac{q}{\gamma\lambda+\lambda-1}} \int_0^t |\delta(s)|ds \right) \\ &\le C\left(|\delta(t)| + (1+\Lambda+\Lambda_\delta)^{\frac{q}{\gamma\lambda+\lambda-1}} \int_0^t |\delta(s)|ds \right), \end{aligned}$$

which completes the proof. □

6.3 Malliavin Differentiability

Gaussianity of the Volterra driver in (6.7) enables a very powerful tool, *Malliavin calculus*, that can be used in a wide variety of ways. Some of such applications will be presented in Chaps. 7 and 8: for example, Malliavin differentiability will allow to establish some key regularity properties of probability laws in financial models based on (6.7) and will serve as a basis for several numerical pricing algorithms. In this section, we will consider a slightly simplified version of the equation (6.7) with $\mathcal{G} \equiv 0$, i.e.

$$Y(t) = y_0(t) + \int_0^t b(s, Y(s))ds + Z(t), \tag{6.19}$$

where $Z(t) = \int_0^t \mathcal{K}(t,s)dB(s)$, $t \in [0, T]$, with all the coefficients satisfying Assumption 6.1 and Assumption 6.3.

Recall also that we work on the canonical Wiener space $(\Omega, \mathcal{F}, \mathbb{P})$ with the Brownian motion $B\colon [0,T] \times \Omega \to \mathbb{R}$ being such that $B\colon [0,T] \times \Omega \to \mathbb{R}$, $B(t,\omega) = \omega(t)$. Let C be the Cameron-Martin space, i.e. the space of all continuous functions $F \in C_0([0,T])$ such that

$$F(\cdot) = \int_0^{\cdot} f(s)ds,$$

where $f \in L^2([0,T])$. By $\mathbb{D}^{1,2}$, we denote the space of Malliavin differentiable functions w.r.t. the L^2-closure (see Appendix A.3).

6.3.1 *First Order Malliavin Differentiability*

As the main tool for establishing the Malliavin differentiability of $Y(t)$, we will utilize the famous result by H. Sugita that we present below without the proof.

Theorem 6.2 **([7], Theorem 3.1)** *A random variable $\eta \in L^2(\Omega, \mathcal{F}_T, \mathbb{P})$ belongs to $\mathbb{D}^{1,2}$ w.r.t. B if and only if the following two conditions are satisfied:*

(i) *η is* ray absolutely continuous, *i.e. for any $F \in C$ there exists a version of the process $\{\eta(\omega + \varepsilon F),\ \varepsilon \geq 0\}$ that is absolutely continuous;*

(ii) *η is* stochastically Gateaux differentiable, *i.e. there exists a random vector*

$$D\eta \in L^2\left(\Omega \times [0,T]\right)$$

such that for any $F = \int_0^{\cdot} f(s)ds \in C$, $f \in L^2([0,T])$,

$$\frac{1}{\varepsilon}\left(\eta\left(\omega+\varepsilon F\right)-\eta(\omega)\right)\xrightarrow{\mathbb{P}}\langle D\eta, f\rangle_{L^2([0,T])},\quad \varepsilon\to 0.$$

In this case the random vector $D\eta$ from (ii) is the Malliavin derivative of η.

In order to apply Theorem 6.2 to obtain the Malliavin differentiability of random variables $Y(t)=Y(\omega,t)$, $t\in[0,T]$, we first need to understand what exactly is

$$Y^{\varepsilon,F}(\omega,t):=Y\left(\omega+\varepsilon F,t\right)\quad t\in[0,T],$$

for any $F=\int_0^{\cdot} f(s)ds\in C$. Using (6.19), we can see that

$$Y^{\varepsilon,F}(\omega,t)=y_0(t)+\int_0^t b\left(s,Y^{\varepsilon,F}(\omega,s)\right)ds+Z^{\varepsilon,F}\left(\omega,t\right),\quad t\in[0,T],\tag{6.20}$$

where $Z^{\varepsilon,F}\left(\omega,t\right):=Z\left(\omega+\varepsilon F,t\right)$. Note that, with probability 1, the equation (6.20) has a unique continuous solution since $y_0(t)+Z^{\varepsilon,F}\left(\omega,t\right)$ is Hölder continuous w.r.t. t a.s.

Since the Brownian motion B satisfies $B(t,\omega)=\omega(t)$, it is intuitively clear that $Z^{\varepsilon,F}\left(\omega,t\right)$ should be obtainable from Z by "*replacing*" the integrator B with $B+\varepsilon F$ – in other words, one should be able to represent $Z^{\varepsilon,F}\left(\omega,t\right)$ as

$$Z^{\varepsilon,F}\left(\cdot,t\right)=Z(\cdot,t)+\varepsilon\int_0^t\mathcal{K}(t,s)f(s)ds,\quad t\in[0,T].$$

This guess is formalized in the next proposition.

Proposition 6.2 *Let $\varepsilon\geq 0$ and $F=\int_0^{\cdot} f(s)ds\in C$. Then there exists $\Omega_{\varepsilon,F}$, $\mathbb{P}(\Omega_{\varepsilon,F})=1$, such that for any $i=1,...,d$ and $\omega\in\Omega_{\varepsilon,F}$*

$$Z^{\varepsilon,F}\left(\omega,t\right)=Z(\omega,t)+\varepsilon\int_0^t\mathcal{K}(t,s)f(s)ds,\quad t\in[0,T].$$

Proof Fix an arbitrary $t\in[0,T]$. By definition,

$$\begin{aligned}Z(\cdot,t)&=L^2\text{-}\lim_{n\to\infty}\sum_{l=0}^{k_n-1}K^{n,l}(t)\left(B(\cdot,t_{n,l+1})-B(\cdot,t_{n,l})\right)\\&=:L^2\text{-}\lim_{n\to\infty}Z^n(\cdot,t),\end{aligned}$$

where $\pi^n = \{0 = t_{n,0} < t_{n,1} < ... < t_{n,k_n} = t\}$, $n \geq 1$, is a system of partitions such that

$$\sup_{l=0,...,k_n-1} (t_{n,l+1} - t_{n,l}) \to 0$$

as $n \to \infty$ and $\mathcal{K}^n(t,s) := \sum_{l=0}^{k_n-1} K^{n,l}(t)\mathbb{1}_{[t_{n,l},t_{n,l+1})}(s)$ are simple functions w.r.t. s such that $\int_0^t (\mathcal{K}(t,s) - \mathcal{K}^n(t,s))^2 ds \to 0$, $n \to \infty$.

By the Riesz theorem, there exist $\Omega'_t \subset \Omega$, $\mathbb{P}(\Omega'_t) = 1$, and a subsequence of partitions that will again be denoted by $\{\pi^n,\ n \geq 1\}$ such that for all $\omega \in \Omega'_t$

$$Z(\omega, t) = \lim_{n\to\infty} Z^n(\omega, t). \tag{6.21}$$

Moreover, for any $\omega \in \Omega = C_0([0, T])$,

$$\begin{aligned} Z^n(\omega, t) &= \sum_{l=0}^{k_n-1} K^{n,l}(t)\left(B(\omega, t_{n,l+1}) - B(\omega, t_{n,l})\right) \\ &= \sum_{l=0}^{k_n-1} K^{n,l}(t)\left(\omega(t_{n,l+1}) - \omega(t_{n,l})\right), \end{aligned}$$

whence, for all $\omega \in \Omega$,

$$Z^n(\omega + \varepsilon F, t) = Z^n(\omega) + \varepsilon \sum_{l=0}^{k_n-1} K^{n,l}(t)\left(F(t_{n,l+1}) - F(t_{n,l})\right)$$

and thus, for all $\omega \in \Omega'_t$,

$$\begin{aligned} \lim_{n\to\infty} Z^n(\omega + \varepsilon F, t) &= \lim_{n\to\infty} Z^n(\omega) + \lim_{n\to\infty} \varepsilon \sum_{l=0}^{k_n-1} K^{n,l}(t)\left(F(t_{n,l+1}) - F(t_{n,l})\right) \\ &= Z_i(\omega, t) + \varepsilon \int_0^t \mathcal{K}(t,s) f(s) ds. \end{aligned} \tag{6.22}$$

Now, denote $\Omega_{\varepsilon,F,t} := \{\omega \in \Omega'_t :\ \omega + \varepsilon F \in \Omega'_t\}$. Then $\mathbb{P}(\Omega_{\varepsilon,F,t}) = 1$ since, by the Cameron-Martin theorem,

$$\mathbb{P}(\Omega_{\varepsilon,F,t}) = \int_\Omega \mathbb{1}_{\Omega'_t}(\omega + \varepsilon F) d\mathbb{P}(\omega) = \mathbb{E}\left[\mathbb{1}_{\Omega'_t} \exp\left\{\varepsilon \mathcal{I}_f - \frac{\varepsilon^2}{2}\|\mathcal{I}_f\|^2_{L^2(\Omega)}\right\}\right] = 1,$$

where $I_f := \int_0^T f(s)dB(s)$, and therefore, taking into account (6.21) and (6.22), for any $\omega \in \Omega_{\varepsilon,F,t}$

$$Z(\omega + \varepsilon F, t) = Z(\omega, t) + \varepsilon \int_0^t \mathcal{K}(t, s) f(s) ds. \tag{6.23}$$

Finally, it remains to notice that, for the fixed $\varepsilon \geq 0$ and $F \in C$, both left- and right-hand sides of (6.23) are continuous w.r.t. t, whence we can put

$$\Omega_{\varepsilon,F} := \bigcap_{t \in \mathbb{Q} \cap [0,T]} \Omega_{\varepsilon,F,t}.$$

□

Now we are ready to prove the key result of this subsection: the Malliavin differentiability of $Y(t)$.

Theorem 6.3 *Let Assumptions 6.1 and 6.3 hold. Then, for all $t \in [0, T]$, $Y(t) \in \mathbb{D}^{1,2}$ and*

$$D_u Y(t) := \left(\mathcal{K}(t, u) + \int_u^t \mathcal{K}(s, u) \frac{\partial b}{\partial y}(s, Y(s)) e^{\int_s^t \frac{\partial b}{\partial y}(v, Y(v)) dv} ds \right) \mathbb{1}_{[0,t]}(u). \tag{6.24}$$

Proof We will split the proof into three steps. First, we verify that $DY(t)$ from (6.24) indeed satisfies the relation

$$\mathbb{E} \int_0^T (D_u Y(t))^2 du < \infty$$

and then check the conditions of Theorem 6.2.

Step 1. Correctness of the form of derivative. By Theorem 3.8, for all $s \in [0, T]$: $(s, Y(s)) \in \mathcal{D}_{\frac{1}{\xi}, \frac{1}{\xi}}$, where

$$\xi := C(1 + \Lambda_\lambda)^{\frac{1}{\gamma\lambda + \lambda - 1}}$$

with $C > 0$ being a deterministic constant that does not depend on the noise Z, $\lambda \in \left(\frac{1}{\gamma+1}, H\right)$, and Λ_λ being such that

$$|Z(t_1) - Z(t_2)| \leq \Lambda_\lambda |t_1 - t_2|^\lambda, \quad t_1, t_2 \in [0, T].$$

Moreover, Λ_λ can be chosen such that $\mathbb{E}[\Lambda_\lambda^r] < \infty$ for all $r \in \mathbb{R}$. Hence, due to Assumption 6.3(**b5**),

$$\left|\frac{\partial b}{\partial y}(s, Y(s))\right| \leq C(1+\xi^q), \quad s \in [0, T].$$

Moreover, by the same Assumption 6.3(**b5**),

$$e^{\int_s^t \frac{\partial b}{\partial y}(v, Y(v))dv} \leq e^{C_b T}.$$

Therefore, for the value $D_u Y(t)$ given by (6.24), we have:

$$\begin{aligned}\mathbb{E}\int_0^T (D_u Y(t))^2\, du &\leq 2\int_0^t \mathcal{K}^2(t,u)du \\ &\quad + 2T\int_u^t \mathcal{K}^2(s,u)\mathbb{E}\left[\left|\frac{\partial b}{\partial y}(s, Y(s))\right|^2 e^{2\int_s^t \frac{\partial b}{\partial y}(v, Y(v))dv}\right]ds \\ &\leq 2\int_0^t \mathcal{K}^2(t,u)du \\ &\quad + 2CT\int_u^t \mathcal{K}^2(s,u)\mathbb{E}\left[(1+\xi^q)^2\right]ds < \infty,\end{aligned}$$

because $\int_0^t \mathcal{K}^2(t,s)ds < \infty$ by Assumption 6.1(**K1**), .and the random variable ξ has moments of all orders.

Step 2. Ray absolute continuity. Let $F \in \mathcal{H}$ be fixed. Then, as explained above, the random process $\{Y^{\varepsilon,F}(t),\ \varepsilon \geq 0\}$ defined for every $\varepsilon \geq 0$ by the equation

$$Y^{\varepsilon,F}(t) = y_0(t) + \int_0^t b(s, Y^{\varepsilon,F}(s))ds + Z(t) + \varepsilon\int_0^t \mathcal{K}(t,s)f(s)ds, \tag{6.25}$$

is a version of the process $\{Y(\omega+\varepsilon F, t),\ \varepsilon \geq 0\}$. Let us prove that this version is Lipschitz w.r.t. ε and hence is absolutely continuous. First, fix $\lambda \in \left(\frac{1}{\gamma+1}, H\right)$, where H and γ are from Assumption 6.3, and take an arbitrary $\varepsilon^* > 0$. Denote $Z^{\varepsilon,F}(t) = Z(t) + \varepsilon\int_0^t \mathcal{K}(t,s)f(s)ds$ and take arbitrary $\varepsilon_1, \varepsilon_2 \in [0, \varepsilon^*]$. Note that the processes $Y^{\varepsilon_i,F}$, $i = 1, 2$ can be viewed as a two-sided sandwiched processes driven by Hölder continuous noises $Z^{\varepsilon_i,F}$, $i = 1, 2$, respectively, and, in addition, for any $\varepsilon \in [0, \varepsilon^*]$ and $0 \leq t_1 \leq t_2 \leq T$,

$$
\begin{aligned}
|Z^{\varepsilon,F}(t_1) - Z^{\varepsilon,F}(t_2)| &\le |Z(t_1) - Z(t_2)| + \varepsilon \left| \int_0^{t_2} (\mathcal{K}(t_2,s) - \mathcal{K}(t_1,s)) f(s)ds \right| \\
&\le \Lambda_\lambda |t_1 - t_2|^\lambda + \varepsilon \left(\int_0^{t_2} (\mathcal{K}(t_2,s) - \mathcal{K}(t_1,s))^2 ds \right)^{\frac{1}{2}} \|f\|_{L^2([0,T])} \\
&\le \left(\Lambda_\lambda + \varepsilon^* \ell_\lambda \|f\|_{L^2([0,T])}\right) |t_1 - t_2|^\lambda .
\end{aligned}
\tag{6.26}
$$

Moreover, it is easy to see that $\delta(t) := Z^{\varepsilon_2,F}(t) - Z^{\varepsilon_1,F}(t) = (\varepsilon_2 - \varepsilon_1) \int_0^t \mathcal{K}(t,s) f(s)ds$ satisfies

$$
|\delta(t_1) - \delta(t_2)| \le (\varepsilon_2 - \varepsilon_1)\ell_\lambda \|f\|_{L^2([0,T])} |t_1 - t_2|^\lambda \le \varepsilon^* \ell_\lambda \|f\|_{L^2([0,T])} |t_1 - t_2|^\lambda .
$$

Therefore, we can apply Theorem 6.1 to establish the existence of a deterministic constant C that does not depend on ε and such that

$$
\begin{aligned}
|Y^{\varepsilon_2,F}(t) - Y^{\varepsilon_1,F}(t)| &\le C \left(|\delta(t)| + \left(1 + \Lambda_\lambda + 2\varepsilon^* \ell_\lambda \|f\|_{L^2([0,T])}\right)^{\frac{q}{\gamma\lambda+\lambda-1}} \int_0^t |\delta(u)| du \right) \\
\le C \Bigg(\left| \int_0^t \mathcal{K}(t,s) f(s) ds \right| \\
&+ \left(1 + \Lambda_\lambda + 2\varepsilon^* \ell_\lambda \|f\|_{L^2([0,T])}\right)^{\frac{q}{\gamma\lambda+\lambda-1}} \int_0^t \int_0^u |\mathcal{K}(u,s) f(s)| \, ds du \Bigg) |\varepsilon_2 - \varepsilon_1|.
\end{aligned}
\tag{6.27}
$$

Therefore, $\{Y^{\varepsilon,F}(t),\ \varepsilon \ge 0\}$ is locally Lipschitz which implies the desired absolute continuity w.r.t. ε.

Step 3. Stochastic Gateaux differentiability. It is sufficient to compute the limit in probability of the form

$$
\mathbb{P}\text{-} \lim_{\varepsilon \to 0} \frac{Y^{\varepsilon,F}(t) - Y^{0,F}(t)}{\varepsilon},
$$

where $Y^{\varepsilon,F}(t)$ is defined by (6.25).

Using the mean-value theorem, we obtain that

$$\begin{aligned}
Y^{\varepsilon,F}(t) - Y^{0,F}(t) &= \int_0^t (b(s, Y^{\varepsilon,F}(s)) - b(s, Y^{0,F}(s))ds \\
&\quad + \varepsilon \int_0^t \mathcal{K}(t,s) f(s) ds \\
&= \int_0^t \Theta_\varepsilon(s)(Y^{\varepsilon,F}(s) - Y^{0,F}(s))ds \\
&\quad + \varepsilon \int_0^t \mathcal{K}(t,s) f(s) ds,
\end{aligned} \tag{6.28}$$

where $\Theta_\varepsilon(s) := \frac{\partial b}{\partial y}\left(s, Y^{0,F}(s) + \theta_\varepsilon(s)(Y^{\varepsilon,F}(s) - Y^{0,F}(s))\right)$ with $\theta_\varepsilon(s)$ being some value between 0 and 1. It is easy to verify that (6.28) implies the representation of the form

$$Y^{\varepsilon,F}(t) - Y^{0,F}(t) = \varepsilon \int_0^t \exp\left\{\int_s^t \Theta_\varepsilon(v)dv\right\} dG(s),$$

where

$$G(t) := \int_0^t \mathcal{K}(t,s) f(s) ds.$$

Note that, just like in Step 1, it follows from Theorem 3.8 and (6.26) that, for all $\varepsilon \in [0,1]$ and $s \in [0,T]$,

$$\varphi(t) + \frac{1}{\xi_1} \le Y^{\varepsilon,F}(s) \le \psi(t) - \frac{1}{\xi_1},$$

where

$$\xi_1 := \frac{1}{C\left(1 + \Lambda_\lambda + \ell_\lambda \|f\|_{L^2([0,T])}\right)^{\frac{1}{\gamma\lambda+\lambda-1}}}$$

and $C > 0$ is some positive constant. This estimate, together with the definition of $\Theta_\varepsilon(s)$ and Assumption 3.5, implies that there exists a finite positive random variable η such that $|\Theta_\varepsilon(s)| \le \eta$ for all $\varepsilon \in [0,1]$ and $s \in [0,T]$. Moreover, the mapping $s \mapsto \exp\left\{\int_s^t \Theta_\varepsilon(v)dv\right\}$, $s \in [0,t]$, is Lipschitz and hence the integrals

$$\int_0^t \exp\left\{\int_s^t \Theta_\varepsilon(v)dv\right\} dG(s)$$

are well-defined as the pathwise limits of Riemann-Stieltjes integral sums due to the classical results of Young (see [8, Section 10], see also [4, Chapter 6] for a more recent overview on the subject).

Now, applying the integration by parts formula, we obtain

$$\begin{aligned}\frac{Y^{\varepsilon,F}(t) - Y^{0,F}(t)}{\varepsilon} &= \int_0^t \exp\left\{\int_s^t \Theta_\varepsilon(v)dv\right\} dG(s) \\ &= \left(G(t) - \int_0^t G(s)d\left(\exp\left\{\int_s^t \Theta_\varepsilon(v)dv\right\}\right)\right) \\ &= \left(G(t) + \int_0^t G(s)\Theta_\varepsilon(s)\exp\left\{\int_s^t \Theta_\varepsilon(v)dv\right\} ds\right),\end{aligned}$$

and, taking into account the definition of G,

$$\frac{Y^{\varepsilon,F}(t) - Y^{0,F}(t)}{\varepsilon} = \int_0^T D(\varepsilon, t, u) f(u)du$$

with

$$D(\varepsilon, t, u) := \left(\mathcal{K}(t, u) + \int_u^t \mathcal{K}(s, u)\Theta_\varepsilon(s)\exp\left\{\int_s^t \Theta_\varepsilon(v)dv\right\} ds\right) \mathbb{1}_{[0,t]}(u).$$

Now, note that for all $s \in [0, T]$ and $\varepsilon \in [0, 1]$

$$|\Theta_\varepsilon(s)| \exp\left\{\int_s^t \Theta_\varepsilon(v)dv\right\} < \eta \exp\{C_b T\},$$

where c is from Assumption 6.3(**b5**), and therefore we can apply the dominated convergence theorem to obtain that

$$\frac{Y^{\varepsilon,F}(t) - Y^{0,F}(t)}{\varepsilon} \to \langle DY(t), f\rangle_{L^2([0,T])} \quad a.s., \quad \varepsilon \to 0,$$

where $DY(t)$ has exactly the form (6.24), which finishes the proof. □

6.3.2 Malliavin Derivatives of Higher Orders

In the previous section, we established the existence of the first-order Malliavin derivative. However, one may be interested in more nuanced properties: for example, we already know that $Y(t) \in \mathbb{D}^{1,2}$, but what about $\mathbb{D}^{1,p}$ for $p > 2$? Furthermore, does $Y(t)$ have Malliavin derivatives of higher orders? As we will see later in Chap. 7, these two questions are not purely theoretical: as it will turn out, they allow to study the implied volatility skews generated by financial models with Y as stochastic volatility! And, since our ultimate goal is to use sandwiched processes as stochastic volatilities, it is very natural to answer these two questions.

First of all, let us remark that Theorem 6.3, and, in particular, Eq. (6.24), gives a good intuition of how the second-order Malliavin derivative should look like. Indeed, one should expect that the following computations are true:

$$
\begin{aligned}
&D_r D_s Y(t) \\
&= D_r \int_s^t \mathcal{K}(u,s) b'_y(u, Y(u)) \exp\left\{ \int_u^t b'_y(v, Y(v)) dv \right\} du \\
&= \int_s^t \mathcal{K}(u,s) D_r \left[b'_y(u, Y(u)) \exp\left\{ \int_u^t b'_y(v, Y(v)) dv \right\} \right] du \\
&= \int_s^t \mathcal{K}(u,s) b'_y(u, Y(u)) D_r \left[\exp\left\{ \int_u^t b'_y(v, Y(v)) dv \right\} \right] du \\
&\quad + \int_s^t \mathcal{K}(u,s) \exp\left\{ \int_u^t b'_y(v, Y(v)) dv \right\} D_r \left[b'_y(u, Y(u)) \right] du \\
&= \int_s^t \mathcal{K}(u,s) b'_y(u, Y(u)) \exp\left\{ \int_u^t b'_y(v, Y(v)) dv \right\} \int_u^t b''_{yy}(v, Y(v)) D_r[Y(v)] dv du \\
&\quad + \int_s^t \mathcal{K}(u,s) b''_{yy}(u, Y(u)) \exp\left\{ \int_u^t b'_y(v, Y(v)) dv \right\} D_r \left[Y(u) \right] du.
\end{aligned}
\tag{6.29}
$$

However, the transitions in (6.29) are not straightforward to justify. For example, under Assumption 6.3, the functions $y \mapsto b'_y(t, y)$ and $y \mapsto b''_{yy}(t, y)$ demonstrate explosive behavior as $y \to \varphi(t)+$ and $y \to \psi(t)-$ for any $t \in [0, T]$. This makes it impossible to use the classical Malliavin chain rules given in Appendix A.3, therefore we have to use special properties of the sandwiched process (6.19) and tailor a version of the Malliavin chain rule specifically for our needs.

Let us start our analysis from the following observation: our "unjustified" computations (6.29) contain the second derivative b''_{yy} of the drift function. Assumption

6.3 does not contain any information about it (we do not even know if this derivative exists!), so we have to put extra conditions on b.

Assumption 6.4 The drift function b: $\mathcal{D}_{0,0} \to \mathbb{R}$ has continuous partial derivative b''_{yy} and

(1) there are constants C_b and $q > 0$ such that

$$|b''_{yy}(t,y)| \le C_b \left(1 + \frac{1}{(y-\varphi(t))^q} + \frac{1}{(\psi(t)-y)^q}\right), \quad (t,y) \in \mathcal{D}_{0,0}; \quad (6.30)$$

(2) there exist two sequences $\{r_n,\ n \ge 1\}$, $\{s_n,\ n \ge 1\}$ such that $r_n, s_n \to \infty$, $n \to \infty$, and

$$\left\{(t,y) \in \mathcal{D}_{0,0} \mid |b''_{yy}(t,y)| \ge s_n\right\} \subset \left\{(t,y) \in \mathcal{D}_{0,0} \mid |b'_y(t,y)| \ge r_n\right\}$$

for all large enough n.

Remark 6.5 Essentially, the second condition of Assumption 6.4 can be read as follows: if the second derivative $b''_{yy}(t,y)$ takes a "large" value at some point (t,y), the first derivative $b'_y(t,y)$ also must be "large" at this point.

Remark 6.6 Without loss of generality, we can assume that the constants C_b and $q > 0$ in (6.30) are the same as in Assumption 6.3(**b5**).

Example 6.2 Let

$$b(t,y) := \frac{\theta_1(t)}{(y-\varphi(t))^{\gamma_1}} - \frac{\theta_2(t)}{(\psi(t)-y)^{\gamma_2}} + a(t,y), \quad (6.31)$$

where

- $\gamma_1 > \frac{1}{H} - 1$, $\gamma_2 > \frac{1}{H} - 1$ with $H \in (0,1)$ being from Assumption 6.1(K2);
- the functions θ_1, θ_2: $[0,T] \to (0,\infty)$ are strictly positive and continuous;
- the function a: $[0,T] \times \mathbb{R} \to \mathbb{R}$ is locally Lipschitz in y uniformly in t, i.e. for any $N > 0$ there exists a constant $C_N > 0$ that does not depend on t such that

$$|a(t,y_2) - a(t,y_1)| \le C_N |y_2 - y_1|, \quad t \in [0,T], \quad y_1, y_2 \in [-N,N];$$

- a: $[0,T] \times \mathbb{R} \to \mathbb{R}$ is two times differentiable w.r.t. the spatial variable y with a, a'_y, a''_{yy} all being continuous on $[0,T] \times \mathbb{R}$.

It is easy to see that this function satisfies Assumption 6.3(**b1**)–(**b5**) as well as (6.30). Let us check the second condition of Assumption 6.4. Put

$$r_n := \min\{\gamma_1 \min_{t\in[0,T]} \theta_1(t) n^{\gamma_1+1}, \gamma_2 \min_{t\in[0,T]} \theta_2(t) n^{\gamma_2+1}\} - \sup_{(t,y)\in\mathcal{D}_{0,0}} |a'_y(t,y)|, \quad (6.32)$$

$$s_n := \gamma_1(\gamma_1+1) \max_{t\in[0,T]} \theta_1(t) n^{\gamma_1+2} + \gamma_2(\gamma_2+1) \max_{t\in[0,T]} \theta_2(t) n^{\gamma_2+2} \quad (6.33)$$

$$+ \sup_{(t,y)\in\mathcal{D}_{0,0}} |a''_{yy}(t,y)| \tag{6.34}$$

and fix an arbitrary $n > \frac{2}{\max_{t\in[0,T]} |\psi(t)-\varphi(t)|}$. Our goal is to prove that, if

$$|b''_{yy}(t,y)| \geq s_n$$

for some point $(t,y) \in \mathcal{D}_{0,0}$, then

$$|b'_y(t,y)| \geq r_n.$$

Indeed, observe that

$$\begin{aligned} b'_y(t,y) &= -\frac{\gamma_1\theta_1(t)}{(y-\varphi(t))^{\gamma_1+1}} - \frac{\gamma_2\theta_2(t)}{(\psi(t)-y)^{\gamma_2+1}} + a'_y(t,y) \\ b''_{yy}(t,y) &= \frac{\gamma_1(\gamma_1+1)\theta_1(t)}{(y-\varphi(t))^{\gamma_1+2}} + \frac{\gamma_2(\gamma_2+1)\theta_2(t)}{(\psi(t)-y)^{\gamma_2+2}} + a''_{yy}(t,y), \end{aligned}$$

so it is easy to see that $|b''_{yy}(t,y)| < s_n$ for all $(t,y) \in \mathcal{D}_{\frac{1}{n},\frac{1}{n}}$. Therefore, the inequality $|b''_{yy}(t,y)| \geq s_n$ can only hold on a subset of $\mathcal{D}_{0,0} \setminus \mathcal{D}_{\frac{1}{n},\frac{1}{n}}$. However, it is straightforward to verify that

$$|b'_y(t,y)| \geq r_n, \quad (t,y) \in \mathcal{D}_{0,0} \setminus \mathcal{D}_{\frac{1}{n},\frac{1}{n}},$$

so the function (6.31) satisfies the second condition of Assumption 6.4.

From now on, let $Y = \{Y(t),\ t \in [0,T]\}$ be the continuous solution to the sandwiched equation (6.19) with the coefficients satisfying Assumptions 6.1, 6.3 and 6.4. We start with a simple proposition.

Proposition 6.3 *There exists a random variable $\xi > 0$ such that*

- *for any $p \geq 1$, $\mathbb{E}[\xi^p] < \infty$;*
- *for any $t \in [0,T]$,*

$$|b(t,Y(t))| + |b'_y(t,Y(t))| + |b''_{yy}(t,Y(t))| < \xi.$$

In particular, for any $p \geq 1$,

$$\mathbb{E}\left[\sup_{t\in[0,T]} \left(|b(t,Y(t))|^p + |b'_y(t,Y(t))|^p + |b''_{yy}(t,Y(t))|^p\right)\right] < \infty.$$

Proof Fix $\lambda \in (0,H)$ and take the corresponding $\widetilde{\Lambda} > 0$ from (6.10). Then by Assumptions 6.3(**b3**), 6.3(**b5**), 6.4 and Theorem 3.8, there exists constant $C > 0$ that depends only on the shape of the deterministic coefficients of (6.19) such that, for all $t \in [0,T]$,

$$|b(t, Y(t))| \le C(1 + \widetilde{\Lambda})^{\frac{\gamma}{\gamma\lambda+\lambda-1}},$$

$$|b'_y(t, Y(t))| \le C(1 + \widetilde{\Lambda})^{\frac{q}{\gamma\lambda+\lambda-1}},$$

$$|b''_{yy}(t, Y(t))| \le C(1 + \widetilde{\Lambda})^{\frac{q}{\gamma\lambda+\lambda-1}}.$$

It remains to notice that $\widetilde{\Lambda}$ has moments of all orders. □

Let us now address the first question posed at the beginning of this subsection and demonstrate that $Y(t) \in \mathbb{D}^{1,p}$ for all $p \ge 2$. To achieve this, we will utilize the following technical lemma, the proof of which can be found in [6, Proposition 1.5.5].

Lemma 6.1 *Let $p > 1$ and $X \in \mathbb{D}^{1,2}$ be such that*

$$\mathbb{E}\left[|X|^p\right] < \infty$$

and

$$\mathbb{E}\left[\left(\int_0^T (D_s X)^2 ds\right)^{\frac{p}{2}}\right] < \infty.$$

Then $X \in \mathbb{D}^{1,p}$.

Proposition 6.4 *For any $t \in [0, T]$ and $p > 1$, $Y(t) \in \mathbb{D}^{1,p}$.*

Proof Note that $Y(t)$ is a bounded random variable which immediately yields that $\mathbb{E}[|Y(t)|^p] < \infty$ for any $p > 1$. Therefore, by Lemma 6.1, it is sufficient to prove that

$$\mathbb{E}\left[\left(\int_0^T (D_s Y(t))^2\, ds\right)^{\frac{p}{2}}\right] < \infty$$

for any $p > 1$. Note that, by Assumption 6.3**(b5)**,

$$\exp\left\{\int_s^t b'_y(v, Y(v))dv\right\} < \exp\left\{C_b T\right\},$$

and, by Proposition 6.3, there exists a random variable ξ having all moments such that

$$\sup_{s\in[0,T]} |b'_y(s, Y(s))| \le \xi.$$

Hence

$$
\begin{aligned}
|D_s Y(t)| &\leq |\mathcal{K}(t,s)| + \int_s^t |\mathcal{K}(u,s)||b_y'(u,Y(u))| \exp\left\{\int_u^t b_y'(v,Y(v))dv\right\} du \\
&\leq |\mathcal{K}(t,s)| + \xi \exp\{C_b T\} \int_s^t |\mathcal{K}(u,s)| du.
\end{aligned}
\tag{6.35}
$$

By Assumption 6.1 and Remark 6.1,

$$
\left(\int_0^T \mathcal{K}^2(t,s)ds\right)^{\frac{p}{2}} < \infty,
$$

therefore

$$
\begin{aligned}
\mathbb{E}\left[\left(\int_0^T (D_s Y(t))^2\, ds\right)^{\frac{p}{2}}\right] &\leq C\left(\int_0^T \mathcal{K}^2(t,s)ds\right)^{\frac{p}{2}} \\
&+ C\mathbb{E}\left[\left(\int_0^T\int_0^t \mathcal{K}^2(u,s)(b_y'(u,Y(u)))^2 \exp\left\{2\int_u^t b_y'(v,Y(v))dv\right\} duds\right)^{\frac{p}{2}}\right] \\
&\leq C\left(\int_0^T \mathcal{K}^2(t,s)ds\right)^{\frac{p}{2}} + C\mathbb{E}\left[\xi^p\right]\exp\{pC_bT\}\left(\int_0^T\int_0^t \mathcal{K}^2(u,s)duds\right)^{\frac{p}{2}} < \infty,
\end{aligned}
\tag{6.36}
$$

which ends the proof. □

Our next goal is to establish the Malliavin chain rule for the random variables $b_y'(t,Y(t))$ and $\exp\left\{\int_u^t b_y'(v,Y(v))dv\right\}$.

Proposition 6.5 *For any* $0 \leq u \leq t \leq T$ *and* $p > 1$,

1) $b_y'(t,Y(t)) \in \mathbb{D}^{1,p}$ *with*

$$
D_s\left[b_y'(t,Y(t))\right] = b_{yy}''(t,Y(t))D_sY(t),
\tag{6.37}
$$

2) $\exp\left\{\int_u^t b_y'(v,Y(v))dv\right\} \in \mathbb{D}^{1,p}$ *with*

$$D_s\left[\exp\left\{\int_u^t b'_y(v, Y(v))dv\right\}\right] = \exp\left\{\int_u^t b'_y(v, Y(v))dv\right\}\int_u^t b''_{yy}(v, Y(v))D_sY(v)dv. \tag{6.38}$$

Proof (1) We shall start from proving that $b'_y(t, Y(t)) \in \mathbb{D}^{1,p}$. Note that b'_y is not a bounded function itself and it does not have bounded derivatives, hence the classical chain rule from Appendix A.3 cannot be applied here in a straightforward manner. For the reader's convenience, we divide the proof into several steps.

Step 0. First of all, observe that $b'(t, Y(t)) \in L^2(\Omega)$ as a direct consequence of Proposition 6.3. Also, for any $p > 1$,

$$\mathbb{E}\left[\left(\int_0^T \left(b''_{yy}(t, Y(t))D_sY(t)\right)^2 ds\right)^{\frac{p}{2}}\right] < \infty.$$

Indeed, again by Proposition 6.3 together with the proof of Proposition 6.4, we have

$$\begin{aligned}
&\mathbb{E}\left[\left(\int_0^T \left(b''_{yy}(t, Y(t))D_sY(t)\right)^2 ds\right)^{\frac{p}{2}}\right] \\
&\le \mathbb{E}\left[\xi^p\left(\int_0^T (D_sY(t))^2 ds\right)^{\frac{p}{2}}\right] \\
&\le \left(\mathbb{E}\left[\xi^{2p}\right]\right)^{\frac{1}{2}}\left(\mathbb{E}\left[\left(\int_0^T (D_sY(t))^2 ds\right)^{p}\right]\right)^{\frac{1}{2}} < \infty.
\end{aligned}$$

Therefore, by Lemma 6.1, it is sufficient to prove that $b'_y(t, Y(t)) \in \mathbb{D}^{1,2}$ with (6.37) being the corresponding Malliavin derivative.

Step 1. Let $\phi \in C^1(\mathbb{R})$ be an arbitrary compactly supported function such that $\phi(x) = x$ whenever $|x| \le 1$ and $|\phi(x)| \le |x|$ for all $|x| > 1$. Fix $t \in [0, T]$ and, for $m \ge 1$, put

$$f_m(y) := m\phi\left(\frac{b'_y(t, y)}{m}\right).$$

Observe that, for any fixed $m \ge 1$ and $t \in [0, T]$, the derivative

$$f_m'(y) = b_{yy}''(t, y)\phi'\left(\frac{b_y'(t, y)}{m}\right)$$

is bounded. Indeed, take the sequences $\{r_n, s_n, n \geq 1\}$ from Assumption 6.4 and let n_m be such that $\frac{r_{n_m}}{m} > \max\{-\inf \operatorname{supp}\phi, \sup\operatorname{supp}\phi\}$. Then, for all $(t, y) \in \mathcal{D}_{0,0}$ such that $|b_{yy}''(t, y)| \geq s_{n_m}$, it holds that $\frac{b_y'(t,y)}{m} \notin \operatorname{supp}\phi$ and hence

$$f_m'(y) = 0.$$

Therefore

$$|f_m'(y)| \leq s_{n_m} \max_{x+in\mathbb{R}} |\phi'(x)| < \infty,$$

and it means that the function f_m satisfies the conditions of the classical Malliavin chain rule [6, Proposition 1.2.3], so $f_m(Y(t)) \in \mathbb{D}^{1,2}$ and, with probability 1 for a.a. $s \in [0, T]$,

$$D_s f_m(Y(t)) = b_{yy}''(t, Y(t))\phi'\left(\frac{b_y'(t, Y(t))}{m}\right) D_s Y(t).$$

Now it remains to prove that

$$f_m(Y(t)) \to b'(t, Y(t))$$

in $L^2(\Omega)$ and

$$Df_m(Y(t)) \to b_{yy}''(t, Y(t))DY(t)$$

in $L^2(\Omega \times [0, T])$ as $m \to \infty$, then the result will follow immediately from the closedness of the Malliavin derivative operator D.

Step 2: $f_m(Y(t)) \to b'(t, Y(t))$ in $L^2(\Omega)$ as $m \to \infty$. Indeed, by the definitions of f_m and ϕ, $f_m(Y(t)) \to b'(t, Y(t))$ a.s., $m \to \infty$. Moreover, with probability 1, $|f_m(Y(t))| \leq |b_y'(t, Y(t))| \in L^2(\Omega)$ and hence the required convergence follows from the dominated convergence theorem.

Step 3: $Df_m(Y(t)) \to b_{yy}''(t, Y(t))DY(t)$ in $L^2(\Omega \times [0, T])$ as $m \to \infty$. By the definitions of f_m and ϕ, with probability 1,

$$\left(b_{yy}''(t, Y(t))\phi'\left(\frac{b_y'(t, Y(t))}{m}\right)\right)^2 \int_0^T (D_s Y(t))^2 ds \to (b_{yy}''(t, Y(t)))^2 \int_0^T (D_s Y(t))^2 ds$$

as $m \to \infty$. Moreover, since ϕ has compact support, $\max_{y\in\mathbb{R}}(\phi'(y))^2 < \infty$, so we can write

$$\int_0^T (D_s f_m(Y(t)))^2 ds = \left(b''_{yy}(t, Y(t))\phi'\left(\frac{b'_y(t, Y(t))}{m}\right)\right)^2 \int_0^T (D_s Y(t))^2 ds$$

$$\leq \max_{y\in\mathbb{R}}(\phi'(y))^2 (b''_{yy}(t, Y(t)))^2 \int_0^T (D_s Y(t))^2 ds \in L^2(\Omega).$$

Therefore, by the dominated convergence theorem,

$$\mathbb{E}\left[\int_0^T \left(D_s f_m(Y(t)) - b''_{yy}(t, Y(t)) D_s Y(t)\right)^2 ds\right] \to 0, \quad m \to \infty.$$

This, together with the convergence established in Step 2 above and the closedness of the Malliavin derivative operator D, proves the first claim of the Proposition.

2) Let us proceed with the second claim and verify that

$$\exp\left\{\int_u^t b'_y(v, Y(v))dv\right\} \in \mathbb{D}^{1,p}$$

with (6.38) being the corresponding Malliavin derivative. Note that, since b'_y is bounded from above, $\exp\left\{\int_u^t b'_y(v, Y(v))dv\right\}$ is also bounded from above and hence is an element of $L^p(\Omega)$ for any $p > 1$. Moreover, by Proposition 6.3, boundedness of $\exp\left\{\int_u^t b'_y(v, Y(v))dv\right\}$ and (6.35), we can write

$$\mathbb{E}\left[\left(\int_0^T \left(\exp\left\{\int_u^t b'_y(v, Y(v))dv\right\} \int_u^t b''_{yy}(v, Y(v)) D_s Y(v) dv\right)^2 ds\right)^{\frac{p}{2}}\right]$$

$$\leq C\mathbb{E}\left[\xi^p \left(\int_0^T \int_u^t (D_s Y(v))^2 dv ds\right)^{\frac{p}{2}}\right]$$

$$\leq C\mathbb{E}\left[\xi^p \left(\int_0^T \int_u^t \mathcal{K}^2(v, s) dv ds\right)^{\frac{p}{2}}\right]$$

$$+ C\exp\{pC_b T\}\mathbb{E}\left[\xi^{2p}\right] \left(\int_0^T \int_u^t \int_s^v \mathcal{K}^2(u, s) du dv ds\right)^{\frac{p}{2}} < \infty,$$

and hence it is sufficient to prove that $\exp\left\{\int_u^t b'_y(v, Y(v))dv\right\} \in \mathbb{D}^{1,2}$.

Since the Malliavin derivative operator D is closed and the expression

$$\int_u^t b''_{yy}(v, Y(v))D_s Y(v)dv$$

is well-defined by Proposition 6.3, combination of Step 1 of the current proof and Hille theorem guaran/tee/s that $\int_u^t b'_y(v, Y(v))Y(v)dv \in \mathbb{D}^{1,2}$ and

$$D_s \int_u^t b'_y(v, Y(v))Y(v)dv = \int_u^t b''_{yy}(v, Y(v))D_s Y(v)dv.$$

Finally, the function $x \mapsto e^x$ satisfies the conditions of the chain rule from Appendix A.3 and hence $\exp\left\{\int_u^t b'_y(v, Y(v))dv\right\} \in \mathbb{D}^{1,2}$ and (6.38) holds. □

For the next step, we will need the following version of the Malliavin product rule.

Lemma 6.2 *Let $X_1, X_2 \in \mathbb{D}^{1,2}$ be such that*

(i) $X_1 X_2 \in L^2(\Omega)$;
(ii) $X_2 D X_1,\ X_1 D X_2 \in L^2(\Omega \times [0, T])$.

Then $X_1 X_2 \in \mathbb{D}^{1,2}$ and

$$D[X_1 X_2] = X_2 D X_1 + X_1 D X_2.$$

If, in addition,

$$\mathbb{E}[|X_1 X_2|^p] < \infty, \quad \mathbb{E}\left[\left(\int_0^T (X_2 D_u X_1 + X_1 D_u X_2)^2 du\right)^{\frac{p}{2}}\right] < \infty$$

for some $p \geq 2$, then $X_1 X_2 \in \mathbb{D}^{1,p}$.

Proof Let $\phi \in C^\infty(\mathbb{R})$ be a compactly supported function such that $\phi(x) = x$ whenever $|x| \leq 1$ and $|\phi(x)| \leq |x|$ for all $|x| > 1$. For $m \geq 1$, put

$$f_m(x_1, x_2) := m^2 \phi\left(\frac{x_1}{m}\right)\phi\left(\frac{x_2}{m}\right)$$

and observe that both partial derivatives

$$\frac{\partial f_m}{\partial x_1}(x_1, x_2) = m\phi'\left(\frac{x_1}{m}\right)\phi\left(\frac{x_2}{m}\right), \quad \frac{\partial f_m}{\partial x_2}(x_1, x_2) = m\phi\left(\frac{x_1}{m}\right)\phi'\left(\frac{x_2}{m}\right)$$

are bounded. Therefore, by the classical chain rule [6, Proposition 1.2.3],

$$Df_m(X_1, X_2) = m\left(\phi'\left(\frac{X_1}{m}\right)\phi\left(\frac{X_2}{m}\right)DX_1 + \phi\left(\frac{X_1}{m}\right)\phi'\left(\frac{X_2}{m}\right)DX_2\right).$$

Now it is sufficient to prove that

$$f_m(X_1, X_2) \to X_1X_2 \tag{6.39}$$

in $L^2(\Omega)$ and

$$m\left(\phi'\left(\frac{X_1}{m}\right)\phi\left(\frac{X_2}{m}\right)DX_1 + \phi\left(\frac{X_1}{m}\right)\phi'\left(\frac{X_2}{m}\right)DX_2\right) \to X_2DX_1 + X_1DX_2 \tag{6.40}$$

in $L^2(\Omega \times [0, T])$ as $m \to \infty$.

Observe that $|f_m(X_1, X_2)| \to X_1X_2$ a.s. as $m \to \infty$ and

$$|f_m(X_1, X_2)| \le X_1X_2 \in L^2(\Omega),$$

so (6.39) holds by the dominated convergence theorem. Next, since ϕ' is bounded, we have that, with probability 1,

$$\begin{aligned} &m\left|\phi'\left(\frac{X_1}{m}\right)\phi\left(\frac{X_2}{m}\right)DX_1 + \phi\left(\frac{X_1}{m}\right)\phi'\left(\frac{X_2}{m}\right)DX_2\right| \\ &\quad\le \max_{x\in\mathbb{R}} |\phi'(x)|\,(|X_2DX_1| + |X_1DX_2|) \in L^2(\Omega \times [0, T]). \end{aligned}$$

Therefore, since $m\phi'\left(\frac{X_1}{m}\right)\phi\left(\frac{X_2}{m}\right) \to X_2$ a.s. and $m\phi\left(\frac{X_1}{m}\right)\phi'\left(\frac{X_2}{m}\right) \to X_1$ a.s. as $m \to \infty$, (6.40) holds by the dominated convergence, which ends the proof of the first claim.

The second claim immediately follows from Lemma 6.1. □

Proposition 6.5 and Lemma 6.2 together allow us to deduce the following corollary.

Corollary 6.1 *For any* $0 \le s < t \le T$ *and* $p > 1$,

$$b'_y(s, Y(s)) \exp\left\{\int_s^t b'_y(v, Y(v))dv\right\} \in \mathbb{D}^{1,p}$$

and

$$
\begin{aligned}
D_u &\left[b'_y(s, Y(s)) \exp\left\{ \int_s^t b'_y(v, Y(v))dv \right\} \right] \\
&= b''_{yy}(s, Y(s)) \exp\left\{ \int_s^t b'_y(v, Y(v))dv \right\} D_u Y(s) \\
&\quad + b'_y(s, Y(s)) \exp\left\{ \int_s^t b'_y(v, Y(v))dv \right\} \int_s^t b''_{yy}(v, Y(v)) D_u Y(v) dv.
\end{aligned} \tag{6.41}
$$

Proof For fixed $0 \le s < t \le T$, denote

$$
X_1 := b'_y(s, Y(s)), \quad X_2 := \exp\left\{ \int_s^t b'_y(v, Y(v))dv \right\}.
$$

By Proposition 6.5 and Lemma 6.2 from the Appendix, it is sufficient to check that for all $p \ge 2$

(i) the product $X_1 X_2 \in L^p(\Omega)$,
(ii) $\mathbb{E}\left[\left(\int_0^T (X_2 D_u X_1)^2 \right)^{\frac{p}{2}} \right] < \infty$ and
(iii) $\mathbb{E}\left[\left(\int_0^T (X_1 D_u X_2)^2 \right)^{\frac{p}{2}} \right] < \infty$.

All conditions (i)–(iii) can be checked in a straightforward manner using Proposition 6.3 and the arguments similar to the proof of Proposition 6.4. □

We are now ready to formulate the main result of this section.

Theorem 6.4 *For any $t \in [0, T]$ and $p \ge 2$,*

(1) $Y(t) \in \mathbb{D}^{2,p}$,
(2) with probability 1 and for a.a. $r, s \in [0, T]$,

$$
\begin{aligned}
D_r D_s Y(t) &= \int_s^t \mathcal{K}(u, s) F_1(t, u) \left(\int_u^t b''_{yy}(v, Y(v)) D_r Y(v) dv \right) du \\
&\quad + \int_s^t \mathcal{K}(u, s) F_2(t, u) D_r Y(u) du,
\end{aligned} \tag{6.42}
$$

where

$$F_1(t,u) := b'_y(u, Y(u)) \exp\left\{ \int_u^t b'_y(v, Y(v)) dv \right\},$$

$$F_2(t,u) := b''_{yy}(u, Y(u)) \exp\left\{ \int_u^t b'_y(v, Y(v)) dv \right\}.$$

Proof Our goal is to prove that $Y(t) \in \mathbb{D}^{2,p}$ and

$$\begin{aligned} D_r D_s Y(t) &= \int_s^t \mathcal{K}(u,s) D_r \left[b'_y(u, Y(u)) \exp\left\{ \int_u^t b'_y(v, Y(v)) dv \right\} \right] du \\ &= \int_s^t \mathcal{K}(u,s) D_r \left[F_1(t,u) \right] du \end{aligned}$$

since in this case (6.42) follows immediately from Corollary 6.1. Recall that

$$D_s Y(t) = \mathcal{K}(t,s) + \int_s^t \mathcal{K}(u,s) F_1(t,u) du.$$

Clearly, for any $0 \le r, s < t \le T$,

$$D_r \mathcal{K}(t,s) = 0,$$

thus, due to the closedness of D and Hille theorem, it suffices to show that

(i) for a.a. $0 \le s \le u < t \le T$, $\mathcal{K}(u,s) F_1(t,u) \in \mathbb{D}^{1,p}$ and
(ii) for a.a. $0 \le s < t \le T$,

$$\begin{aligned} &\int_0^T \left(\mathbb{E}\left[\left(\int_0^T (D_r[\mathcal{K}(u,s) F_1(t,u)])^2 dr \right)^{\frac{p}{2}} \right] \right)^{\frac{1}{p}} du \\ &\quad = \int_0^T \mathcal{K}(u,s) \left(\mathbb{E}\left[\left(\int_0^T (D_r[F_1(t,u)])^2 dr \right)^{\frac{p}{2}} \right] \right)^{\frac{1}{p}} du < \infty. \end{aligned}$$

Item (i) above follows immediately from Corollary 6.1. As for item (ii), observe that, by Proposition 6.3, (6.35) as well as the boundedness of

$$\exp\left\{\int_u^t b'_y(v, Y(v))dv\right\},$$

we have

$$\begin{aligned}
&(D_r[F_1(t,u)])^2 \\
&\quad \le C\Bigg((b''_{yy}(u, Y(u)))^2 \exp\left\{2\int_u^t b'_y(v, Y(v))dv\right\}(D_rY(u))^2 \\
&\qquad + (b'_y(u, Y(u)))^2 \exp\left\{2\int_u^t b'_y(v, Y(v))dv\right\}\int_u^t (b''_{yy}(v, Y(v))D_rY(v))^2 dv\Bigg) \\
&\quad \le C\left(\xi^2 (D_rY(u))^2 + \xi^4 \int_u^t (D_rY(v))^2 dv\right) \\
&\quad \le C\xi^2\left(\mathcal{K}^2(u,r) + \int_r^u \mathcal{K}^2(z,r)dz\right) \\
&\qquad + C\xi^4\left(\int_u^t \mathcal{K}^2(v,r)dv + \int_u^t\int_r^v \mathcal{K}^2(z,r)dzdv\right).
\end{aligned}$$

Hence, for any $p \ge 2$, Remark 6.1 implies

$$\begin{aligned}
&\int_0^T (D_r[F_1(t,u)])^2 dr \\
&\quad \le C\xi^2\left(\int_0^T \mathcal{K}^2(u,r)dr + \int_0^T\int_r^u \mathcal{K}^2(z,r)dzdr\right) \\
&\qquad + C\xi^4\left(\int_0^T\int_u^t \mathcal{K}^2(v,r)dvdr + \int_0^T\int_u^t\int_r^v \mathcal{K}^2(z,r)dzdvdr\right) \\
&\quad \le C\left(\xi^2 + \xi^4\right),
\end{aligned}$$

so, since ξ has moments of all orders, (ii) holds, which finalizes the proof. □

Finally, denote $\mathbb{L}^{2,p} := L^p([0,T]; \mathbb{D}^{2,p})$. We complete the section with the following result.

Corollary 6.2 *For any* $p \ge 2$, $Y \in \mathbb{L}^{2,p}$.

Proof By the definition of the $\|\cdot\|_{2,p}$-norm from Appendix A.3, it is sufficient to check that

$$\int_0^T \mathbb{E}[|Y(t)|^p] < \infty, \tag{6.43}$$

$$\int_0^T \mathbb{E}\left[\left(\int_0^T (D_s Y(t))^2 ds\right)^{\frac{p}{2}}\right] dt < \infty \tag{6.44}$$

and

$$\int_0^T \mathbb{E}\left[\left(\int_0^T \int_0^T (D_r D_s Y(t))^2 ds dr\right)^{\frac{p}{2}}\right] dt < \infty. \tag{6.45}$$

It is clear that (6.43) holds automatically due to the bounded nature of $Y(t)$. Next, (6.44) can be easily deduced from (6.36). Finally, using Proposition 6.3 and the boundedness of $\exp\left\{\int_u^t b_y'(v, Y(v))dv\right\}$, it is easy to establish the upper bounds similar to (6.36) for

$$\mathbb{E}\left[\left(\int_0^T \int_0^T (D_r D_s Y(t))^2 ds dr\right)^{\frac{p}{2}}\right],$$

which implies (6.45). This completes the proof. □

6.4 Markovian Approximation

The ability of the stochastic process (6.19) to incorporate memory into dynamics is simultaneously its weakness: the absence of Markovianity creates multiple technical problems from a computational perspective. Therefore, it may be convenient to have some type of a *Markovian approximation* of (6.19) that can be used in numerical algorithms.

Our general approach to the approximation is quite simple. It is evident that the original kernel $\mathcal{K}$ can be approximated in L^2 by *degenerate* Volterra kernels and

$$\mathcal{K}_m(t, s) = \sum_{i=0}^m e_{m,i}(t) f_{m,i}(s) \mathbb{1}_{s<t},$$

where $\int_0^T e_{m,i}^2(t)dt < \infty$, $\int_0^T f_{m,i}^2(t)dt < \infty$, $i = 0, ..., m$, $m \geq 0$. Then one can attempt to define the approximation of the sandwiched equation (6.19) as

$$Y_m(t) = y_0(t) + \int_0^t b(s, Y_m(s))ds + Z_m(t), \tag{6.46}$$

where Z_m is given by

$$Z_m(t) := \int_0^t \mathcal{K}_m(t,s)dB(s) = \sum_{i=0}^m e_{m,i}(t) \int_0^t f_{m,i}(s)dB(s), \quad t \in [0,T]. \tag{6.47}$$

In this case, the $(m+2)$-dimensional process $(Y_m, U_{m,0}, ..., U_{m,m})$, where

$$U_{m,i}(t) = \int_0^t f_{m,i}(s)dB(s), \quad i = 0, 1, ..., m, \tag{6.48}$$

is Markovian w.r.t. the filtration $\mathbb{F} = \{\mathcal{F}_t,\ t \in [0,T]\}$ generated by B and, *intuitively*, should approximate the original non-Markovian SVV model if $\mathcal{K}_m$ is close enough to $\mathcal{K}$. However, such intuition is not simple to back up analytically due to the explosive behavior of the drift b:

- Assumption 6.3(**b4**) demands $\gamma > \frac{1}{H} - 1$, i.e. the order of the drift explosion in Assumption 6.3(**b3**) is coupled with the Hölder regularity of the driving Volterra noise. Therefore, changing the kernel may violate (**b4**) and hence (6.46) may not have a solution.
- It is not clear why Y_m approximates Y in L^2 in the first place. Keep in mind that b is non-Lipschitz, so the standard reasoning via direct application of Gronwall inequality is not applicable in this situation.

The goal of this section is to overcome the problems mentioned above and study the convergence of approximations (6.46).

6.4.1 Admissible Approximating Kernels

As noted above, we want to replace the kernel $\mathcal{K}$ with its appropriate L^2-approximations $\mathcal{K}_m$, $m \geq 1$, and then consider stochastic processes $Z_m = \{Z_m(t),\ t \in [0,T]\}$ and $Y_m = \{Y_m(t),\ t \in [0,T]\}$, defined by (6.47) and (6.46), respectively. However, we need to make sure that the solution to (6.46) exists, i.e. the conditions of Assumptions 6.1 and 6.3 hold for all $m \geq 1$. In order to ensure this, we have to introduce additional restrictions on the approximating kernels.

Assumption 6.5 The sequence of kernels $\{\mathcal{K}_m, m \geq 1\}$ is such that each function $\mathcal{K}_m \colon [0, T]^2 \to \mathbb{R}$ is a degenerate Volterra kernel, i.e.

$$\mathcal{K}_m(t, s) = \sum_{i=0}^{m} e_{m,i}(t) f_{m,i}(s) \mathbb{1}_{s<t},$$

where $\int_0^T e_{m,i}^2(t)dt < \infty$, $\int_0^T f_{m,i}^2(t)dt < \infty$, $i = 0, \ldots, m$, $m \geq 0$, and

(Km1) for each $m \geq 1$, $\mathcal{K}_m$ is square integrable, i.e.

$$\int_0^T \int_0^T \mathcal{K}_m^2(t, s) ds dt = \int_0^T \int_0^t \mathcal{K}_m^2(t, s) ds dt < \infty;$$

(Km2) for each $m \geq 1$ and $\lambda \in (0, H)$ with H being from Assumption 6.1(**K2**),

$$\int_0^t (\mathcal{K}_m(t, u) - \mathcal{K}_m(s, u))^2 \, du \leq \ell_\lambda^2 (t - s)^{2\lambda}, \quad 0 \leq s \leq t \leq T,$$

where $\ell_\lambda > 0$ is a constant that possibly depends on λ but does not depend on m.

Example 6.3 (Hölder continuous kernels) Assume that

$$\mathcal{K}(t, s) = \mathcal{K}(t - s) \mathbb{1}_{s<t}$$

with $\mathcal{K}(0) = 0$ and $\mathcal{K}$ is an H-Hölder continuous function of order $H \in (0, 1)$, i.e.

$$|\mathcal{K}(t) - \mathcal{K}(s)| \leq C_H |t - s|^H, \quad s, t \in [0, T]. \tag{6.49}$$

Let $\mathcal{K}_m$ be the corresponding Bernstein polynomial of order m defined as

$$\begin{aligned} \mathcal{K}_m(t) &= \frac{1}{T^m} \sum_{i=0}^{m} \mathcal{K}\left(\frac{Ti}{m}\right) \binom{m}{i} t^i (T - t)^{m-i} \\ &= \sum_{i=0}^{m} \left(\frac{1}{T^i} \sum_{j=0}^{i} (-1)^{i-j} \mathcal{K}\left(\frac{Tj}{m}\right) \binom{m}{j} \binom{m - j}{i - j} \right) t^i \\ &=: \sum_{i=0}^{m} \kappa_{m,i} t^i, \end{aligned} \tag{6.50}$$

and let

$$
\begin{aligned}
\mathcal{K}_m(t,s) = \mathcal{K}_m(t-s)\mathbb{1}_{s<t} &= \sum_{i=0}^{m} \kappa_{m,i} \left(\sum_{j=0}^{i} \binom{i}{j} (-1)^{i-j} t^j s^{i-j} \right) \mathbb{1}_{s<t} \\
&= \sum_{j=0}^{m} t^j \left(\sum_{i=j}^{m} \kappa_{m,i} \binom{i}{j} (-1)^{i-j} s^{i-j} \right) \mathbb{1}_{s<t} =: \sum_{i=0}^{m} e_{m,i}(t) f_{m,i}(s) \mathbb{1}_{s<t}.
\end{aligned}
\tag{6.51}
$$

By [5, Proposition 2],

$$
|\mathcal{K}_m(t) - \mathcal{K}_m(s)| \le C_H |t-s|^H, \quad s,t \in [0,T],
$$

where C_H is the same as in (6.49) for all $m \ge 1$. Since $\mathcal{K}(0) = 0$, $\mathcal{K}_m(0) = \kappa_{m,0} = 0$ and we can write for any $0 \le s \le t \le T$

$$
\begin{aligned}
\int_0^t (\mathcal{K}_m(t-u) - \mathcal{K}_m(s-u))^2 du + \int_s^t \mathcal{K}_m^2(t-u) du \\
&\le C_H^2 T |t-s|^{2H} + \int_0^{t-s} \mathcal{K}_m^2(v) dv \\
&\le C_H^2 T |t-s|^{2H} + C_H^2 \int_0^{t-s} v^{2H} dv \\
&\le C_H^2 T |t-s|^{2H} + \frac{C_H^2}{2H+1} |t-s|^{2H+1} \\
&\le C |t-s|^{2H},
\end{aligned}
$$

where C does not depend on m, i.e. Assumption 6.5 holds. Moreover, [5, Theorem 1] gives the estimate

$$
\sup_{t \in [0,T]} |\mathcal{K}(t) - \mathcal{K}_m(t)| \le C m^{-\frac{H}{2}},
$$

which implies that

$$
\|\mathcal{K} - \mathcal{K}_m\|_{L^2([0,T])} \le C m^{-\frac{H}{2}} \to 0, \quad m \to \infty. \tag{6.52}
$$

Remark 6.7 Note that each process

$$
Z_m(t) = \int_0^t \mathcal{K}_m(t-s) dB(s) = \sum_{i=0}^{m} \kappa_{m,i} \int_0^t (t-s)^i dB(s), \quad t \in [0,T],
$$

defined in Example 6.3 is actually Hölder continuous of *any* order $\lambda \in (0,1)$, i.e. the solution to the SDE

$$Y_m(t) = Y(0) + \int_0^t b(s, Y_m(s))ds + Z_m(t), \quad t \in [0, T],$$

exists and is unique for any value $\gamma > 0$ in Assumption 6.3(**b4**) provided that the bounds φ and ψ are both Lipschitz.

Example 6.4 (Fractional kernels) Let $H \in \left(0, \frac{1}{2}\right)$ and

$$\mathcal{K}(t, s) = \mathcal{K}(t - s)\mathbb{1}_{s<t} = \frac{(t - s)^{H-\frac{1}{2}}}{\Gamma\left(H + \frac{1}{2}\right)}\mathbb{1}_{s<t}. \tag{6.53}$$

For such a kernel, an appropriate sequence $\{\mathcal{K}_m,\ m \geq 1\}$ of degenerate L^2-approximations satisfying Assumption 6.5 is presented in [1, Section 3]. Namely, let μ be a measure on $\mathbb{R}_+$ defined as

$$\mu(d\alpha) := \frac{\alpha^{-H-\frac{1}{2}}}{\Gamma\left(H + \frac{1}{2}\right)\Gamma\left(\frac{1}{2} - H\right)}d\alpha,$$

$$\sigma_{m,i} := \int_{\tau_{m,i-1}}^{\tau_{m,i}} \mu(d\alpha), \quad \alpha_{m,i} := \frac{1}{\sigma_{m,i}} \int_{\tau_{m,i-1}}^{\tau_{m,i}} \alpha\mu(d\alpha), \quad i = 1, ..., m,$$

where $0 = \tau_{m,0} < \tau_{m,1} < ... < \tau_{m,m}$ is such that

$$\tau_{m,m} \to \infty, \quad \sum_{i=1}^{m} \int_{\tau_{m,i-1}}^{\tau_{m,i}} (\alpha_{m,i} - \alpha)^2\mu(d\alpha) \to 0, \quad m \to \infty. \tag{6.54}$$

Denote for $m \geq 1$

$$\begin{aligned} \mathcal{K}_m(t - s)\mathbb{1}_{s<t} &= \sum_{i=1}^{m} \sigma_{m,i}e^{-\alpha_{m,i}(t-s)}\mathbb{1}_{s<t} \\ &= \sum_{i=1}^{m} \left(\sigma_{m,i}e^{-\alpha_{m,i}t}\right) e^{\alpha_{m,i}s}\mathbb{1}_{s<t} \\ &=: \sum_{i=1}^{m} e_{m,i}(t) f_{m,i}(s)\mathbb{1}_{s<t}. \end{aligned} \tag{6.55}$$

By [1, Lemma 5.2], the sequence $\{\mathcal{K}_m,\ m \geq 1\}$ satisfies Assumption 6.5 and, moreover, by [1, Proposition 3.3],

$$\|\mathcal{K} - \mathcal{K}_m\|_{L^2([0,T])} \to 0, \quad m \to \infty.$$

Note that the values $0 = \tau_{m,0} < \tau_{m,1} < ... < \tau_{m,m}$ satisfying (6.54) can be chosen in multiple ways. For example, as suggested in [1, Subsection 3.2], one can put

$$\tau_{m,i} := \frac{1}{T}\left(\frac{\sqrt{10}(1-2H)}{5-2H}\right)^{\frac{2}{5}} i m^{-\frac{1}{5}}, \quad i = 0, 1, 2, ..., m. \tag{6.56}$$

In this case,

$$\|\mathcal{K} - \mathcal{K}_m\|_{L^2([0,T])} \le C_H m^{-\frac{4H}{5}}, \tag{6.57}$$

where C_H is a positive constant that depends only on the Hurst parameter $H \in \left(0, \frac{1}{2}\right)$.

Remark 6.8 Putting $e_{m,i}(t) := \sigma_{m,i} e^{-\alpha_{m,i} t}$, $f_{m,i}(s) := e^{\alpha_{m,i} s}$ as suggested in Example 6.4, we obtain that each process of the form

$$e_{m,i}(t) U_{m,i}(t) = \sigma_{m,i} \int_0^t e^{-\alpha_{m,i}(t-s)} dB_1(s) =: V_{m,i}(t), \quad t \in [0, T],$$

from (6.48) is an Ornstein-Uhlenbeck process satisfying the SDE

$$V_{m,i}(t) = -\alpha_{m,i} \int_0^t V_{m,i}(s) ds + \sigma_{m,i} B_1(t), \quad t \in [0, T].$$

It is important to note that Assumption 6.5 guarantees that each Z_m has a continuous modification that satisfies the λ-Hölder property for all $\lambda \in (0, H)$. In fact, **(Km2)** allows to deduce a stronger statement: the Hölder seminorms of Z_m, $m \ge 1$, are uniformly bounded in $L^r(\mathbb{P})$ for any $r \ge 1$. The corresponding result is presented in the next lemma.

Lemma 6.3 *Let $\mathcal{K}$ satisfy Assumption 6.1 and the sequence $\{\mathcal{K}_m, m \ge 1\}$ satisfy Assumption 6.5. Then, for any constant $\lambda \in (0, H)$, there exist positive random variables Λ_m, $m \ge 1$, such that for each $m \ge 1$, $r \ge 1$ and $s, t \in [0, T]$*

$$|Z_m(t) - Z_m(s)| \le \Lambda_m |t - s|^\lambda,$$

and

$$\sup_{m \ge 1} \mathbb{E}\Lambda_m^r < \infty. \tag{6.58}$$

Proof Fix an arbitrary $\lambda \in (0, H)$, choose $\alpha > 1$ such that $\lambda + \frac{2}{\alpha} < H$ and denote

$$\Lambda_m := A_{\lambda,\alpha} \left(\int_0^T \int_0^T \frac{|Z_m(t) - Z_m(s)|^\alpha}{|t-s|^{\lambda\alpha+2}} ds dt\right)^{\frac{1}{\alpha}}, \quad m \ge 1,$$

with $A_{\lambda,\alpha} := 2^{3+\frac{2}{\alpha}}\left(\frac{\lambda\alpha+2}{\lambda\alpha}\right)$. By the Corollary A.1, for any $m \geq 1$ and $s, t \in [0, T]$,

$$|Z_m(t) - Z_m(s)| \leq \Lambda_m |t - s|^\lambda,$$

and, moreover, as explained in Appendix A.1, $\mathbb{E}\Lambda_m^r < \infty$ for each $r > 0$. It remains to show that $\sup_{m\geq 1} \mathbb{E}\Lambda_m^r < \infty$. Take any $r > \alpha$, and observe that Minkowski integral inequality, Gaussian distribution of $Z_m(t) - Z_m(s)$ and Assumption 6.5(**Km2**) yield

$$
\begin{aligned}
(\mathbb{E}[\Lambda_m^r])^{\frac{\alpha}{r}} &\leq A_{\lambda,\alpha}^{\alpha} \int_0^T\int_0^T \frac{(E[|Z_m(t) - Z_m(s)|^r])^{\frac{\alpha}{r}}}{|t-s|^{\lambda\alpha+2}} dsdt \\
&= 2^{\frac{\alpha}{2}}\left(\frac{\Gamma\left(\frac{r+1}{2}\right)}{\sqrt{\pi}}\right)^{\frac{\alpha}{r}} A_{\lambda,\alpha}^{\alpha} \int_0^T\int_0^T \frac{\left(\int_0^{t\vee s} (\mathcal{K}_m(t,u) - \mathcal{K}_m(s,u))^2 du\right)^{\frac{\alpha}{2}}}{|t-s|^{\lambda\alpha+2}} dsdt \\
&\leq 2^{\frac{\alpha}{2}}\left(\frac{\Gamma\left(\frac{r+1}{2}\right)}{\sqrt{\pi}}\right)^{\frac{\alpha}{r}} A_{\lambda,\alpha}^{p} \ell_{\lambda+\frac{2}{\alpha}}^{\alpha} \int_0^T\int_0^T \frac{|t-s|^{\lambda\alpha+2}}{|t-s|^{\lambda\alpha+2}} dsdt \\
&\leq 2^{\frac{\alpha}{2}} T^2 \left(\frac{\Gamma\left(\frac{r+1}{2}\right)}{\sqrt{\pi}}\right)^{\frac{\alpha}{r}} A_{\lambda,\alpha}^{\alpha} \ell_{\lambda+\frac{2}{\alpha}}^{\alpha},
\end{aligned}
$$

and thus

$$\sup_{m\geq 1} E[\Lambda_m^r] \leq 2^{\frac{r}{2}} T^{\frac{2r}{\alpha}} \frac{\Gamma\left(\frac{r+1}{2}\right)}{\sqrt{\pi}} A_{\lambda,\alpha}^{r} \ell_{\lambda+\frac{2}{\alpha}}^{r} < \infty,$$

which completes the proof. □

6.4.2 *Approximation of Two-Sided Sandwiched Processes*

Next, consider a family of sandwiched process $Y_m = \{Y_m(t),\ t \in [0, T]\}$, $m \geq 1$, defined by equations of the form (6.46) with the kernels $\mathcal{K}_m$ satisfying Assimption 6.5. Note that these processes are well-defined since each of the equations satisfies Assumptions 6.1 and 6.3. Moreover, by Theorem 3.8, one can choose a deterministic constant $C > 0$ that does not depend on m such that

$$\varphi(t) + \frac{1}{C(1+\Lambda_m)^{\frac{1}{\gamma\lambda+\lambda-1}}} < Y_m(t) < \psi(t) - \frac{1}{C(1+\Lambda_m)^{\frac{1}{\gamma\lambda+\lambda-1}}}, \quad t \in [0, T], \tag{6.59}$$

$m \geq 1$, where the choice of Λ_m is described in Lemma 6.3. This, as well as Theorem 6.1, allow to deduce the following result.

Theorem 6.5 *Let the processes Y and Y_m, $m \geq 1$, be given by equations* (6.19) *and* 6.46, *respectively, with the coefficients satisfying Assumptions 6.1, 6.3 and 6.5. Then, for each $m \geq 1$, there exist a deterministic constant $C > 0$ that does not depend on m or t, and a random variable ξ_m that does not depend on t such that, for any $r > 0$,*

$$\sup_{m\geq 1} \mathbb{E}[\xi_m^r] < \infty, \tag{6.60}$$

and, for any $t \in [0, T]$,

$$|Y(t) - Y_m(t)| \leq C \left(|Z(t) - Z_m(t)| + \xi_m \int_0^t |Z(u) - Z_m(u)| du \right).$$

Proof First, pick any $\lambda \in \left(\frac{1}{1+\gamma}, H\right)$. Since

$$\begin{aligned} |(Z(t_1) - Z_m(t_1)) - (Z(t_2) - Z_m(t_2))| &\leq |Z(t_1) - Z(t_2)| + |Z_m(t_1) - Z_m(t_2)| \\ &\leq (\Lambda + \Lambda_m)|t_1 - t_2|^{\lambda}, \quad t_1, t_2 \in [0, T], \end{aligned}$$

and

$$Y_m(t) = y_0(t) + \int_0^t b(s, Y_m(s))ds + Z(t) + (Z_m(t) - Z(t)), \tag{6.61}$$

we can use Theorem 6.1 and immediately obtain that

$$|Y(t) - Y_m(t)| \leq C \left(|Z(t) - Z_m(t)| + \xi_m \int_0^t |Z(u) - Z_m(u)| du \right),$$

where $\xi := (1 + \Lambda + \Lambda_m)^{\frac{q}{\gamma\lambda+\lambda-1}}$. The rest of the result follows from (6.58) in Lemma 6.3. □

Given that $Z(t) = \int_0^t \mathcal{K}(t, s)dB(s)$ and $Z_m(t) = \int_0^t \mathcal{K}_m(t, s)dB(s)$, Theorem 6.5 yields the following obvious corollary.

Corollary 6.3 *Under conditions of Theorem 6.5,*

$$\mathbb{E}\left[\int_0^T |Y(t) - Y_m(t)|^2 dt \right] \leq C \|\mathcal{K} - \mathcal{K}_m\|^2_{L^2([0,T]^2)}.$$

References

1. Abi Jaber, E., El Euch, O.: Multifactor approximation of rough volatility models. SIAM J. Financ. Math. **10**(2), 309–349 (2019). https://doi.org/10.1137/18m1170236
2. Di Nunno, G., Mishura, Y., Yurchenko-Tytarenko, A.: Option pricing in sandwiched Volterra volatility model. SIAM J. Financ. Math. **15**(3), 824–882 (2024). https://doi.org/10.1137/22m1521328
3. Di Nunno, G., Yurchenko-Tytarenko, A.: Sandwiched Volterra volatility model: Markovian approximations and hedging. Financ. Stoch. **30**(1), 277–325 (2026). https://doi.org/10.1007/s00780-025-00584-2
4. Friz, P.K., Victoir, N.B.: Multidimensional Stochastic Processes as Rough Paths: Theory and Applications. Cambridge Studies in Advanced Mathematics. Cambridge University Press (2010). https://doi.org/10.1017/CBO9780511845079
5. Mathé, P.: Approximation of Hölder continuous functions by Bernstein polynomials. Am. Math. Mon. Off. J. Math. Assoc. Am. **106**(6), 568–574 (1999). https://doi.org/10.1080/00029890.1999.12005085
6. Nualart, D.: The Malliavin calculus and related topics. Springer-Verlag, Berlin/Heidelberg (2006)
7. Sugita, H.: On a characterization of the Sobolev spaces over an abstract Wiener space. J. Math. Kyoto Univ. **25**(4), 717–725 (1985). https://doi.org/10.1215/kjm/1250521019
8. Young, L.C.: An inequality of the Hölder type, connected with Stieltjes integration. Acta Mathematica **67**, 251–282 (1936). https://doi.org/10.1007/bf02401743

Chapter 7
The SVV Model of Financial Market

Abstract This chapter introduces the Sandwiched Volterra Volatility (SVV) model—a financial market framework in which stochastic volatility is represented by a two-sided sandwiched process driven by a Hölder-continuous Gaussian Volterra noise. We demonstrate that this model, which was originally introduced in [7], supports efficient numerical approximation schemes and allows for a complete characterization of equivalent local martingale measures. In addition, we further investigate the implied volatility surfaces produced by the SVV model and demonstrate how the choice of the Gaussian Volterra driver can combine features of both roughness and long-memory.

7.1 Introduction

So far, in Chaps. 3–6, we have discussed purely theoretical aspects of a class of stochastic integro-differential equations with explosive drift, driven by noises with memory. It is time to put this theory into practice: in this and the following chapters, we will consider a financial market model with stochastic volatility represented by two-sided sandwiched processes

$$Y(t) = y_0(t) + \int_0^t b(s, Y(s))ds + \int_0^t \mathcal{G}(t, s)a(s, Y(s))ds + Z(t). \tag{7.1}$$

In this context, sandwiched processes offer several advantages that we aim to exploit. First, recall that the explosive component of the drift ensures that the solution Y to the two-sided sandwiched equation above remains confined between two functions, φ and ψ. By setting the lower bound $\varphi > 0$, we guarantee that the trajectories of Y remain strictly positive with probability 1, which is an essential and very natural requirement for a volatility model. Furthermore, this ensures that the integral $\int_0^T \frac{1}{Y^2(s)}ds$ is well-defined and finite; as we will see later, this property allows us to provide a clear description of all equivalent local martingale measures in the market. The upper bound ψ, in turn, guarantees that there are no moment explosions in the

G. Di Nunno et al., *Volterra Volatility Models*, Springer Finance,
https://doi.org/10.1007/978-3-032-26576-0_7

price process S. Specifically, as we will see later, under the two-sided sandwiched volatility,

$$\mathbb{E}\left[\sup_{t\in[0,T]} S^r(t)\right] < \infty,$$

for any $r \in \mathbb{R}$. This prevents several technical and modeling issues: for example, if $\mathbb{P}$ and $\mathbb{Q}$ are the physical and martingale measures respectively, then $S(T) \notin L^2(\mathbb{P})$ implies that $(S(T) - K)_+ \notin L^2(\mathbb{P})$ and hence it is generally not clear whether the option price

$$\mathbb{E}_{\mathbb{Q}}\left[(S(T) - K)_+\right] = \mathbb{E}_{\mathbb{P}}\left[\frac{d\mathbb{Q}}{d\mathbb{P}}(S(T) - K)_+\right] < \infty.$$

In addition to that, the infinite second moment of $(S(T) - K)_+$ entirely excludes the application of quadratic hedging techniques, and this limitation impacts stochastic volatility models that usually produce incomplete markets.

Second, note that the driving noise Z in (7.1) can be *any* Hölder-continuous process with arbitrarily low regularity. In particular, as discussed in Chap. 6, we can work with *any* Hölder-continuous Gaussian Volterra process. This class of drivers is very rich and allows to capture various market phenomena: for example, even a simple linear combination of two fractional Brownian motions B^{H_1} and B^{H_2} with $H_1 > 1/2$ and $H_2 < 1/2$ can be sufficient to get both roughness/power law **and** long memory simultaneously. We can also go one step further and introduce more complex covariance structures in the noise. For instance, one may consider a *multifractional Brownian motion* $Z(t) = \int_0^t \mathcal{K}(t, s)dB(s)$ with

$$\mathcal{K}(t, s) := (t - s)^{h(t)-\frac{1}{2}}\mathbb{1}_{s<t},$$

where $h\colon [0, T] \to (0, 1)$ is some Hölder continuous function. In fact, such a choice is supported by econometric evidence: in [4, Section 2.2], the regularity of SPX volatility is found to vary over time between 0.1 and 0.9.

In this chapter, which contains the reworked material from our papers [7, 8], we introduce a financial market model based on (7.1). Since its defining feature is *volatility* modeled by a *sandwiched* process driven by Gaussian *Volterra* noise, we refer to it as the *Sandwiched Volterra Volatility model*—or, in short, *SVV*. Right now, we are interested in more straightforward matters, namely the specifications of the model, some basic properties of the resulting price as well as some immediate questions related to the structure of the set of martingale measures and the shape of the implied volatility surface. Later on, in Chaps. 8 and 9, we will explore more specialized topics related to pricing and hedging.

7.2 Specifications of the Model

Let $(\Omega, \mathcal{F}, \mathbb{P})$ be the canonical $2d$-dimensional Wiener space, i.e.

$$\Omega := C_0([0, T]; \mathbb{R}^{2d}),$$

with C_0 denoting the space of continuous functions starting from zero, $\mathbb{P}$ is the classical Wiener measure, $\mathcal{F}$ is the $\mathbb{P}$-augmented Borel sigma-algebra. Let also W be the corresponding $2d$-dimensional Brownian motion, i.e. $W: [0, T] \times \Omega \to \mathbb{R}^{2d}$, $W(t, \omega) = \omega(t)$, and $\mathbb{F} = \{\mathcal{F}_t,\ t \in [0, T]\}$ be the $\mathbb{P}$-augmented filtration generated by W. In what follows, to simplify the notation, we will omit ω in brackets when writing $W(t, \omega)$.

Now, let $\Sigma = (\sigma_{i,j})_{i,j=1}^{2d}$ be an arbitrary non-degenerate $2d \times 2d$ real correlation matrix, i.e. Σ is symmetric, positive definite, $\sigma_{i,j} \in [-1, 1]$ and $\sigma_{i,i} = 1$ for all $i = 1, ..., 2d$. Since Σ is symmetric and positive definite, it admits a Cholesky decomposition of the form

$$\Sigma = \mathcal{L}\mathcal{L}^T, \tag{7.2}$$

where $\mathcal{L} = (\ell_{i,j})_{i,j=1}^{2d}$ is a lower triangular matrix with strictly positive diagonal entries. Define

$$B(t) = (B_1(t), ..., B_{2d}(t))^T := \mathcal{L}W(t) \tag{7.3}$$

and note that components of the $2d$-dimensional stochastic process $\{B(t),\ t \in [0, T]\}$ are 1-dimensional Brownian motions with

$$\mathbb{E}\left[B_i(t)B_j(s)\right] = (t \wedge s)\sigma_{i,j}, \quad t, s \in [0, T],\ i, j = 1, ..., 2d.$$

In what follows, the first d components of this new $2d$-dimensional Gaussian process B will generate the "outer" noise for the price, whereas the remaining d components will generate the noise for the volatility. To reflect this difference, we introduce the notation

$$\begin{aligned} B^S(t) &= \left(B_1^S(t), ..., B_d^S(t)\right)^T := (B_1(t), ..., B_d(t))^T, \\ B^Y(t) &= \left(B_1^Y(t), ..., B_d^Y(t)\right)^T := (B_{d+1}(t), ..., B_{2d}(t))^T. \end{aligned} \tag{7.4}$$

Remark 7.1 Since the matrix $\mathcal{L}$ is non-degenerate, the filtration generated by the $2d$-dimensional process B coincides with $\mathbb{F}$.

7.2.1 Stochastic Volatility Processes

Consider now d measurable Volterra kernels $\mathcal{K}_i\colon [0,T]^2 \to \mathbb{R}$, $i = 1, \dots, d$. From now on, we assume that each $\mathcal{K}_i$ satisfies Assumption 6.1 with a given regularity $H_i \in (0,1)$ (which, in principle, can depend on i). This allows to define a d-dimensional Gaussian Volterra process $Z = (Z_1, \dots, Z_d)$ such that

$$Z_i(t) := \int_0^t \mathcal{K}_i(t,s)dB_i^Y(s), \quad t \in [0,T]. \tag{7.5}$$

As discussed previously in Chap. 6, Assumption 6.1(**K2**) ensures that each Z_i has a modification with paths that are a.s. λ-Hölder continuous for each $\lambda \in (0, H_i)$. In what follows, by Z_i, we *always* mean this modification.

Now, let $\varphi_i, \psi_i\colon [0,T] \to \mathbb{R}$, $i = 1, \dots, d$, be d pairs of functions such that

(1) φ_i, ψ_i are Hölder continuous up to the order H_i (where H_i is the same as the order of regularity of the corresponding Z_i);
(2) for each $i = 1, \dots, d$ and $t \in [0,T]$:

$$0 < \varphi_i(t) < \psi_i(t). \tag{7.6}$$

In addition, for any $\varepsilon_1, \varepsilon_2 \in \left[0, \frac{1}{2}\min_{i=1,\dots,d}\min_{t\in[0,T]}(\psi_i(t) - \varphi_i(t))\right)$, we denote

$$\mathcal{D}^i_{\varepsilon_1,\varepsilon_2} := \{(t,y) \in [0,T] \times \mathbb{R}_+,\ y \in (\varphi_i(t) + \varepsilon_1, \psi_i(t) - \varepsilon_2)\}. \tag{7.7}$$

For each $i = 1, \dots, d$, we also choose an H_i-Hölder continuous deterministic function $y_0^i : [0,T] \to \mathbb{R}$ such that $\varphi_i(0) < y_0^i(0) < \psi_i(0)$, a Volterra kernel $\mathcal{G}_i : [0,T]^2 \to \mathbb{R}$, and functions $a_i : [0,T] \times \mathbb{R} \to \mathbb{R}$, $b_i : \mathcal{D}^i_{0,0} \to \mathbb{R}$, and assume that $y_0^i, \mathcal{G}_i, a_i, b_i$ collectively satisfy the *two-sided* Assumption 6.3 for the chosen order of regularity H_i. Under such conditions, according to Chap. 6, for each $i = 1, \dots, d$, the equation

$$Y_i(t) = y_0^i(t) + \int_0^t b_i(s, Y_i(s))ds + \int_0^t \mathcal{G}_i(t,s)a_i(s, Y_i(s))ds + Z_i(t) \tag{7.8}$$

has a unique continuous solution that, in addition, has the property

$$\varphi_i(t) < Y_i(t) < \psi_i(t), \quad t \in [0,T], \quad a.s.$$

Definition 7.1 Under the conditions specified above, we refer to the process (7.8) as a d-dimensional sandwiched volatility process.

In the previous chapters, we have already discussed various properties of the sandwiched process (7.8). In particular, when $\mathcal{G}_i \equiv 0$, the results of Section 6.3

allow us to conclude that the d-dimensional volatility process $(Y_1, ..., Y_d)$ is Malliavin differentiable with respect to the initial $2d$-dimensional Brownian motion W. The exact result is provided below. It will be very handy in the next few sections, when we explore the absolute continuity of the price distribution as well as the shape of the implied volatility surface.

Theorem 7.1 *Assume that* $\mathcal{G}_i \equiv 0$ *for all* $i = 1, ..., d$*, and all other coefficients of* (7.8) *collectively satisfy Assumption 6.3 for each* $i = 1, ..., d$*. Then, for all* $i = 1, ..., d$ *and* $t \in [0, T]$, $Y_i(t) \in \mathbb{D}^{1,2}(W)$ *and*

$$DY_i(t) = (D^1 Y_i(t), ..., D^{2d} Y_i(t)) \in L^2\left(\Omega \times [0, T]; \mathbb{R}^{2d}\right),$$

$$D_u^j Y_i(t) := \ell_{d+i,j} \left(\mathcal{K}_i(t,u) + \int_u^t \mathcal{K}_i(s,u) \frac{\partial b_i}{\partial y}(s, Y_i(s)) e^{\int_s^t \frac{\partial b_i}{\partial y}(v, Y_i(v)) dv} ds \right) \mathbb{1}_{[0,t]}(u), \tag{7.9}$$

with $\ell_{d+i,j}$, $i = 1, ..., d$, $j = 1, ..., 2d$, *being the corresponding elements of the matrix* $\mathcal{L}$ *defined by* (7.2)–(7.3).

Proof The claim follows from the proof of Theorem 6.3. □

7.2.2 *Price Process*

Having the sandwiched volatility process in place, let us now define the price $S = (S_1, ..., S_d)$ as the solution to the system of stochastic differential equations

$$S_i(t) = S_i(0) + \int_0^t \mu_i(s) S_i(s) ds + \int_0^t Y_i(s) S_i(s) dB_i^S(s), \quad t \in [0, T], \quad i = 1, ..., d, \tag{7.10}$$

where

- B_i^S, $i = 1, ..., d$, are defined by (7.4);
- $S_i(0) > 0$, $i = 1, ..., d$, are deterministic constants;
- Y_i, $i = 1, ..., d$ are sandwiched volatility processes defined above;
- μ_i: $[0, T] \to \mathbb{R}$, $i = 1, ..., d$, are H_i-Hölder continuous functions, where each H_i is the regularity of the driving noise Z_i of the corresponding volatility process Y_i.

Remark 7.2 As a numéraire, we will use the function $e^{\int_0^t \nu(s) ds}$, $t \in [0, T]$, where $\nu : [0, T] \to \mathbb{R}_+$ denotes a Hölder continuous function of order $\min_{i=1,...,d} H_i$ representing an instantaneous interest rate. The discounted price process denoted as

$$\widetilde{S}_i(t) := e^{-\int_0^t \nu(s)ds} S_i(t), \quad t \in [0, T], \quad i = 1, ..., d,$$

has thus the dynamics of the form

$$\widetilde{S}_i(t) = S_i(0) + \int_0^t \widetilde{\mu}_i(s)\widetilde{S}_i(s)ds + \int_0^t Y_i(s)\widetilde{S}_i(s)dB_i^S(s), \quad t \in [0, T], \quad i = 1, ..., d,$$

where $\widetilde{\mu}_i := \mu_i - \nu$.

Definition 7.2 The market model given by equations (7.8), (7.10) is called *Sandwiched Volterra Volatility* (SVV) model.

The existence and uniqueness of the solution to (7.10) are rather straightforward and follow from the standard and well-known results from stochastic analysis. In fact, thanks to the linear nature of (7.10), we can even write the explicit form of the solution: quite unsurprisingly, for each $i = 1, ..., d$,

$$S_i(t) = S_i(0) \exp\left\{\int_0^t \left(\mu_i(s) - \frac{Y_i^2(s)}{2}\right) ds + \int_0^t Y_i(s)dB_i^S(s)\right\}. \tag{7.11}$$

What is a bit more interesting is the integrability properties of S_i: it turns out that the random variable $\sup_{t\in[0,T]} S_i(t)$ has moments of *all* orders, i.e., the price with the two-sided sandwiched volatility does not have moment explosions! All these observations are summarized and proved in the following theorem.

Theorem 7.2 *For any $i = 1, ..., d$, equation* (7.10) *has a unique solution that has the form* (7.11). *Furthermore, for any $r \in \mathbb{R}$:*

$$\mathbb{E}\left[\sup_{t\in[0,T]} S_i^r(t)\right] < \infty.$$

Proof Each Y_i, $i = 1, ..., d$, is a bounded process which yields the existence and uniqueness of the solution to the corresponding equation. The explicit form of the solution can be checked straightforwardly via Itô's formula.

Next, using the explicit form of S_i as well as boundedness of Y_i one can see that there exists a constant $C > 0$ such that

$$
\begin{aligned}
S_i^r(t) &= S_i^r(0)\exp\left\{ r\int_0^t \mu_i(s)ds - \frac{r}{2}\int_0^t Y_i^2(s)ds + r\int_0^t Y_i(s)dB_i^S(s)\right\}\\
&\le S_i^r(0)\exp\left\{|r|T\max_{s\in[0,T]}|\mu_i(s)| + \frac{|r|T}{2}\max_{s\in[0,T]}\psi_i^2(s)\right\}\exp\left\{r\int_0^t Y_i(s)dB_i^S(s)\right\}\\
&= S_i^r(0)\exp\left\{|r|T\max_{s\in[0,T]}|\mu_i(s)| + \frac{|r|T}{2}\max_{s\in[0,T]}\psi_i^2(s)\right\}\exp\left\{\frac{r^2}{2}\int_0^t Y_i^2(s)ds\right\}\times\\
&\quad\times\exp\left\{\int_0^t rY(s)dB_i^S(s) - \frac{1}{2}\int_0^t (rY(s))^2 ds\right\}\\
&\le c_i\exp\left\{\int_0^t rY_i(s)dB_i^S(s) - \frac{1}{2}\int_0^t (rY_i(s))^2 ds\right\}\\
&:= c_i\mathcal{E}_t\{rY_i\cdot B_i^S\},
\end{aligned}
\tag{7.12}
$$

where

$$
c_i := S_i^r(0)\exp\left\{|r|T\max_{s\in[0,T]}|\mu_i(s)| + \frac{|r|(|r|+1)T}{2}\max_{s\in[0,T]}\psi_i^2(s)\right\}
$$

and

$$
\mathcal{E}_t\{rY_i\cdot B_i^S\} := \exp\left\{\int_0^t rY_i(s)dB_i^S(s) - \frac{1}{2}\int_0^t (rY_i(s))^2 ds\right\},\quad t\in[0,T].
$$

Note that the Novikov's criterion immediately yields that the process $\mathcal{E}_t\{rY_i\cdot B_i^S\}$, $t\in[0,T]$, is a uniformly integrable martingale such that

$$
\mathbb{E}\left[\mathcal{E}_t\{rY_i\cdot B_i^S\}\right] = 1,\quad t\in[0,T],
$$

and, moreover,

$$
\mathcal{E}_t\{rY_i\cdot B_i^S\} = 1 + \int_0^t \mathcal{E}_s\{rY_i\cdot B_i^S\}rY_i(s)dB_i^S(s),\quad t\in[0,T].
$$

By the Burkholder–Davis–Gundy inequality, there exists a constant $C_1 > 0$ such that

$$\mathbb{E}\left[\sup_{t\in[0,T]} \mathcal{E}_t\{rY_i \cdot B_i^S\}\right] = 1 + \mathbb{E}\left[\sup_{t\in[0,T]} \int_0^t \mathcal{E}_s\{rY_i \cdot B_i^S\} rY_i(s) dB_i^S(s)\right]$$
$$\leq 1 + C_1 \mathbb{E}\left[\left(\int_0^T \mathcal{E}_s^2\{rY_i \cdot B_i^S\} Y_i^2(s) ds\right)^{\frac{1}{2}}\right]$$
$$\leq 1 + C_1 \sup_{s\in[0,T]} \psi_i(s) \left(\int_0^T \mathbb{E}\left[\mathcal{E}_s^2\{rY_i \cdot B_i^S\}\right] ds\right)^{\frac{1}{2}}. \tag{7.13}$$

By Novikov's criterion, $\mathbb{E}\left[\mathcal{E}_s(2rY_i \cdot B_i^S)\right] = 1$, so

$$\mathbb{E}\left[\mathcal{E}_s^2\{rY_i \cdot B_i^S\}\right] = \mathbb{E}\left[\mathcal{E}_s\{2rY_i \cdot B_i^S\} \exp\left\{\int_0^s (rY_i(u))^2 du\right\}\right]$$
$$\leq \exp\left\{r^2 T \max_{s\in[0,T]} \psi_i^2(s)\right\} \mathbb{E}\left[\mathcal{E}_s\{2rY_i \cdot B_i^S\}\right]$$
$$= \exp\left\{r^2 T \max_{s\in[0,T]} \psi_i^2(s)\right\} < \infty.$$

Hence, taking into account (7.12), we obtain that

$$\mathbb{E}\left[\sup_{t\in[0,T]} S_i^r(t)\right] \leq c_i \left(1 + C_1\sqrt{T} \sup_{s\in[0,T]} \psi_i(s) \exp\left\{\frac{r^2 T}{2} \max_{s\in[0,T]} \psi_i^2(s)\right\}\right),$$

which yields the required result. □

Our next topic is an investigation of the Malliavin differentiability of the price processes $(S_1, ..., S_d)^T$. An intuitive way to obtain it would be to notice that each S_i has the form

$$S_i(t) = e^{X_i(t)}, \quad t \in [0, T],$$

where X_i is the log-price process which, due to Theorem 7.2, can be written as

$$X_i(t) := \log S_i(t) = X_i(0) + \int_0^t \left(\mu_i(s) - \frac{Y_i^2(s)}{2}\right) ds + \int_0^t Y_i(s) dB_i^S(s) \tag{7.14}$$

with $X_i(0) := \log S_i(0)$. Hence, in order to prove that $S_i(t) \in \mathbb{D}^{1,2}(W)$, one can first establish that $X_i(t) \in \mathbb{D}^{1,2}(W)$ and then apply an appropriate chain rule. The first step of this plan is actually relatively simple: one can simply say that the Malliavin differentiability of $X_i(t)$ a direct corollary of Theorem 7.1 and the explicit shape (7.14) of $X_i(t)$.

Corollary 7.1 *Under assumptions of Theorem 7.1, for any $i = 1, ..., d$ and $t \in [0, T]$, $X_i(t) \in \mathbb{D}^{1,2}(W)$ and its Malliavin derivative*

$$DX_i(t) = (D^1 X_i(t), ..., D^{2d} X_i(t))$$

has the form

$$D_u^j X_i(t) = \left(- \int_0^t Y_i(s) D_u^j Y_i(s) ds + \ell_{i,j} Y_i(u) + \int_0^t D_u^j Y_i(s) d B_i^S(s) \right) \mathbb{1}_{[0,t]}(u).$$

Proof First, note that if $u > t$, then the claim is true due to [13, Corollary 1.2.1]. Let now $u \leq t$. By the definition of the Malliavin derivative,

$$D_u^j \left[\int_0^t \mu_i(s) ds \right] = 0. \tag{7.15}$$

Next, note that, for any $s \in [0, T]$, $Y_i^2(s) \in \mathbb{D}^{1,2}(W)$, where W is defined in Sect. 7.2. Indeed, consider a function $f_i \in C^{(1)}(\mathbb{R})$ such that both f_i and f_i' are bounded and $f_i(x) = x^2$ for all $x \in \left(-\max_{t\in[0,T]} \varphi(t), \max_{t\in[0,T]} \varphi(t)\right)$. Then f_i satisfies the condition of the classical Malliavin chain rule from Appendix A.3 and, moreover, $f_i(Y_i(s)) = Y_i^2(s)$ since $Y(s) \in \left(-\max_{t\in[0,T]} \varphi(t), \max_{t\in[0,T]} \varphi(t)\right)$, which implies that $Y_i^2(s) \in \mathbb{D}^{1,2}(W)$ and $D_u^j Y_i^2(s) = 2Y_i(s) D_u^j Y_i(s)$ for all $j = 1, ..., 2d$.

Next, consider the integral $\frac{1}{2} \int_0^t D_u^j Y_i^2(s) ds = \int_0^t Y_i(s) D_u^j Y_i(s) ds$, $j = 1, ..., 2d$. Using the chain rule and the Hille theorem for the closed operator D (see e.g. [17, Theorem 1.2.2] or [12, Theorem 1.2.4]), we can write

$$D_u^j \left[\frac{1}{2} \int_0^t Y_i^2(s) ds \right] = \int_0^t Y_i(s) D_u^j Y_i(s) ds, \quad j = 1, ..., 2d. \tag{7.16}$$

Finally, by adaptedness and boundedness of Y_i and [13, Proposition 1.3.2],

$$D_u^j\left[\int_0^t Y_i(s)dB_i^S(s)\right] = D_u^j\left[\sum_{k=1}^{2d}\int_0^t \ell_{i,k}Y_i(s)dW_k(s)\right]$$
$$= \ell_{i,j}Y_i(u) + \sum_{k=1}^{2d}\ell_{i,k}\int_0^t D_u^jY_i(s)dW_k(s) \qquad (7.17)$$
$$= \ell_{i,j}Y_i(u) + \int_0^t D_u^jY_i(s)dB_i^S(s),$$

as required. □

The next step, namely the application of the chain rule to $S_i(t) = e^{X_i(t)}$, is a bit more problematic. The classical chain rule results require either the derivatives of the transformation to be bounded (which is obviously not true for $f(x) = e^x$) or the law of the initial random variable to be absolutely continuous (which we have not established yet for the log-price $X_i(t)$). Therefore, below, we give a version of the Malliavin chain rule that fits our needs.

Proposition 7.1 *Let $\eta = (\eta_1, ..., \eta_k) \in (\mathbb{D}^{1,2})^k$ and $f \in C^{(1)}(\mathbb{R}^k; \mathbb{R})$ be a real-valued function such that*

(1) if $\frac{\partial f}{\partial x_j}$ is unbounded, then there exist sequences $\{r_n,\ n \ge 1\}$, $\{s_{j,n},\ n \ge 1\}$, $r_n, s_{j,n} \to \infty$ as $n \to \infty$, such that

$$\left|\frac{\partial f}{\partial x_j}(x_1, ..., x_k)\right| \ge s_{j,n}$$

implies that $|f(x_1, ..., x_k)| \ge r_n$;

(2) $\mathbb{E}\left[|f(\eta)|^2 + \left\|\sum_{j=1}^k \frac{\partial f}{\partial x_j}(\eta)D\eta_j\right\|^2_{L^2([0,T];\mathbb{R}^{2d})}\right] < \infty$.

Then $f(\eta) \in \mathbb{D}^{1,2}$ and $Df(\eta) = \sum_{j=1}^k \frac{\partial f}{\partial x_j}(\eta)D\eta_j$.

Proof According to [13, Proposition 1.2.3], if all partial derivatives $\frac{\partial f}{\partial x_j}$ are bounded, then the claim of our Proposition is true and the chain rule holds. For the general case, let $\phi \in C^\infty(\mathbb{R})$ be compactly supported function such that $\phi(x) = x$ whenever $|x| \le 1$ and $|\phi(x)| \le |x|$ for all $x \in \mathbb{R}$.

For any $m \ge 1$ define $f_m(x) := m\phi\left(\frac{f(x)}{m}\right)$ and observe that f_m has bounded partial derivatives. Indeed, if $\frac{\partial f}{\partial x_j}$ is bounded, so is $\frac{\partial f_m}{\partial x_j}$. Otherwise, let n^* be such that $\frac{r_{n^*}}{m} > \max\{-\inf \operatorname{supp}\phi, \sup\operatorname{supp}\phi\}$. Then, for all $(x_1, ..., x_k) \in \mathbb{R}^k$ such that $|\frac{\partial f}{\partial x_j}(x_1, ..., x_k)| \ge s_{j,n^*}$, it holds that $\frac{f(x_1,...,x_k)}{m} \notin \operatorname{supp}\phi$ and therefore

$$\frac{\partial f_m}{\partial x_j}(x_1, ..., x_k) = \frac{\partial f}{\partial x_j}(x_1, ..., x_k)\phi'\left(\frac{f(x_1, ..., x_k)}{m}\right) = 0.$$

Hence,

$$\left|\frac{\partial f_m}{\partial x_j}(x_1, ..., x_k)\right| = \left|\frac{\partial f}{\partial x_j}(x_1, ..., x_k)\right|\left|\phi'\left(\frac{f(x_1, ..., x_k)}{m}\right)\right| \leq s_{j,n^*} \max_{z\in\mathbb{R}} |\phi'(z)|.$$

Note that $|f_m(\eta)| \leq |f(\eta)|$ for all $m \geq 1$ and $\lim_{m\to\infty} f_m(\eta) = f(\eta)$ a.s. Moreover,

$$|Df_m(\eta)| = \left|\sum_{j=1}^{k} \frac{\partial f_m}{\partial x_j}(\eta)D\eta_j\right| \leq \max_{z\in\mathbb{R}} |\phi'(z)| \left|\sum_{j=1}^{k} \frac{\partial f}{\partial x_j}(\eta)D\eta_j\right|$$

for all $m \geq 1$, and

$$\lim_{m\to\infty} Df_m(\eta) = \sum_{j=1}^{k} \frac{\partial f}{\partial x_j}(\eta)D\eta_j \quad a.s.$$

Therefore, by the dominated convergence theorem,

$$\lim_{m\to\infty} \mathbb{E}\left[|f_m(\eta) - f(\eta)|^2 + \left\|Df_m(\eta) - \sum_{j=1}^{k} \frac{\partial f}{\partial x_j}(\eta)D\eta_j\right\|^2_{L^2([0,T];\mathbb{R}^{2d})}\right] = 0,$$

and the result follows by the closedness of the operator D. □

Corollary 7.2 *Let assumptions of Theorem 7.1 hold for any $i = 1, ..., d$ and $t \in [0, T]$ and, in addition, assume that*

$$\sup_{s\in[0,T]} \int_0^T \mathcal{K}_i^2(u, s)du < \infty \tag{7.18}$$

for all $i = 1, ..., d$. Then $S_i(t) \in \mathbb{D}^{1,2}(W)$ and its Malliavin derivative $DS_i(t) = (D^1 S_i(t), ..., D^{2d} S_i(t))$ has the form

$$D_u^j S_i(t) = S_i(t)D_u^j X_i(t).$$

Proof The function $f(x) = e^x$ satisfies the assumptions of Proposition 7.1, so it is sufficient to check that

$$\mathbb{E}\left[S_i^2(t) + \|S_i(t)DX_i(t)\|^2_{L^2([0,T];\mathbb{R}^{2d})}\right] < \infty,$$

and, since the bound

$$\mathbb{E}\left[S_i^2(t)\right] < \infty$$

follows immediately from Theorem 7.2, it remains to check that

$$\int_0^T \mathbb{E}\left[S_i^2(t)(D_u^j X_i(t))^2\right] du < \infty$$

for all $j = 1, ..., 2d$.

By Hölder inequality and Theorem 7.2,

$$\int_0^T \mathbb{E}\left[S_i^2(t)(D_u^j X_i(t))^2\right] du \le \int_0^T \left(\mathbb{E}\left[S_i^4(t)\right]\right)^{\frac{1}{2}} \left(\mathbb{E}\left[(D_u^j X_i(t))^4\right]\right)^{\frac{1}{2}} du$$

$$\le C \int_0^T \left(\mathbb{E}\left[(D_u^j X_i(t))^4\right]\right)^{\frac{1}{2}} du$$

$$= C \int_0^t \left(\mathbb{E}\left[\left(\ell_{i,j} Y_i(u) - \int_0^t Y_i(s) D_u^j Y_i(s) ds + \int_0^t D_u^j Y_i(s) d B_i^S(s)\right)^4\right]\right)^{\frac{1}{2}} du$$

and, thanks to the boundedness of Y_i,

$$\int_0^T \mathbb{E}\left[S_i^2(t)(D_u^j X_i(t))^2\right] du$$

$$\le C \int_0^t \left(\mathbb{E}\left[1 + \left(\int_0^t |D_u^j Y_i(s)| ds\right)^4 + \left(\int_0^t D_u^j Y_i(s) d B_i^S(s)\right)^4\right]\right)^{\frac{1}{2}} du. \tag{7.19}$$

Note that, by Theorem 3.8, for all $s \in [0, T]$: $(s, Y_i(s)) \in \mathcal{D}_{\frac{1}{\xi_i}, \frac{1}{\xi_i}}$, where

$$\xi_i := C(1 + \Lambda_i)^{\frac{1}{\gamma_i \lambda_i + \lambda_i - 1}}$$

with $C > 0$ being a deterministic constant that does not depend on the noise Z_i, $\lambda_i \in \left(\frac{1}{\gamma_i + 1}, H_i\right)$ and $\Lambda_i - \Lambda_i(\lambda_l)$ being such that

$$|Z_i(t_1) - Z_i(t_2)| \le \Lambda_i |t_1 - t_2|^{\lambda_i}, \quad t_1, t_2 \in [0, T].$$

We also take Λ_i such that $\mathbb{E}[\Lambda_i^r] < \infty$ for all $r \in \mathbb{R}$. By Assumption 6.3(**b5**),

$$\left|\frac{\partial b_i}{\partial y}(s, Y_i(s))\right| \leq C(1 + \xi_i^{q_i}), \quad s \in [0, T],$$

and, moreover, by the same Assumption 6.3(**b5**),

$$e^{\int_s^t \frac{\partial b_i}{\partial y}(v, Y_i(v))dv} \leq e^{C_{b_i} T}.$$

Therefore, for any $s \in [0, T]$, $i = 1, ..., d$, $j = 1, ..., 2d$,

$$\begin{aligned}|D_u^j Y_i(s)| &= \ell_{d+i,j} \left| \mathcal{K}_i(s,u) + \int_u^s \mathcal{K}_i(v,u) \frac{\partial b_i}{\partial y}(v, Y_i(v)) e^{\int_v^s \frac{\partial b_i}{\partial y}(w, Y_i(w))dw} dv \right| \mathbb{1}_{[0,s]}(u) \\ &\leq C \left(|\mathcal{K}_i(s,u)| + (1 + \xi_i^{q_i}) \int_u^s |\mathcal{K}_i(v,u)| dv \right) \mathbb{1}_{[0,s]}(u),\end{aligned}$$

where $\xi_i^{q_i}$ has moments of all orders.

Using this estimation as well as the Burkholder-Davis-Gundy inequality and assumption (7.18),

$$\begin{aligned}\mathbb{E}\left[\left(\int_0^t D_u^j Y_i(s) dB_i^S(s)\right)^4\right] &\leq C\mathbb{E}\left[\left(\int_0^t (D_u^j Y_i(s))^2 ds\right)^2\right] \\ &\leq C\mathbb{E}\left[\left(\int_u^t \mathcal{K}_i^2(s,u)ds + (1 + \xi_i^{q_i})^2 \int_u^t \int_u^s \mathcal{K}_i^2(v,u)dvds\right)^2\right] \\ &\leq C\mathbb{E}\left[\left(\int_u^t \mathcal{K}_i^2(s,u)ds + (1 + \xi_i^{q_i})^2 \int_u^t \int_u^s \mathcal{K}_i^2(v,u)dvds\right)^2\right] \\ &\leq C < \infty\end{aligned}$$

for some deterministic constant C. Similarly, again due to (7.18),

$$\begin{aligned}\mathbb{E}\left[\left(\int_0^t |D_u^j Y_i(s)| ds\right)^4\right] &\leq C\mathbb{E}\left[\left(\int_u^t \left(|\mathcal{K}_i(s,u)| + (1 + \xi_i^{q_i}) \int_u^s |\mathcal{K}_i(v,u)| dv\right) ds\right)^4\right] \\ &\leq C < \infty.\end{aligned}$$

Therefore, taking into account (7.19), we obtain that

$$\int_0^T \mathbb{E}\left[S_i^2(t)(D_u^j X_i(t))^2\right] du < \infty$$

for any $t \in [0, T]$, $i = 1, ..., d$, $j = 1, ..., 2d$, which yields the required result. □

So far, we have discussed Malliavin differentiability with respect to the "initial" Brownian motion W, but we can easily obtain Malliavin differentiability with respect to linear transformations of W as well. In fact, this proves to be quite convenient for certain applications: for example, it is well known (see e.g. [13, Theorem 2.1.3]) that a sufficient condition for the absolute continuity of the law of $X_i(t)$ w.r.t. the Lebesgue measure is

$$\|D_\cdot X_i(t)\|_{L^2([0,T];\mathbb{R}^{2d})} > 0 \quad a.s., \tag{7.20}$$

but the form of $DX_i(t)$ obtained above is not very convenient for this matter. However, (7.20) becomes trivial if we calculate the derivative with respect to a certain transformation of W. Namely, recall that all Itô integrals within our model are written with respect a $2d$-dimensional Gaussian martingale $B = (B_1, ..., B_{2d})^T$ that is obtained from W by a linear transformation

$$B(t) = \begin{pmatrix} B^S(t) \\ B^Y(t) \end{pmatrix} = \mathcal{L}W(t),$$

where $\mathcal{L} = (\ell_{i,j})_{i,j=1}^{2d}$ is a lower-triangular matrix such that $\mathcal{L}\mathcal{L}^T = \Sigma = (\sigma_{i,j})_{i,j=1}^{2d}$. Now, define $\mathcal{U} = (u_{i,j})_{i,j=1}^{2d}$ an *upper* $2d \times 2d$-triangular matrix with positive values on the diagonal such that

$$\widetilde{W} := \mathcal{U}^{-1}B = \mathcal{U}^{-1}\mathcal{L}W$$

is a $2d$-dimensional Brownian motion. Let now $\widetilde{D} = (\widetilde{D}^1, ..., \widetilde{D}^{2d})$ be the Malliavin derivative with respect to the new Brownian motion $\widetilde{W}$ and $\widetilde{\mathbb{D}}^{1,2}$ be its domain. Clearly, the matrix $\mathcal{U}^{-1}\mathcal{L}$ is non-degenerate and hence W and $\widetilde{W}$ generate the same filtration $\mathbb{F}$ and, moreover, it is easy to see that $\widetilde{\mathbb{D}}^{1,2}$ coincides with $\mathbb{D}^{1,2}$. Finally, by construction, $\widetilde{W}_1$ is independent of the processes $B_2^S, ..., B_d^S, B_1^Y, ..., B_d^Y$ and therefore the following result holds.

Theorem 7.3 *Under assumptions of Theorem 7.1, for any $t \in (0, T]$,*

$$\widetilde{D}_v^1 X_i(t) = \begin{cases} u_{1,1} Y_1(v) \mathbb{1}_{[0,t]}(v), & \text{if } i = 1, \\ 0, & \text{if } i \neq 1, \end{cases}$$

and, if, in addition, (7.18) *holds,*

$$\widetilde{D}_v^1 S_i(t) = \begin{cases} u_{1,1} S_1(t) Y_1(v) \mathbb{1}_{[0,t]}(v), & \text{if } i = 1, \\ 0, & \text{if } i \neq 1, \end{cases}$$

where $u_{1,1}$ is the upper left element of $\mathcal{U}$. In particular,

$$\|\widetilde{D}.X_1(t)\|_{L^2([0,T];\mathbb{R}^{2d})} > 0 \quad a.s.$$

Proof The expression for $\widetilde{D}^1 X_i(t)$ is obtained straightforwardly by differentiating the right-hand side of (7.14) and taking into account that

$$B_1^S(t) = \sum_{k=1}^{2d} u_{1,k}\widetilde{W}_k(t)$$

and $\widetilde{D}^1 Y_i(t) = 0$ due to independence of $\widetilde{W}_1$ and B_i^Y. The expression for $\widetilde{D}^1 S_i(t)$ can be easily obtained using the chain rule from Proposition 7.1. □

Corollary 7.3 *Under assumptions of Theorem 7.1, for each $i = 1, ..., d$ and $t \in (0, T]$, the law of $X_i(t)$ (and therefore of $S_i(t) = e^{X_i(t)}$) is absolutely continuous w.r.t. the Lebesgue measure.*

Proof By Theorem 7.3,

$$\|\widetilde{D}.X_1(t)\|_{L^2([0,T];\mathbb{R}^{2d})} > 0 \quad a.s.$$

and, moreover, $\widetilde{D}X_1(t)$ is adapted to the filtration generated by $\widetilde{B}^i$. Thus the result for $X_1(t)$ follows, for example, from [13, Theorem 2.1.3]. □

Theorem 7.4 *For all $t \in (0, T]$, the law of the random vector $X(t)$ is absolutely continuous with respect to the Lebesgue measure on $\mathbb{R}^d$.*

Proof Our strategy is to check that, for any $f \in C^\infty(\mathbb{R}^d; \mathbb{R})$, such that f and all its partial derivatives are bounded, the following relation holds:

$$\left|\mathbb{E}\left[\frac{\partial f}{\partial x_i}(X(t))\right]\right| \le c_i \sup_{x\in\mathbb{R}^d} |f(x)|, \quad i = 1, ..., d.$$

Then the result will follow from [13, Lemma 2.1.1].

Note that, by applying sequentially the chain rule for the Malliavin derivative and the duality relationship between Malliavin derivative and the Skorokhod integral, we can obtain

$$\begin{aligned}\left|\mathbb{E}\left[\frac{\partial f}{\partial x_1}(X(t))\right]\right| &= \frac{1}{t}\left|\mathbb{E}\left[\int_0^t \frac{\partial f}{\partial x_1}(X(t))\widetilde{D}_u^1 X_1(t)\frac{1}{\widetilde{D}_u^1 X_1(t)}du\right]\right| \\ &= \frac{1}{t}\left|\mathbb{E}\left[\int_0^t \widetilde{D}_u^1[f(X(t))]\frac{1}{\widetilde{D}_u^1 X_1(t)}du\right]\right|\end{aligned}$$

$$
\begin{aligned}
&= \frac{1}{t}\left|\mathbb{E}\left[f(X(t))\int_0^t \frac{1}{u_{1,1}Y_1(u)}d\widetilde{W}_1(u)\right]\right| \\
&\le \frac{1}{tu_{1,1}}\mathbb{E}\left[\left|\int_0^t \frac{1}{Y_i(u)}d\widetilde{W}_1(u)\right|\right]\sup_{x\in\mathbb{R}^d}|f(x)| \\
&=: c_1 \sup_{x\in\mathbb{R}^d}|f(x)|.
\end{aligned}
$$

So, the required property holds for $\frac{\partial f}{\partial x_1}(X(t))$. As for $\frac{\partial f}{\partial x_i}(X(t))$ with $i = 2, ..., d$, one can reorder the elements of B so that by swapping B_i becomes the first one and then apply the same arguments. □

7.3 Martingale Measures in SVV Models

The next question we aim to answer is the absence of *arbitrage* in the SVV model. To begin, let us quickly outline what we want to see. First, we seek to establish that the SVV model is arbitrage-free. Indeed, if it is not the case, the presence of arbitrage opportunities would raise significant concerns about the model's principal applicability in a financial context. Since the issue of arbitrage naturally leads to a discussion of equivalent local martingale measures, our initial objective is to prove they *exist*. Once that is established, the second logical question is whether the SVV market is *complete*, that is, whether the local martingale measure is *unique*. We can already make an educated guess on this point: since the SVV model is a stochastic volatility model driven by multiple sources of noise, we expect the market to be incomplete, meaning that there exist multiple equivalent local martingale measures. This leads us to the final question we aim to address in this section: can we fully characterize the set of local martingale measures produced by the SVV model?

Fortunately for us, the answer is yes. Thanks to the structure of the SVV model and its nice integrability properties, we can rely on the results of Biagini, Guasoni, and Pratelli [3], which apply to general stochastic volatility models. The objective of this section is therefore to adapt those results to our specific case.

Recall that our model is driven by the $2d$-dimensional Gaussian process

$$
\begin{pmatrix} B^S(t) \\ B^Y(t) \end{pmatrix} = \mathcal{L}W(t), \quad t \in [0, T],
$$

where $W = \{W(t),\ t \in [0, T]\}$ is a standard $2d$-dimensional Brownian motion and $\mathcal{L} = (\ell_{i,j})_{i,j=1}^{2d}$ is nondegenerate lower triangular $2d \times 2d$ real matrix with positive entries on a diagonal such that $\mathcal{L}\mathcal{L}^T = \Sigma$. Let also

$$\mathcal{L} = \begin{pmatrix} \mathcal{L}_{11} & \mathcal{L}_{12} \\ \mathcal{L}_{21} & \mathcal{L}_{22} \end{pmatrix},$$

where by $\mathcal{L}_{ij}$ we denote the $d \times d$ blocks of matrix $\mathcal{L}$. Note that $\mathcal{L}_{12}$ is a zero matrix and $\mathcal{L}_{11}$ is a lower triangular matrix with positive elements on the diagonal. In particular, $\mathcal{L}_{11}$ is invertible, so we can define its inverse by $\mathcal{L}_{11}^{-1} = \left(\ell_{j,k}^{(-1)}\right)_{j,k=1}^{d}$. Finally, for a given semimartingale X, we denote the corresponding Doléans-Dade exponential by

$$\mathcal{E}_t\left\{X\right\} := \exp\left\{X_t - X_0 - \frac{1}{2}[X]_t\right\}.$$

Lemma 7.1 *Let M be a positive local $\mathbb{P}$-martingale w.r.t. $\mathbb{F}$ with $M(0) = 1$ a.s. Then $M\widetilde{S}_i$, $i = 1, ..., d$, are all local martingales if and only if*

$$\begin{aligned} M(t) &= \mathcal{E}_t\left\{-\sum_{j=1}^{d}\int_0^{\cdot}\left(\sum_{k=1}^{d}\ell_{j,k}^{(-1)}\frac{\widetilde{\mu}_k(s)}{Y_k(s)}\right)dW_j(s)\right\}\mathcal{E}_t\left\{\sum_{l=d+1}^{2d}\int_0^{\cdot}\xi_l(s)dW_l(s)\right\} \\ &= \mathcal{E}_t\left\{-\sum_{j=1}^{d}\int_0^{\cdot}\left(\sum_{k=1}^{d}\ell_{j,k}^{(-1)}\frac{\widetilde{\mu}_k(s)}{Y_k(s)}\right)dW_j(s) + \sum_{l=d+1}^{2d}\int_0^{\cdot}\xi_l(s)dW_l(s)\right\}, \end{aligned} \tag{7.21}$$

where $\left(\ell_{j,k}^{(-1)}\right)_{j,k=1}^{d}$ are the elements of $\mathcal{L}_{11}^{-1}$ and ξ_l, $l = d+1, ..., 2d$, are any predictable processes such that $\int_0^{\cdot}\xi_l(s)dW_l(s)$ are well-defined local martingales.

Proof By the virtue of the local martingale representation theorem, M is a local $\mathbb{P}$-martingale w.r.t. $\mathbb{F}$ if and only if there exist predictable processes $\zeta_j = \{\zeta_j(t),\ t \in [0, T]\}$, $j = 1, ..., 2d$, such that

$$\mathbb{P}\left(\int_0^T \zeta_j^2(s)ds < \infty\right) = 1, \quad j = 1, ..., 2d,$$

and

$$M(t) = 1 + \sum_{j=1}^{2d}\int_0^t \zeta_j(s)dW_j(s). \tag{7.22}$$

By Itô's formula,

$$\begin{aligned} d(M(t)\widetilde{S}_i(t)) = &\left(\tilde{\mu}_i(t)M(t) + Y_i(t)\sum_{j=1}^{2d}\ell_{i,j}\zeta_j(t)\right)\widetilde{S}_i(t)dt \\ &+ \sum_{j=1}^{2d}\left(\ell_{i,j}M(t)Y_i(t) + \zeta_j(t)\right)\widetilde{S}_i(t)dW_j(t), \end{aligned}$$

and thus $M\widetilde{S}_i$ is a local martingale if and only if

$$\int_0^t \left(\tilde{\mu}_i(s)M(s) + Y_i(s) \sum_{j=1}^{2d} \ell_{i,j}\zeta_j(s) \right) \widetilde{S}_i(s)ds = 0, \quad t \in [0, T].$$

This, in turn, implies that

$$\sum_{j=1}^{2d} \ell_{i,j}\zeta_j(t) = -M(t)\frac{\tilde{\mu}_i(t)}{Y_i(t)}, \quad i = 1, ..., d,$$

almost everywhere w.r.t. $dt \otimes d\mathbb{P}$. Since $\mathcal{L}$ is lower-triangular and each $Y_i(t)$ is positive a.s., we have that

$$\mathcal{L}_{11} \begin{pmatrix} \zeta_1(t) \\ \vdots \\ \zeta_d(t) \end{pmatrix} = -M(t) \begin{pmatrix} \frac{\tilde{\mu}_1(t)}{Y_1(t)} \\ \vdots \\ \frac{\tilde{\mu}_d(t)}{Y_d(t)} \end{pmatrix} \tag{7.23}$$

almost everywhere w.r.t. $dt \otimes d\mathbb{P}$, and hence

$$\begin{pmatrix} \zeta_1(t) \\ \vdots \\ \zeta_d(t) \end{pmatrix} = \mathcal{L}_{11}^{-1} \begin{pmatrix} -M(t)\frac{\tilde{\mu}_1(t)}{Y_1(t)} \\ \vdots \\ -M(t)\frac{\tilde{\mu}_d(t)}{Y_d(t)} \end{pmatrix} \tag{7.24}$$

almost everywhere w.r.t. $dt \otimes d\mathbb{P}$. Using notation $\mathcal{L}_{11}^{-1} = \left(\ell_{j,k}^{(-1)}\right)_{j,k=1}^{d}$ and substituting (7.24) to (7.22), we see that the process M should satisfy the SDE of the form

$$dM(t) = -M(t)\sum_{j=1}^{d}\sum_{k=1}^{d} \ell_{j,k}^{(-1)} \frac{\widetilde{\mu}_k(t)}{Y_k(t)} dW_j(t) + \sum_{l=d+1}^{2d} \zeta_l(t)dW_l(t).$$

This SDE has a unique solution of the form

$$M(t) = \mathcal{E}_t\left\{-\sum_{j=1}^{d}\int_0^{\cdot}\left(\sum_{k=1}^{d}\ell_{j,k}^{(-1)}\frac{\widetilde{\mu}_k(s)}{Y_k(s)}\right)dW_j(s)\right\}$$
$$\times\left(1+\sum_{l=d+1}^{2d}\int_0^{t}\mathcal{E}_u^{-1}\left\{-\sum_{j=1}^{d}\int_0^{\cdot}\left(\sum_{k=1}^{d}\ell_{j,k}^{(-1)}\frac{\widetilde{\mu}_k(s)}{Y_k(s)}\right)dW_j(s)\right\}\zeta_l(u)dW_l(u)\right)$$
$$=: \mathcal{E}_t\left\{-\sum_{j=1}^{d}\int_0^{\cdot}\left(\sum_{k=1}^{d}\ell_{j,k}^{(-1)}\frac{\widetilde{\mu}_k(s)}{Y_k(s)}\right)dW_j(s)\right\}\left(1+\sum_{l=d+1}^{2d}\int_0^{t}\widetilde{\zeta}_l(u)dW_l(u)\right). \tag{7.25}$$

Furthermore, since the process M is positive, the continuous process $\widetilde{M}$ defined by

$$\widetilde{M}(t) := 1+\sum_{l=d+1}^{2d}\int_0^{t}\widetilde{\zeta}_l(s)dW_l(s)$$

must also be positive, thus it can be represented as

$$\widetilde{M}(t) = \mathcal{E}_t\left(\sum_{l=d+1}^{2d}\int_0^{\cdot}\xi_l(s)dW_l(s)\right) \tag{7.26}$$

with $\xi_l, l = d+1, \ldots, 2d$, being some predictable processes such that the corresponding stochastic integrals are well-defined. It remains to notice that $\widetilde{\zeta}_l, l = d+1, \ldots, 2d$, can be arbitrary predictable stochastic processes such that the corresponding stochastic integrals are well-defined, thus $\xi_l, l = d+1, \ldots, 2d$, can also be chosen arbitrarily. Taking into account (7.25) and (7.26), we get the representation (7.21). □

In the literature, positive Radon-Nikodym densities $M(T)$ associated with local $\mathbb{P}$-martingales M such that $M\widetilde{S}_i$ all become local martingales are commonly called *strict local martingale densities* (see [14] for more detail), and Lemma 7.1 thus gives the description of the set of all strict local martingale densities on the market. However, note that the measure defined by $d\mathbb{Q} := M(T)d\mathbb{P}$ is not necessarily a probability measure and one has to additionally check that $\mathbb{Q}(\Omega) = \mathbb{E}[M(T)] = 1$ (or, equivalently, that M is a uniformly integrable martingale). Moreover, exploiting the Bayes' theorem for conditional expectations, it is easy to see that if $\mathbb{Q}$ defined above is a probability measure, then $\widetilde{S}_i$, $i = 1, \ldots, d$, are local $\mathbb{Q}$-martingales, which would imply the no-arbitrage property of the market. It turns out that the market model in consideration indeed has this property which is formulated in the next theorem.

Theorem 7.5 *The SVV market is arbitrage-free and incomplete.*

Proof Each Y_k, $k = 1, \ldots, d$, is bounded away from zero by the corresponding $\min_{t\in[0,T]}\varphi_k(t) > 0$. Therefore, by Novikov's condition,

$$M(t) = \mathcal{E}_t \left\{ -\sum_{j=1}^{d} \int_0^{\cdot} \left(\sum_{k=1}^{d} \ell_{j,k}^{(-1)} \frac{\widetilde{\mu}_k(s)}{Y_k(s)} \right) dW_j(s) \right\}$$

is a uniformly integrable martingale (this case corresponds to $\xi_l \equiv 0$ in (7.21), $l = d+1, ..., 2d$). Therefore the associated $\mathbb{Q}$ is a probability measure, i.e. the market is arbitrage free.

Furthermore, again by Novikov's condition, any ξ_l such that

$$\mathbb{E}\left[\exp\left\{ \frac{1}{2} \sum_{l=d+1}^{2d} \int_0^T \xi_l^2(t)dt \right\} \right] < \infty$$

leads to M being a uniformly integrable martingale, thus the market is incomplete. □

Remark 7.3 The martingale measure with the Radon-Nikodym density

$$M(T) = \mathcal{E}_T \left\{ -\sum_{j=1}^{d} \int_0^{\cdot} \left(\sum_{k=1}^{d} \ell_{j,k}^{(-1)} \frac{\widetilde{\mu}_k(s)}{Y_k(s)} \right) dW_j(s) \right\} \tag{7.27}$$

discussed in Theorem 7.5 is called the *minimal martingale measure* (see e.g. [15, 18]). Further, in Section 8.2.3, we will use it in our option pricing algorithms since this measure has the simplest structure. However, there are other equivalent martingale probability measuers that can be interesting for certain applications. One such example could be the *variance optimal martingale measure* that is important in mean-variance hedging; see [5, 16]. Note that the SVV price is continuous and, using the same arguments as in Theorem 7.2, it is easy to prove that the SVV minimal martingale density $M(T)$ in (7.27) is square integrable. Therefore, by [5, Theorem 1.3] the optimal variance measure exists on the SVV market.

7.4 Simulation of the SVV Model

Since the volatility processes in the SVV model exhibit a fairly complex structure in both drift and driving noises, it would be naive to expect compact closed-form expressions for option prices. Therefore, to *use* the model effectively, we must rely on numerical methods, specifically, algorithms for simulating the paths of both price and volatility.

In fact, we have already completed the most challenging part of this process, the simulation of the volatility paths, in Chap. 5 by introducing the drift-implicit Euler scheme. Let us briefly recall the algorithm.

Consider a uniform partition $\{0 = t_0 < t_1 < ... < t_N = T\}$ of $[0, T]$, $t_k := \frac{Tk}{N}$, $k = 0, 1, ..., N$, with the mesh $\Delta_N := \frac{T}{N}$ such that

$$\Delta_N \max_{i=1,\dots,d} \sup_{(t,y)\in\mathcal{D}^i_{0,0}} \frac{\partial b_i}{\partial y}(t,y) < 1. \tag{7.28}$$

Then, for each $i = 1, \dots, d$, define $\widehat{Y}_i(t)$ as

$$\begin{aligned}
\widehat{Y}_i(0) &= Y_i(0),\\
\widehat{Y}_i(t_{k+1}) &= \widehat{Y}_i(t_k) + b_i(t_{k+1}, \widehat{Y}_i(t_{k+1}))\Delta_N + (Z_i(t_{k+1}) - Z_i(t_k)),\\
\widehat{Y}_i(t) &= \widehat{Y}_i(t_k), \quad t \in [t_k, t_{k+1}),
\end{aligned} \tag{7.29}$$

where $Z_i(t) := \int_0^t \mathcal{K}_i(t,s)dB_i^Y(s)$. As explained in detail in Chap. 5, the second expression in (7.29) should be viewed as an equation with respect to $\widehat{Y}_i(t_{k+1})$ and, under Assumptions 6.1 and 6.3, it has a unique solution such that $\widehat{Y}_i(t_{k+1}) \in (\varphi_i(t_{k+1}), \psi_i(t_{k+1}))$. For this $\widehat{Y}_i$, according to Theorem 5.2, for an arbitrary $r \ge 1$ and $\lambda_i \in (0, H_i)$, where H_i is from Assumption 6.3 corresponding to Y_i, there exists a constant $C > 0$ that does not depend on the partition such that

$$\mathbb{E}\left[\sup_{t\in[0,T]} |Y_i(t) - \widehat{Y}_i(t)|^r\right] \le C\Delta_N^{\lambda_i r}.$$

Moreover, since we assume that

$$\varphi_* := \min_{i=1,\dots,d} \min_{t\in[0,T]} \varphi_i(t) > 0,$$

it is easy to see that

$$\begin{aligned}
\mathbb{E}\left[\sup_{t\in[0,T]} \left|\frac{1}{Y_i(t)} - \frac{1}{\widehat{Y}_i(t)}\right|^r\right] &= \mathbb{E}\left[\sup_{t\in[0,T]} \frac{|\widehat{Y}_i(t) - Y_i(t)|^r}{Y_i^r(t)\widehat{Y}_i^r(t)}\right]\\
&\le \mathbb{E}\left[\sup_{t\in[0,T]} \frac{|\widehat{Y}_i(t) - Y_i(t)|^r}{\varphi_*^{2r}}\right] \le C\Delta_N^{\lambda_i r}.
\end{aligned} \tag{7.30}$$

With a numerical scheme for the volatility processes Y_i, $i = 1, \dots, d$, the simulation of the price processes S_i, $i = 1, \dots, d$, turns out to be quite straightforward. Indeed, we can utilize the explicit shape (7.11) of S_i and define

$$\widehat{S}_i(t) := e^{\widehat{X}_i(t)},$$

where

$$\widehat{X}_i(t_n) = X_i(0) + \frac{T}{N}\sum_{k=0}^{n-1}\left(\mu_i(t_k) - \frac{\widehat{Y}_i^2(t_k)}{2}\right) + \sum_{k=0}^{n-1} \widehat{Y}_i(t_k)\left(B_i^S(t_{k+1}) - B_i^S(t_k)\right),$$

with B_i^S, $i = 1, ..., d$, being the correlated Brownian motions defined by (7.4), $X_i(0) := \log S_i(0)$ and $\widehat{Y}_i$ being the drift-implicit approximation (7.29) of Y_i. For the time indices t outside of the partition, we can put

$$\widehat{X}_i(t) = \widehat{X}_i(t_n), \quad t \in (t_n, t_{n+1}).$$

Remark 7.4 Note that the approximating processes $\widehat{Y}_i$, $\widehat{X}_i$ and $\widehat{S}_i$ are *not* continuous on $[0, T]$, despite the fact that the "original" processes Y_i, X_i and S_i are. In priniple, we could have used linear interpolation between the points of the partition: for example, for each $t \in [t_k, t_{k+1})$, we could have put

$$\widehat{Y}_i(t) = \frac{t_{k+1} - t}{t_{k+1} - t_k}\widehat{Y}_i(t_k) + \frac{t - t_k}{t_{k+1} - t_k}\widehat{Y}_i(t_{k+1}),$$
$$\widehat{X}_i(t) = \frac{t_{k+1} - t}{t_{k+1} - t_k}\widehat{X}_i(t_k) + \frac{t - t_k}{t_{k+1} - t_k}\widehat{X}_i(t_{k+1}),$$

and such a definition would preserve continuity of the approximation. However, as we will see later, continuity of the approximating processes does not play any important role for our purposes. Instead, we will be *much more* concerned with the *adaptability* of the filtering approximations generated by the original $2d$-dimensional Brownian motion W. The "discontinuous" version of $\widehat{Y}_i$, $\widehat{X}_i$ and $\widehat{S}_i$ does have this property, so we will use it in what follows.

We have already discussed the convergence properties of the discretized volatilities $\widehat{Y}_i$, $i = 1, ..., d$, so let us fill the remaining gap and provide the convergence results for both $\widehat{X}_i$ and $\widehat{S}_i$.

Theorem 7.6 *Let Assumptions 6.1 and 6.3 hold and the mesh of the partition* Δ_N *satisfy* (7.28)*. Fix* $i = 1, ..., d$*, take an arbitrary* $\lambda_i \in (0, H_i)$*, where* H_i *is from Assumption 6.3 corresponding to* Y_i*, and fix* $r \geq 1$*. Then there exists a constant* $C > 0$ *that does not depend on the partition such that*

$$\mathbb{E}\left[\sup_{n=0,1,...,N} |X_i(t_n) - \widehat{X}_i(t_n)|^r\right] \leq C\Delta_N^{r\lambda_i}.$$

Proof Observe that, for any $n = 1, ..., N$,

$$\begin{aligned}|X_i(t_n) - \widehat{X}_i(t_n)|^r &\leq C\left(\sum_{k=0}^{n-1}\int_{t_k}^{t_{k+1}} |\mu_i(s) - \mu_i(t_k)|ds\right)^r \\ &+ C\left(\int_0^{t_n} |Y_i^2(s) - \widehat{Y}_i^2(s)|ds\right)^r \\ &+ C\left|\int_0^{t_n} (Y_i(s) - \widehat{Y}_i(s))dB_i^S(s)\right|^r\end{aligned}$$

$$
\begin{aligned}
&\leq C\left(\sum_{k=0}^{N-1}\int_{t_k}^{t_{k+1}}|s-t_k|^{\lambda_i}ds\right)^r \\
&+C\left(\int_0^T|Y_i(s)+\widehat{Y}_i(s)||Y_i(s)-\widehat{Y}_i(s)|ds\right)^r \\
&+C\left(\sup_{t\in[0,T]}\left|\int_0^t(Y_i(s)-\widehat{Y}_i(s))dB_i^S(s)\right|\right)^r \\
&\leq C\Delta_N^{r\lambda_i}+C\sup_{t\in[0,T]}|Y_i(t)-\widehat{Y}_i(t)|^r \\
&+C\left(\sup_{t\in[0,T]}\left|\int_0^t(Y_i(s)-\widehat{Y}_i(s))dB_i^S(s)\right|\right)^r.
\end{aligned}
$$

Hence, by the Burkholder-Davis-Gundy inequality and Theorem 5.3,

$$
\begin{aligned}
\mathbb{E}\left[\sup_n|X_i(t_n)-\widehat{X}_i(t_n)|^r\right] &\leq C\Delta_N^{r\lambda_i}+C\mathbb{E}\left[\sup_{t\in[0,T]}|Y_i(t)-\widehat{Y}_i(t)|^r\right] \\
&+C\mathbb{E}\left[\left(\sup_{t\in[0,T]}\left|\int_0^t(Y_i(s)-\widehat{Y}_i(s))dB_i^S(s)\right|\right)^r\right] \\
&\leq C\Delta_N^{r\lambda_i}+C\mathbb{E}\left[\left(\int_0^T(Y_i(s)-\widehat{Y}_i(s))^2ds\right)^{\frac{r}{2}}\right] \\
&\leq C\Delta_N^{r\lambda_i}.
\end{aligned}
$$

□

Proposition 7.2 *Let Assumptions 6.1 and 6.3 hold and the mesh of the partition* Δ_N *satisfy* (7.28). *Then, for any* $r\in\mathbb{R}$

$$
\sup_{N\geq 1}\sup_{n=0,1,\ldots,N}\mathbb{E}\left[\widehat{S}_i^r(t_n)\right]<\infty.
$$

Proof Since both μ_i and $\widehat{Y}_i$ are bounded and the bound does not depend on the partition, it is sufficient to prove that

$$
\sup_{N\geq 1}\sup_{n=0,1,\ldots,N}\mathbb{E}\left[\exp\left\{r\int_0^{t_n}\widehat{Y}_i(s)dB_i^S(s)\right\}\right]<\infty.
$$

Consider a stochastic process

$$\exp\left\{r\int_0^t \widehat{Y}_i(s)dB_i^S(s)\right\}, \quad t \in [0, T].$$

Then, since $\widehat{Y}_i$ is bounded, Novikov's condition implies that

$$\mathbb{E}\left[\exp\left\{\int_0^t r\widehat{Y}_i(s)dB_i^S(s) - \frac{r^2}{2}\int_0^t \widehat{Y}_i^2(s)ds\right\}\right] = 1$$

and hence there exist a constant C that does not depend on the partition such that

$$\begin{aligned}
&\mathbb{E}\left[\exp\left\{r\int_0^t \widehat{Y}_i(s)dB_i^S(s)\right\}\right] \\
&= \mathbb{E}\left[\exp\left\{r\int_0^t \widehat{Y}_i(s)dB_i^S(s) - \frac{r^2}{2}\int_0^t \widehat{Y}_i^2(s)ds\right\}\exp\left\{\frac{r^2}{2}\int_0^t \widehat{Y}_i^2(s)ds\right\}\right] \\
&\le C\mathbb{E}\left[\exp\left\{\int_0^t r\widehat{Y}_i(s)dB_i^S(s) - \frac{r^2}{2}\int_0^t \widehat{Y}_i^2(s)ds\right\}\right] = C < \infty,
\end{aligned}$$

which implies the required result. □

Theorem 7.7 *Let Assumptions 6.1 and 6.3 hold and the mesh of the partition Δ_N satisfy (7.28). Fix $i = 1, \dots, d$, take an arbitrary $\lambda_i \in (0, H_i)$, where H_i is from Assumption 6.3 corresponding to Y_i, and fix $r \ge 1$. Then there exists a constant $C > 0$ that does not depend on the partition such that*

$$\sup_{N\ge 1}\sup_{n=0,1,\dots,N}\mathbb{E}\left[|S_i(t_n) - \widehat{S}_i(t_n)|^r\right] \le C\Delta_N^{r\lambda_i}.$$

In particular,

$$\mathbb{E}\left[|S_i(T) - \widehat{S}_i(T)|^r\right] \le C\Delta_N^{r\lambda_i}.$$

Proof For any $n = 0, 1, \dots, N$, we use Theorem 7.2, Proposition 7.2 and Theorem 7.6 to write:

$$\begin{aligned}
\mathbb{E}\left[|S_i(t_n) - \widehat{S}_i(t_n)|^r\right] &= \mathbb{E}\left[|e^{X_i(t_n)} - e^{\widehat{X}_i(t_n)}|^r\right] \\
&\le \mathbb{E}\left[\left(e^{X_i(t_n)} + e^{\widehat{X}_i(t_n)}\right)^r |X_i(t_n) - \widehat{X}_i(t_n)|^r\right] \\
&\le \left(\mathbb{E}\left[\left(e^{X_i(t_n)} + e^{\widehat{X}_i(t_n)}\right)^{2r}\right]\right)^{\frac{1}{2}} \left(\mathbb{E}\left[|X_i(t_n) - \widehat{X}_i(t_n)|^{2r}\right]\right)^{\frac{1}{2}}
\end{aligned}$$

$$\leq C \left(\mathbb{E}[e^{2rX_i(t_n)}] + \mathbb{E}[e^{2r\widehat{X}_i(t_n)}]\right)^{\frac{1}{2}} \left(\mathbb{E}\left[|X_i(t_n) - \widehat{X}_i(t_n)|^{2r}\right]\right)^{\frac{1}{2}}$$
$$\leq C\Delta_N^{r\lambda_i},$$

where C does not depend on the partition, which implies the required result. □

7.5 SVV Implied Volatility Surfaces

As discussed in Chap. 1, one of the accepted benchmarks for the quality of a financial model is the shape of the implied volatility generated by it. Therefore, our objective for the remaining part of the chapter is to examine implied volatility surfaces generated by the SVV model and convince our reader that we are able to replicate multiple real-life phenomena.

For simplicity, in this section we will stick to a constant interest rate $\nu = \text{const}$ and a slightly more specific SVV model of the form

$$S(t) = S(0) + \nu \int_0^t S(s)ds + \int_0^t Y(s)S(s) \left(\sqrt{1-\rho^2}dB_1(s) + \rho dB_2(s)\right),$$
$$Y(t) = Y(0) + \int_0^t b(s, Y(s))ds + \int_0^t \mathcal{G}(t, s)a(s, Y(s))ds + \int_0^t \mathcal{K}(t, s)dB_2(s), \tag{7.31}$$

where B_1, B_2 are independent Brownian motions, $\rho \in (-1, 1)$, $\mathcal{K}$ satisfies Assumption 6.1 and the coefficients in the volatility equation satisfy Assumption 6.3 so that

$$0 < \min_{t\in[0,T]} \varphi(t) < \varphi(t) < Y(t) < \psi(t) < \max_{t\in[0,T]} \psi(t) < \infty. \tag{7.32}$$

In this case, the discounted price process

$$\widetilde{S} = \{\widetilde{S}(t),\ t \in [0, T]\} = \{e^{-\nu t}S(t),\ t \in [0, T]\}$$

satisfies the SDE

$$\widetilde{S}(t) = S(0) + \int_0^t Y(s)\widetilde{S}(s) \left(\sqrt{1-\rho^2}dB_1(s) + \rho dB_2(s)\right),$$

and hence is a *martingale*. In other words, for the model (7.31), the physical measure $\mathbb{P}$ coincides with the martingale measure $\mathbb{Q}$. This means that, in the model (7.31), the non-arbitrage price Π of a European call-option with maturity T and payoff K is simply equal to

$$\Pi_{SVV}(T, K) = e^{-\nu T}\mathbb{E}\left[(S(T) - K)_+\right].$$

As announced in the beginning of the section, our main object of interest will be the implied volatility surface generated by the model (7.31), so let us briefly recall the necessary definition. Consider a standard Black-Scholes price of a European call-option with payoff K and maturity T at a given (constant) volatility level $\sigma > 0$:

$$\Pi_{BS}(T, K; \sigma) := S(0)\Phi\left(\frac{\log \frac{e^{\nu T}S(0)}{K} + \frac{\sigma^2}{2}T}{\sigma\sqrt{T}}\right) - Ke^{-\nu T}\Phi\left(\frac{\log \frac{e^{\nu T}S(0)}{K} - \frac{\sigma^2}{2}T}{\sigma\sqrt{T}}\right),$$

where Φ is the CDF of the standard normal distribution. The implied volatility surface generated by the SVV model is defined as a value $\widehat{\sigma}(T, K)$ such that

$$\Pi_{BS}(T, K; \widehat{\sigma}(T, K)) = \Pi_{SVV}(T, K).$$

It is easy to check that $\widehat{\sigma}(T, K)$ exists and is uniquely defined for all T, $K > 0$ since the function $\sigma \mapsto \Pi_{BS}(T, K; \sigma)$ is strictly increasing for all T, $K > 0$, and, moreover,

$$\begin{aligned}\lim_{\sigma\to 0+} \Pi_{BS}(T, K; \sigma) &= (S(0) - e^{-\nu T}K)_+ \\ &= (\mathbb{E}\left[e^{-\nu T}S(T)\right] - e^{-\nu T}K)_+ \\ &\le \mathbb{E}\left[(e^{-\nu T}S(T) - e^{-\nu T}K)_+\right] = \Pi_{SVV}(T, K).\end{aligned}$$

Remark 7.5 1. In what follows, we will frequently use a parametrization of the implied volatility surface in terms of *log-moneyness* $\kappa := \log \frac{K}{e^{\nu T}S(0)}$:

$$\widehat{\sigma}_{\text{log-m}}(T, \kappa) = \widehat{\sigma}\left(T, e^{\nu T+\kappa}S(0)\right).$$

It is a bit more convenient in the discussion of the implied volatility *at-the-money* when $T \to 0$: with the log-moneyness parametrization, we will be interested in the value of $\widehat{\sigma}_{\text{log-m}}$ at $\kappa = 0$.

2. In addition to the "strike" and "log-moneyness" parametrization of the implied volatility surface, a third approach can also be found in the literature. This parametrization expresses implied volatility as a function of the log-price for a given strike K:

$$\widehat{\sigma}_{\text{log-price}}(T, x) = \widehat{\sigma}_{\text{log-m}}\left(T, \log\frac{K}{e^{\nu T}} - x\right).$$

This representation, as seen for example in [1], is useful when discussing implied volatility skews.

The first observation about the shape of the SVV implied volatility surface stems directly from the boundedness (7.32) of the volatility process Y.

Theorem 7.8 *Let* $(T, K) \mapsto \widehat{\sigma}(T, K)$, $T, K > 0$, *be the implied volatility surface generated by the SVV model* (7.31). *Then, for all* $T, K > 0$,

$$\min_{t\in[0,T]} \varphi(t) \le \widehat{\sigma}(T, K) \le \max_{t\in[0,T]} \psi(t).$$

Proof By the seminal result [2] of Avellaneda, Levy and Parás, the bounds (7.32) imply that

$$\Pi_{BS}(T, K; \min_{t\in[0,T]} \varphi(t)) \le \Pi_{SVV}(T, K) \le \Pi_{BS}(T, K; \max_{t\in[0,T]} \psi(t)), \tag{7.33}$$

which immediately yields the claim. □

Remark 7.6 The bounds (7.33) on the implied volatility provide a hint about the appropriate choice of the upper bound function ψ in the SVV model (7.31) that is consistent with the market data. Namely, one should choose ψ such that $\max_{t\in[0,T]} \psi(t)$ exceeds the maximum value of the realized implied volatility observed on the market.

Under normal market conditions it is usually sufficient to take

$$\max_{t\in[0,T]} \psi(t) = 1.$$

The plan for the remaining part of this section is as follows: first, we will gain some intuition by looking at the simulated implied volatility surfaces. With this goal in mind, we can immediately see the value of the results presented in Sect. 7.4: after all, we will have to evaluate the SVV option prices $\Pi_{SVV}(T, K)$, so the numerical algorithms can be very handy for the Monte Carlo method. After that, we will delve into a more theoretical subject: the behavior of the implied volatility skew. We will rely on the connection between the implied volatility skew and Malliavin derivative established in [1] as well as our own Malliavin differentiability results obtained in the previous chapters.

7.5.1 *Simulation Study of SVV Implied Volatilitites*

To visualize the SVV implied volatility surfaces, let us fix the exact SVV model that will be simulated:

$$S(t) = 1 + \int_0^t Y(s)S(s)\left(\sqrt{0.75}dB_1(s) - 0.5dB_2(s)\right),$$

$$Y(t) = 0.5 + \int_0^t \left(\frac{0.005}{(Y(s)-0.05)^5} - \frac{0.005}{(1-Y(s))^5} + 0.05(0.5 - Y(s))\right) ds$$

$$+ 0.3 \int_0^t \mathcal{K}(t,s)dB_2(s), \tag{7.34}$$

where the choice of $\mathcal{K}$ will be different in different examples. To slightly simplify the computations, we also put the interest rate $\nu = 0$. To evaluate the implied volatility surface, we first need to estimate the SVV option price $\Pi_{SVV}(T, K)$ using the Monte Carlo method. To do this, we utilize the algorithm described in Sect. 7.4 and simulate $M = 1500000$ realizations $\widehat{S}^{(m)}(T)$, $m = 1, ..., M$, of the random variable $\widehat{S}(T)$. Then, we put

$$\widehat{\Pi}_{SVV}(T, K) := \frac{1}{M}\sum_{m=1}^{M} \widehat{S}^{(m)}(T),$$

and calculate the implied volatility $\widehat{\sigma}(T, K)$ by solving the equation

$$\Pi_{BS}(T, K; \sigma) = \widehat{\Pi}_{SVV}(T, K).$$

with respect to σ. Such a procedure is repeated for maturities $T_i = \frac{i}{200}$, $i = 1, ..., 200$, and strikes $K_j = 0.5 + j/100$, $j = 0, ..., 150$; as a result, we obtain the values of the implied volatility surface $(T, K) \mapsto \widehat{\sigma}(T, K)$ on the mesh (T_i, K_j), $i = 1, ..., 200$, $j = 0, ..., 150$.

Before we dive into simulations, let us outline the things we will be looking at and set what we expect to see.

- *The volatility smile*. While the *presence* of the smile in some form is guaranteed due to the stochastic volatility in the SVV model, it is interesting to look at the "curvature" of the smile at different maturities T. The literature on the topic (see e.g. [10, 11]) allows to make an educated guess about what we may see: a kernel $\mathcal{K}$ that has some "long memory" component—e.g. something similar to a fractional kernel with $H > 1/2$—will induce a surface with more pronounced smiles for larger values of T.
- *The volatility skew*. As explained in Chap. 1, the real-life implied volatility surface tends to become "steeper" at-the-money as $T \to 0$, with a rule-of-thumb behavior

$$\left|\frac{\widehat{\sigma}_{\text{emp}}(T,\kappa) - \widehat{\sigma}_{\text{emp}}(T,\kappa')}{\kappa - \kappa'}\right| \propto T^{-\frac{1}{2}+H}, \quad \kappa, \kappa' \approx 0, \quad H \in \left(0, \frac{1}{2}\right), \tag{7.35}$$

where $\widehat{\sigma}_{\text{emp}}$ denotes the implied volatility surface calculated from the empirical data. To reproduce this phenomenon within the SVV model, we want the SVV implied volatility $\widehat{\sigma}_{\text{log-m}}$ to have the property

$$\left| \frac{\partial \widehat{\sigma}_{\text{log-m}}}{\partial \kappa}(T, 0) \right| = O(T^{-\frac{1}{2}+H}), \quad T \to 0. \tag{7.36}$$

The derivative on the left-hand side of (7.36) is commonly referred to as the *implied volatility skew*, or more specifically, the *at-the-money implied volatility skew* (emphasizing that $\kappa = 0$), whereas the overall asymptotic behavior (7.36) is known as the *power law*. Based on the insights from the literature on rough volatility, we can expect that the kernels $\mathcal{K}$ which behave similarly to a fractional kernel with $H < 1/2$ yield (7.36). Right now, our goal is to see such a behavior visually; the theoretical background for it will be explored in the next section.

Example 7.1 As the first example, consider the SVV model (7.34) with a rough fractional kernel

$$\mathcal{K}(t, s) := \frac{1}{\Gamma(0.7)}(t - s)^{-0.3} \mathbb{1}_{s<t}. \tag{7.37}$$

The corresponding SVV implied volatility surface is depicted on Fig. 7.1. Figure 7.1(b) visually demonstrates that the absolute value of the at-the-money implied volatility skew

$$\left| \frac{\partial \widehat{\sigma}_{\text{log-m}}}{\partial \kappa}(T, 0) \right| \tag{7.38}$$

increases as $T \to 0$. Figure 7.2 contains the variation of (7.38) with respect to T on the standard (Fig. 7.2(a)) and logarithmic (Fig. 7.2(b)) scales and shows that (7.38) indeed seems to follow the power law. Again, we will leave the theoretical proof of this for the next section.

As a final remark, observe that the points on Fig. 7.2(b) are not located on a single line; in fact, the linear fit performed over the short maturities substantially differs from the fit over long maturities (see Fig. 7.3). Such behavior agrees with the recent empirical study [6] reporting the same phenomenon for SPX implied volatility skews (c.f. [6, Fig. 2]).

Example 7.2 The pure rough fractional kernel gives the power law behavior of the skew for short maturities, but Fig. 7.1(b) shows that the smiles flatten out rather quickly. Let us now modify the rough kernel with an extra “long-memory” component, i.e. put

$$\mathcal{K}(t, s) = \left(\frac{\sqrt{0.4}}{\Gamma(0.7)}(t - s)^{-0.3} + \frac{\sqrt{1.8}}{\Gamma(1.4)}(t - s)^{0.4} \right) \mathbb{1}_{s<t}. \tag{7.39}$$

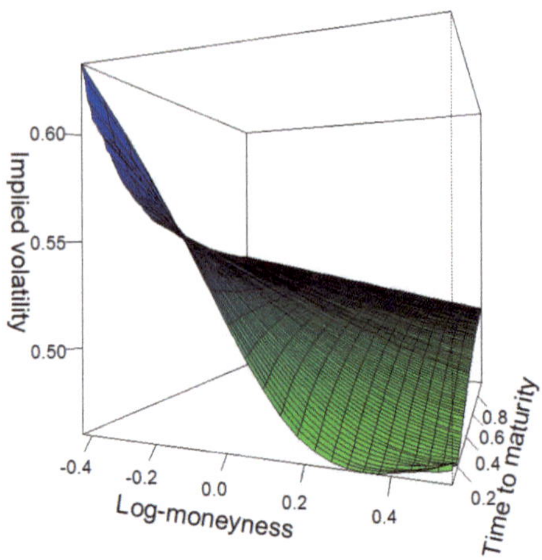

(a) Implied volatility surface

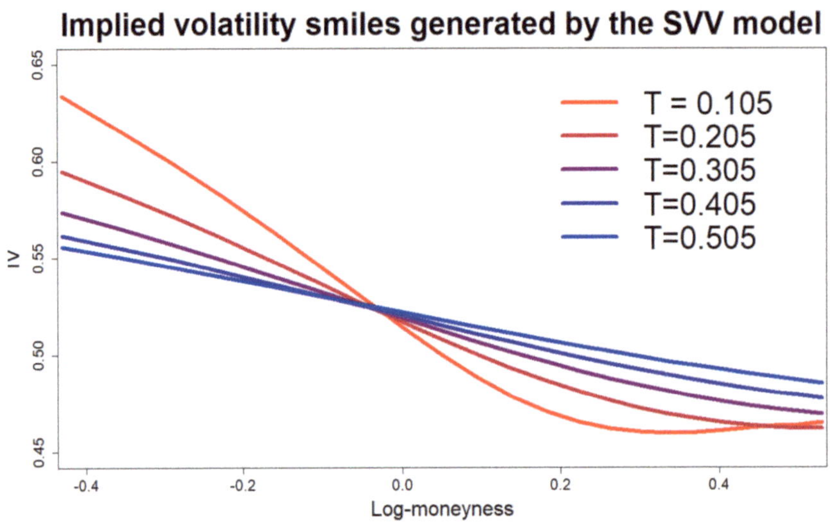

(b) Implied volatility smiles

Fig. 7.1 Implied volatility surface (**a**) and implied volatility smiles (**b**) generated by the SVV model (7.34) with the rough kernel (7.37). Note that the smile becomes steeper as the time to maturity $T \to 0$, which reproduces a similar effect happening on real markets (for more details, see e.g. [6, 9]

The corresponding implied volatility surface and implied volatility smiles are presented on Fig. 7.4. Comparing the results with the purely rough case from Example 7.1, we note that the presence of a "long memory" component $(t - s)^{0.4}$ in

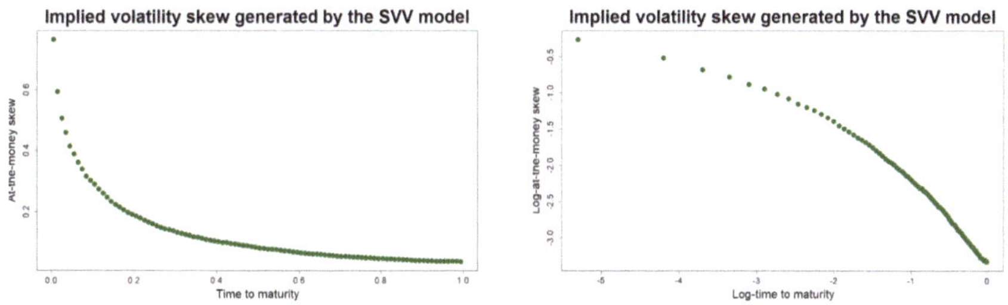

(a) Standard scale (b) Logarithmic scale

Fig. 7.2 Absolute value of the at-the-money implied volatility skew (7.38) generated by the SVV model (7.34) with the rough kernel (7.37) on standard (**a**) and logarithmic (**b**) scales

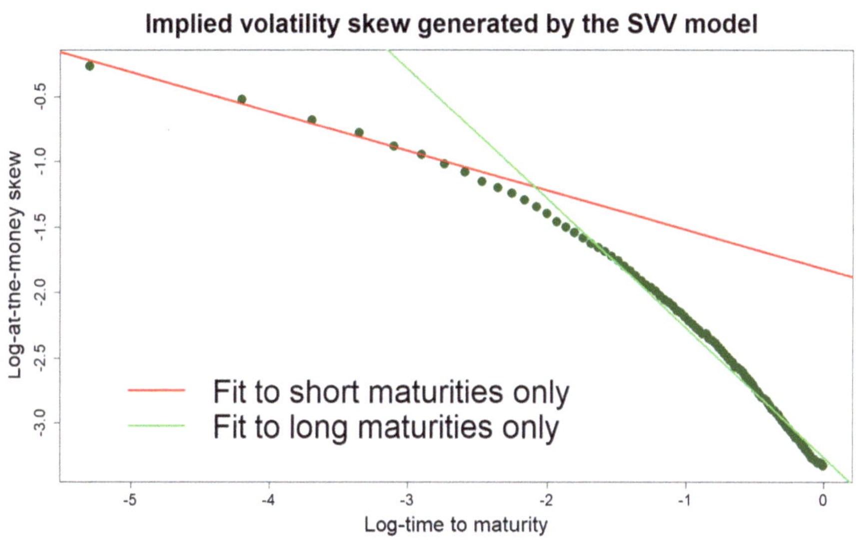

Fig. 7.3 Absolute value of the at-the-money implied volatility skew (7.38) generated by the SVV model (7.34) with the rough kernel (7.37) on a logarithmic scale. The lines depict two linear fits: over the short maturities (red line, the slope is -0.3006433) and over the long maturities (green line, the slope is -0.9920939). Such a behavior is consistent with the recent paper [6]

(7.39) keeps the smiles "less flat" and more pronounced for larger maturities, which is consistent with the results of [10, 11].

In the same time, just like in the purely rough case from Example 7.1, the explosive behavior of the skew as $T \to 0$ is preserved, see Fig. 7.5. In other words, the mixed kernel indeed allows to simultaneously keep the power law of the at-the-money skew as $T \to 0$, and increase the curvature of the smile for larger values of T.

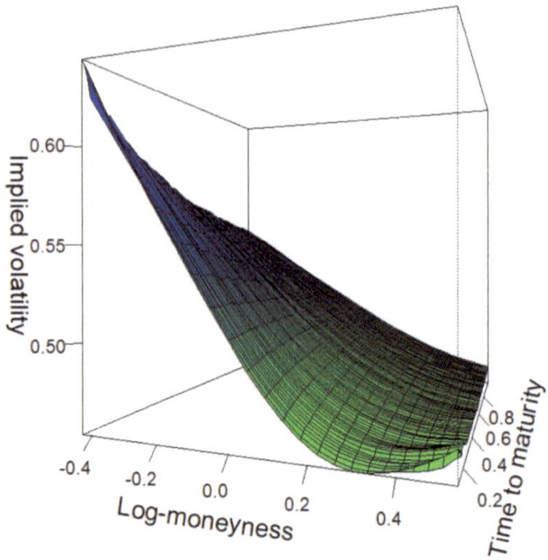

(a) Implied volatility surface

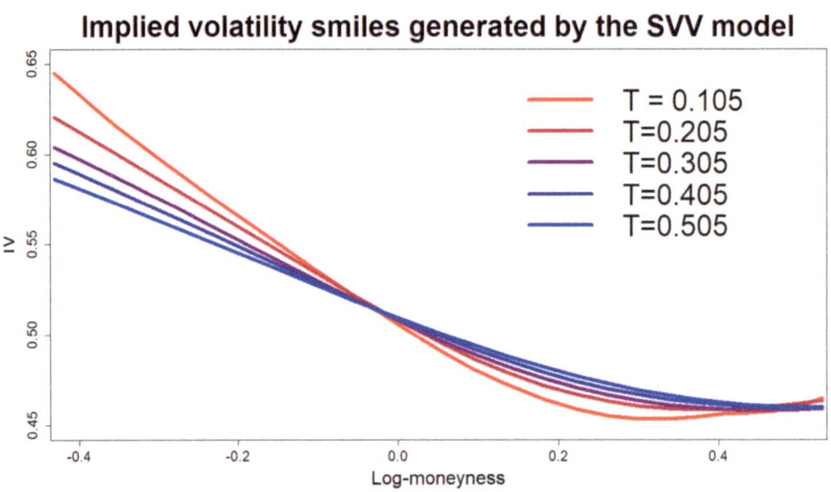

(b) Implied volatility smiles

Fig. 7.4 Implied volatility surface (**a**) and implied volatility smiles (**b**) generated by the SVV model (7.34) with the mixed fractional kernel (7.39). Just as in the purely rough case, the absolute value of the at-the-money implied volatility skew increases for small maturities, which is consistent with the real-life market behavior. Note that the presence of the term $(t-s)^{0.4}$ in the kernel ensures that the smile "flattens out" for larger maturities much slower than in the purely rough case, which agrees with the results of [10, 11]

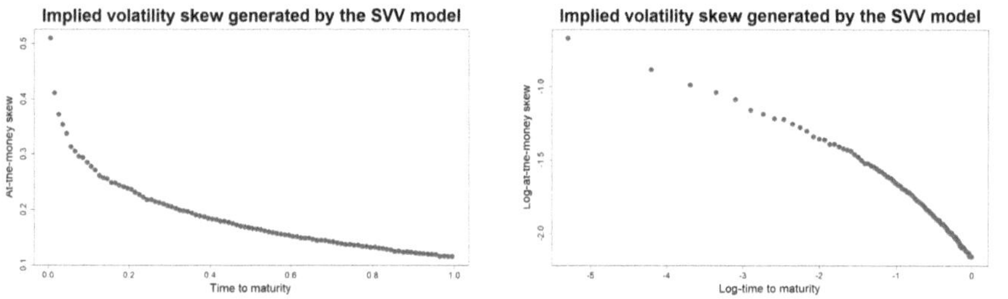

(a) Standard scale (b) Logarithmic scale

Fig. 7.5 Absolute at-the-money implied volatility skew (7.38) generated by the SVV model (7.34) with the mixed kernel (7.39) on standard (**a**) and logarithmic (**b**) scales

7.5.2 *Implied Volatility Skew*

Both Examples 7.1 and 7.2 demonstrate that the SVV model seems to be capable of replicating the power law behavior of at-the-money implied volatility skews:

$$\left|\frac{\partial\widehat{\sigma}_{\text{log-m}}}{\partial\kappa}(T,0)\right| = O(T^{-\frac{1}{2}+H}), \quad T\to 0. \tag{7.40}$$

Let us make this claim more substantial and provide a theoretical result that *guarantees* the power law. Luckily, having the second-order Malliavin differentiability in place, we now possess all the necessary tools to analyze the behavior of implied volatility skew of the SVV model. Let us assume for simplicity that $\mathcal{G}\equiv 0$ and consider a simplified SVV market model with the price process $S=\{S(t),\ t\in[0,T]\}$ of the form

$$\begin{aligned}
S(t) &= e^{X(t)},\\
X(t) &= x_0 + \nu t - \frac{1}{2}\int_0^t Y^2(s)ds + \int_0^t Y(s)\left(\rho dB_1(s) + \sqrt{1-\rho^2}dB_2(s)\right),\\
Y(t) &= y_0 + \int_0^t b(s,Y(s))ds + \int_0^t \mathcal{K}(t,s)dB_1(s),
\end{aligned} \tag{7.41}$$

where B_1, B_2 are two independent Brownian motions, $X=\{X(t),\ t\in[0,T]\}$ denotes the (risk-free) log-price of an asset starting from some level $x_0\in\mathbb{R}$, ν is a constant instantaneous interest rate, and $\rho\in(-1,1)$ is a correlation coefficient that accounts for the *leverage* effect. Throughout this section, we also assume that the drift function b satisfies Assumption 6.4, in addition to the running Assumptions 6.1 and 6.3. Our main tool will be the fundamental result [1, Theorem 6.3] which connects the shape of the skew with the Malliavin derivative of the volatility. We provide a slightly adjusted version of it below.

Theorem 7.9 *Consider a risk-free log-price*

$$X(t) = x_0 + rt - \frac{1}{2}\int_0^t \sigma^2(s)ds + \int_0^t \sigma(s)\left(\rho dB_1(s) + \sqrt{1-\rho^2}dB_2(s)\right), \tag{7.42}$$

where B_1, B_2 are two independent Brownian motions, $x_0 \in \mathbb{R}$ is a deterministic initial value, r is an instantaneous interest rate, $\rho \in (-1, 1)$ is a correlation coefficient and $\sigma = \{\sigma(t),\ t \in [0, T]\}$ is some square-integrable stochastic process with right-continuous trajectories adapted to the filtration $\mathcal{F} = \{\mathcal{F}_t,\ t \in [0, T]\}$ generated by B_1.

Assume that

(H1) $\sigma \in \mathbb{L}^{2,4} := L^4([0, T]; \mathbb{D}^{2,4})$ with respect to B_1;
(H2) there exists a constant $\varphi_* > 0$ such that, with probability 1, $\sigma(t) > \varphi_*$ for all $t \in [0, T]$;
(H3) there exists a constant $H \in \left(0, \frac{1}{2}\right)$ such that, with probability 1, for any $0 < s < t < T$,

$$\mathbb{E}\left[(D_s\sigma(t))^2\right] \le \frac{C}{(t-s)^{1-2H}}, \tag{7.43}$$

$$\mathbb{E}\left[(D_rD_s\sigma(t))^2\right] \le C\left(\frac{t-r}{t-s}\right)^{1-2H}, \tag{7.44}$$

where $C > 0$ is some constant;
(H4) σ has a.s. right-continuous trajectories;
(H5) $\sup_{r,s,t\in[0,\tau]} \mathbb{E}\left[(\sigma(s)\sigma(t) - \sigma^2(r))^2\right] \to 0$ when $\tau \to 0+$.

Finally, assume that there exists a constant $K_\sigma > 0$ such that, with probability 1,

$$\frac{1}{\tau^{\frac{3}{2}+H}}\int_0^\tau\int_s^\tau \mathbb{E}\left[D_s\sigma(t)\right]dtds - K_\sigma \to 0, \quad \tau \to 0+. \tag{7.45}$$

Then, with probability 1,

$$\lim_{\tau\to 0} \tau^{\frac{1}{2}-H}\frac{\partial\widehat{\sigma}_{\text{log-price}}}{\partial x}(\tau, x)\bigg|_{x=\log\frac{K}{e^{r\tau}}} = -\frac{\rho}{\sigma(0)}K_\sigma.$$

Remark 7.7 Following the notation in the original paper [1], we formulated Theorem 7.9 using the "log-price" parametrization of the implied volatility (see Remark 7.5 above). With this parametrization,

$$\frac{\partial \widehat{\sigma}_{\text{log-price}}(T, x)}{\partial x} = -\frac{\partial \widehat{\sigma}_{\text{log-m}}\left(T, \log \frac{K}{e^{\nu T}} - x\right)}{\partial \kappa},$$

and the power law (7.40) is equivalent to

$$\left|\frac{\partial \widehat{\sigma}_{\text{log-price}}}{\partial x}(T, x)\right|_{x=\log \frac{K}{e^{rT}}} = O(T^{-\frac{1}{2}+H}), \quad T \to 0.$$

Remark 7.8 The original formulation of [1, Theorem 6.3] is slightly more general than Theorem 7.9 above in the sense that

(1) in [1, Theorem 6.3], the log-price X is allowed to have jumps;
(2) the result in [1] is formulated for the *future* implied volatility surfaces $\widehat{\sigma}_{\text{log-price}}(t_0, T, X(t_0))$, $t_0 \ge 0$.

Since we are interested in the continuous model (7.41), we removed the jump component in (7.42) and, for the simplicity of notation, we put $t_0 = 0$.

Our ultimate aim in this subsection is to prove the following result.

Theorem 7.10 *Let Assumptions 6.1, 6.3 and 6.4 hold with $H \in \left(\frac{1}{6}, \frac{1}{2}\right)$. Assume also that $\rho \neq 0$ in* (7.41) *and that the Volterra kernel $\mathcal{K}$ is such that for any $0 \le s < t \le T$*

$$|\mathcal{K}(t, s)| \le C|t - s|^{-\frac{1}{2}+H}$$

for some constant $C > 0$ and

$$\frac{1}{\tau^{\frac{3}{2}+H}} \int_0^\tau \int_s^\tau \mathcal{K}(t, s)dtds \to K_Y, \quad \tau \to 0+, \tag{7.46}$$

for some finite constant K_Y. Then, with probability 1, the SVV implied volatility $\widehat{\sigma}$ produced by (7.41) *exhibits the property*

$$\lim_{\tau\to 0} \tau^{\frac{1}{2}-H} \frac{\partial \widehat{\sigma}_{log\text{-}m}}{\partial \kappa}(\tau, \kappa) = -\lim_{\tau\to 0} \tau^{\frac{1}{2}-H} \frac{\partial \widehat{\sigma}_{log\text{-}price}}{\partial x}(\tau, x)\Bigg|_{x=\log \frac{K}{e^{\nu\tau}}} = \frac{\rho}{y_0} K_Y.$$

In particular, if $\rho K_Y \neq 0$, the SVV model (7.41) *reproduces the power law* (7.40) *of the at-the-money implied volatility skew.*

Remark 7.9 Let us briefly take a step back and take a look at certain aspects of Theorem 7.10. First, let us discuss the condition $H > \frac{1}{6}$. At the first glance, it may seem a bit restrictive, but it is, in fact, coherent with the recent empirical estimate $H \approx 0.19$ for the SPX implied volatility obtained in [6]. Second, as we have seen in Chap. 1, the behavior of empirically observed implied volatilities shows that realistic market models should produce $\widehat{\sigma}_{\text{log-m}}$ with

$$\frac{\partial \widehat{\sigma}_{\text{log-m}}}{\partial \kappa}(\tau, \kappa)\bigg|_{\kappa=0} < 0. \tag{7.47}$$

In the SVV setting (7.41), Theorem 7.10 guarantees that (7.47) holds for all small enough maturities τ provided that $\rho K_Y < 0$.

As noted above, our main tool to prove Theorem 7.10 will be Theorem 7.9 and, in fact, the SVV model (7.41) automatically satisfies a number of assumptions of Theorem 7.9. Indeed, (H2) holds with $\varphi^* := \min_{t\in[0,T]} \varphi(t) > 0$ due to the construction of the volatility process. The a.s. continuity of Y, in turn, implies (H4). Finally, the results of Chap. 6, Corollary 6.2 in particular, immediately yield (H1). In addition, by the continuity of Y,

$$\sup_{r,s,t\in[0,\tau]} (Y(s)Y(t) - Y^2(r))^2 \to 0 \quad a.s., \quad \tau \to 0,$$

and, moreover, for all $\tau < T$:

$$\sup_{r,s,t\in[0,\tau]} (Y(s)Y(t) - Y^2(r))^2 \le (\max_{t\in[0,T]} \psi^2(t) - \min_{t\in[0,T]} \varphi^2(t))^2.$$

Therefore, the condition (H5) simply follows from the dominated convergence theorem.

Our next step is to handle (7.45).

Proposition 7.3 *Let Assumptions 6.1, 6.3 and 6.4 hold and the Volterra kernel $\mathcal{K}$ satisfy* (7.46) *for some finite constant K_Y. Then, with probability 1,*

$$\frac{1}{\tau^{\frac{3}{2}+H}} \int_0^\tau \int_s^\tau \mathbb{E}[D_s Y(t)]\, dt ds - K_Y \to 0, \quad \tau \to 0+.$$

Proof Denote

$$F_1(t, u) := b'_y(u, Y(u)) \exp\left\{ \int_u^t b'_y(v, Y(v)) dv \right\},$$

and recall that, by Proposition 6.3,

$$|F_1(t, u)| \le e^{C_b T} \xi, \tag{7.48}$$

where C_b is from Assumption 6.3(**b5**). Then, using the explicit shape of $D_s Y(t)$ given in Theorem 7.1, we can write

$$\frac{1}{\tau^{\frac{3}{2}+H}} \int_0^\tau \int_s^\tau \mathbb{E}\left[D_s Y(t) \right] dt ds$$

$$= \frac{1}{\tau^{\frac{3}{2}+H}} \int_0^\tau \int_s^\tau \mathcal{K}(t, s) dt ds$$

$$+ \frac{1}{\tau^{\frac{3}{2}+H}} \int_0^\tau \int_s^\tau \int_s^t \mathcal{K}(u, s) \mathbb{E}\left[F_1(t, u) \right] du dt ds$$

$$= \frac{1}{\tau^{\frac{3}{2}+H}} \int_0^\tau \int_s^\tau \mathcal{K}(t, s) dt ds$$

$$+ \frac{1}{\tau^{\frac{3}{2}+H}} \int_0^\tau \int_s^\tau \mathcal{K}(u, s) \left(\int_u^\tau \mathbb{E}\left[F_1(t, u) \right] dt \right) du ds.$$

The term $\frac{1}{\tau^{\frac{3}{2}+H}} \int_0^\tau \int_s^\tau \mathcal{K}(t, s) dt ds$ converges to K_Y by (7.46). As for the second term, it should be noted that with probability 1 for any $u \in [0, T]$,

$$\int_u^T \left| \mathbb{E}\left[F_1(t, u) \right] \right| dt \le C \mathbb{E}[\xi] T$$

and hence, given (7.46), we get that with probability 1,

$$\frac{1}{\tau^{\frac{3}{2}+H}} \int_0^\tau \int_s^\tau \mathcal{K}(u, s) \left(\int_u^T \mathbb{E}\left[F_1(t, u) \right] dt \right) du ds \to 0, \quad \tau \to 0+,$$

which completes the proof. □

Finally, let us deal with (H3).

Proposition 7.4 *Let Assumptions 6.1, 6.3 and 6.4 hold with $H \in \left(\frac{1}{6}, \frac{1}{2}\right)$ and the Volterra kernel $\mathcal{K}$ be such that for any $0 \le s < t \le T$*

$$|\mathcal{K}(t, s)| \le C|t - s|^{-\frac{1}{2}+H} \tag{7.49}$$

for some constant $C > 0$. Then the hypothesis (H3) from Theorem 7.9 holds for the sandwiched volatility process $\sigma = Y$.

Proof Fix $0 < r, s < t$ and recall that, by Theorem 6.4,

$$D_r D_s Y(t) = \int_s^t \mathcal{K}(u,s) F_1(t,u) \left(\int_u^t b''_{yy}(v, Y(v)) D_r Y(v) dv \right) du$$
$$+ \int_s^t \mathcal{K}(u,s) F_2(t,u) D_r Y(u) du,$$

where

$$F_1(t,u) := b'_y(u, Y(u)) \exp\left\{ \int_u^t b'_y(v, Y(v)) dv \right\},$$
$$F_2(t,u) := b''_{yy}(u, Y(u)) \exp\left\{ \int_u^t b'_y(v, Y(v)) dv \right\}.$$

Then, taking into account (7.48), with probability 1,

$$\begin{aligned} |D_s Y(t)| &\le |\mathcal{K}(t,s)| + \int_s^t |\mathcal{K}(u,s)||F_1(t,u)| du \\ &\le C \left(|t-s|^{-\frac{1}{2}+H} + \xi \int_s^t |u-s|^{-\frac{1}{2}+H} du \right) \\ &\le C(1+T\xi)|t-s|^{-\frac{1}{2}+H} \\ &=: \zeta |t-s|^{-\frac{1}{2}+H}, \end{aligned} \tag{7.50}$$

which immediately implies (7.43). Next, by Proposition 6.3,

$$|b''_{yy}(v, Y(v))| \le \xi$$

for any $v \in [0,T]$ and, for any $0 \le u \le t \le T$,

$$\begin{aligned} |F_2(t,u)| &= \left| b''_{yy}(u, Y(u)) \exp\left\{ \int_u^t b'_y(v, Y(v)) dv \right\} \right| \\ &\le e^{C_b T} \xi \end{aligned}$$

with C_b taken from Assumption 6.3(**b5**). So, we can write

$$
\begin{aligned}
|D_r D_s Y(t)| &\le \int_s^t |\mathcal{K}(u,s)||F_1(t,u)| \left(\int_u^t |b''_{yy}(v, Y(v))||D_r Y(v)| dv \right) du \\
&\quad + \int_s^t |\mathcal{K}(u,s)||F_2(t,u)||D_r Y(u)| du \\
&\le C \left(\xi^2 \int_s^t |\mathcal{K}(u,s)| \left(\int_u^t |D_r Y(v)| dv \right) du + \xi \int_s^t |\mathcal{K}(u,s)||D_r Y(u)| du \right) \\
&= C \left(\xi^2 \int_s^t |\mathcal{K}(u,s)| \left(\int_{u \vee r}^t |D_r Y(v)| dv \right) du + \xi \int_{r \vee s}^t |\mathcal{K}(u,s)||D_r Y(u)| du \right).
\end{aligned}
$$

Taking into account (7.49) and (7.50), we proceed with the following inequalities:

$$
\begin{aligned}
|D_r D_s Y(t)| &\le C \Bigg(\xi^2 \zeta \int_s^t |u-s|^{-\frac{1}{2}+H} \left(\int_{u \vee r}^t |v-r|^{-\frac{1}{2}+H} dv \right) du \\
&\quad + \xi \zeta \int_{r \vee s}^t |u-s|^{-\frac{1}{2}+H} |u-r|^{-\frac{1}{2}+H} du \Bigg) \\
&\le C \Bigg(\xi^2 \zeta \int_s^t |u-s|^{-\frac{1}{2}+H} |t-r|^{\frac{1}{2}+H} du \\
&\quad + \xi \zeta \int_{r \vee s}^t |u-s|^{-\frac{1}{2}+H} |u-r|^{-\frac{1}{2}+H} du \Bigg).
\end{aligned}
$$

Note that

$$
\begin{aligned}
\int_s^t |u-s|^{-\frac{1}{2}+H} |t-r|^{\frac{1}{2}+H} du &\le C |t-r|^{\frac{1}{2}+H} |t-s|^{\frac{1}{2}+H} \\
&\le C \left(\frac{t-r}{t-s} \right)^{\frac{1}{2}-H}.
\end{aligned}
$$

As for the integral $\int_{r \vee s}^t |u-s|^{-\frac{1}{2}+H} |u-r|^{-\frac{1}{2}+H} du$, we have two cases:

- if $0 < r \le s < t$, we can write

$$\int_s^t |u-s|^{-\frac{1}{2}+H}|u-r|^{-\frac{1}{2}+H}du \le \int_s^t |u-s|^{-1+2H}du$$
$$\le C(t-s)^{2H} \le C\left(\frac{t-r}{t-s}\right)^{\frac{1}{2}-H};$$

- similarly, if $0 < s < r < t$ and given that $H > \frac{1}{6}$, we have

$$\int_r^t |u-s|^{-\frac{1}{2}+H}|u-r|^{-\frac{1}{2}+H}du \le \int_r^t |u-r|^{-1+2H}du$$
$$\le C(t-r)^{2H} \le C\left(\frac{t-r}{t-s}\right)^{\frac{1}{2}-H}.$$

In any case,

$$|D_r D_s Y(t)| \le C\xi\zeta(\xi+1)\left(\frac{t-r}{t-s}\right)^{\frac{1}{2}-H},$$

where ξ and ζ are random variables having all moments, and hence (7.44) holds. □

Taking into account all the above results, we are ready to prove the main result of this section, Theorem 7.10.

Proof of Theorem 7.10 The results above show that the SVV model satisfies conditions (H1)–(H5) of Theorem 7.9. Therefore, Theorem 7.10 follows immediately from Theorem 7.9.

Example 7.3 Let $\frac{1}{6} < H_0 < H_1 < \ldots < H_n < 1$ be such that $H_0 < \frac{1}{2}$ and $\alpha_k > 0$, $k = 0, \ldots, n$. Then the kernel

$$\mathcal{K}(t,s) = \left(\sum_{k=0}^{n} \alpha_k (t-s)^{H_k - \frac{1}{2}}\right) \mathbb{1}_{s<t}$$

satisfies the assumptions of Theorem 7.10, and so the corresponding SVV model generates power law (7.40) with $H = H_0$ provided that $\rho < 0$ in (7.41).

References

1. Alòs, E., León, J.A., Vives, J.: On the short-time behavior of the implied volatility for jump-diffusion models with stochastic volatility. Finance Stochast. **11**(4), 571–589 (2007). https://doi.org/10.1007/s00780-007-0049-1
2. Avellaneda, M., Levy, A., Parás, A.: Pricing and hedging derivative securities in markets with uncertain volatilities. Appl. Math. Financ. **2**(2), 73–88 (1995). https://doi.org/10.1080/13504869500000005
3. Biagini, F., Guasoni, P., Pratelli, M.: Mean-variance hedging for stochastic volatility models. Math. Financ. **10**(2), 109–123 (2000). https://doi.org/10.1111/1467-9965.00084
4. Corlay, S., Lebovits, J., Lévy Véhel, J.: Multifractional stochastic volatility models. Math. Finance. Int. J. Math., Stat. Financ. Econ. **24**(2), 364–402 (2014). https://doi.org/10.1111/mafi.12024
5. Delbaen, F., Schachermayer, W.: The variance-optimal martingale measure for continuous processes. Bernoulli **2**(1), 81 (1996). https://doi.org/10.2307/3318570
6. Delemotte, J., De Marco, S., Segonne, F.: Yet another analysis of the SP500 at-the-money skew: Crossover of different power-law behaviours. SSRN Electron. J. (2023). https://doi.org/10.2139/ssrn.4428407
7. Di Nunno, G., Mishura, Y., Yurchenko-Tytarenko, A.: Option pricing in sandwiched Volterra volatility model. SIAM J. Financ. Math. **15**(3), 824–882 (2024). https://doi.org/10.1137/22m1521328
8. Di Nunno, G., Yurchenko-Tytarenko, A.: Power law in sandwiched Volterra volatility model. Mod. Stoch. Theory Appl. **11**(2), 169–194 (2024). https://doi.org/10.15559/24-vmsta246
9. Fouque, J.P., Papanicolaou, G., Sircar, R., Sølna, K.: Maturity cycles in implied volatility. Financ. Stoch. **8**(4) (2004). https://doi.org/10.1007/s00780-004-0126-7
10. Funahashi, H., Kijima, M.: Does the Hurst index matter for option prices under fractional volatility? Ann. Finance **13**(1), 55–74 (2017). https://doi.org/10.1007/s10436-016-0289-1
11. Funahashi, H., Kijima, M.: A solution to the time-scale fractional puzzle in the implied volatility. Fractal Fract. **1**(1), 14 (2017). https://doi.org/10.3390/fractalfract1010014
12. Hytönen, T., van Neerven, J., Veraar, M., Weis, L.: Analysis in Banach spaces. Volume I: Martingales and Littlewood-Paley theory, 1 edn. Springer International Publishing, Cham, Switzerland (2016)
13. Nualart, D.: The Malliavin Calculus and Related Topics. Springer-Verlag, Berlin/Heidelberg (2006)
14. Schweizer, M.: Martingale densities for general asset prices. J. Math. Econ. **21**(4), 363–378 (1992). https://doi.org/10.1016/0304-4068(92)90014-X. https://www.sciencedirect.com/science/article/pii/030440689290014X
15. Schweizer, M.: On the minimal martingale measure and the Föllmer-Schweizer decomposition. Stoch. Anal. Appl. **13**(5), 573–599 (1995). https://doi.org/10.1080/07362999508809418
16. Schweizer, M.: A guided tour through quadratic hedging approaches. In: E. Jouini, J. Cvitanic, Musiela (eds.) Handbooks in Mathematical Finance: Option Pricing, Interest Rates and Risk Management, pp. 538–574. Cambridge University Press (2001). https://doi.org/10.1017/CBO9780511569708.016
17. Vrabie, I.I.: C_0-Semigroups and Applications. JAI Press, North-Holland Mathematics Studies (2003)
18. Černý, A., Kallsen, J.: On the structure of general mean-variance hedging strategies. Ann. Probab. **35**(4), 1479–1531 (2007). https://doi.org/10.1214/009117906000000872

Chapter 8
Numerical Techniques for Option Pricing in the SVV Model

Abstract This chapter based on the material originally published in [4] develops numerical techniques for option pricing within the Sandwiched Volterra Volatility (SVV) framework. Using the stochastic differentiability of the SVV model and the Malliavin integration by parts formula, we introduce a pricing algorithm that is particularly well-suited for options with discontinuous payoffs. The results are illustrated by simulations.

8.1 Introduction

As we have seen in Chap. 7, the SVV model successfully captures several complex behaviors of the implied volatility observed in real-life market and, additionally, it yields some good integrability properties for the resulting price processes. However, these benefits come at a cost: the dynamics of the volatility process is quite complex and hence it is very difficult to obtain closed-form expressions for derivative prices $\mathbb{E}_{\mathbb{Q}}[f(S(T))]$ (here, f denotes the payoff function of the derivative contract and $\mathbb{Q}$ is a martingale measure).

In Sect. 7.5, we have already introduced a potential approach to this challenge. Specifically, we can use the discretization scheme described in Sect. 7.4 to simulate the drift-implicit Euler approximations $\widehat{S}(T)$ of the asset prices $S(T)$, and then apply the Monte Carlo method to estimate $\mathbb{E}_{\mathbb{Q}}[f(\widehat{S}(T))]$. However, it is important to recognize that this estimation involves *two* distinct sources of error:

- (I) the error between the "theoretical" option price $\mathbb{E}_{\mathbb{Q}}[f(S(T))]$ and its "discretized" counterpart $\mathbb{E}_{\mathbb{Q}}[f(\widehat{S}(T))]$ which arises from the difference between $S(T)$ and $\widehat{S}(T)$;
- (II) the Monte Carlo error coming from approximating the expectation $\mathbb{E}_{\mathbb{Q}}[f(\widehat{S}(T))]$ using a finite number of simulations.

Naturally, the error of type (I) can be reduced by refining the time partition used in simulating $\widehat{S}(T)$; after all, the closer $\widehat{S}(T)$ is to $S(T)$, the smaller this error becomes. However, there is another critical factor that can significantly exacerbate the situation: the nature of the payoff function f.

G. Di Nunno et al., *Volterra Volatility Models*, Springer Finance,
https://doi.org/10.1007/978-3-032-26576-0_8

If f is "good", there is not too much to worry about: for example, if

$$|f(x) - f(y)| \le L_f |x - y|, \quad x, y \in \mathbb{R},$$

one can write

$$|\mathbb{E}_{\mathbb{Q}}[f(S(T))] - \mathbb{E}_{\mathbb{Q}}[f(\widehat{S}(T))]| \le \mathbb{E}_{\mathbb{Q}}[|f(S(T)) - f(\widehat{S}(T))|] \le L_f \mathbb{E}_{\mathbb{Q}}[|S(T) - \widehat{S}(T)|],$$

i.e. the rate of convergence of $\mathbb{E}_{\mathbb{Q}}[f(\widehat{S}(T))]$ to $\mathbb{E}_{\mathbb{Q}}[f(S(T))]$ is the same as the one of $\widehat{S}(T)$. However, what if f happens to be discontinuous, as in the case of digital option payoffs? The answer to this question is rather grim: according to [2], any discontinuities in f lead to a *substantial* deterioration in convergence speed. For example, if

$$\mathbb{E}_{\mathbb{Q}}[|S(T) - \widehat{S}(T)|^2] \le C\Delta_N^{2\lambda},$$

where Δ_N is the diameter of the partition, we can only guarantee that

$$\mathbb{E}[|f(S(T)) - f(\widehat{S}(T))|^2] \le C\Delta_N^{\lambda}.$$

This is particularly problematic because Euler-type schemes for stochastic processes are already relatively slow, especially when the volatility process is driven by a rough noise. In other words, a further reduction in convergence due to the "poor" regularity of f is a price we should strive to avoid.

Luckily, there exists an elegant algorithm, originally developed in [1] and further expanded in [3, 4], that allows to preserve the order of convergence, even if the payoff f is discontinuous. The aim of this chapter is to explore this method in the context of the SVV model and to provide several practical techniques for improving convergence even further.

8.2 Smoothing Payoff Functionals by Malliavin Integration by Parts

8.2.1 One-Dimensional Case

Let us start from a simpler one-dimensional case. Let φ and ψ be H-Hölder continuous functions for some $H \in (0, 1)$ and satisfy the condition

$$0 < \varphi(t) < \psi(t), \quad t \in [0, T].$$

Consider the SVV-model of the form

$$\begin{aligned} S(t) &= S(0) + \int_0^t \mu(s)S(s)ds + \int_0^t Y(s)S(s)dW_1(s), \\ Y(t) &= Y(0) + \int_0^t b(s, Y(s))ds + \int_0^t \mathcal{K}(t,s)\left(\ell_{2,1}dW_1(s) + \ell_{2,2}dW_2(s)\right), \end{aligned} \tag{8.1}$$

where W_1, W_2 are independent Brownian motions, $\mathcal{K}$ satisfies Assumption 6.1 with order of regularity H, b satisfies Assumption 6.3 with φ, ψ serving as the corresponding bounds, μ is an arbitrary H-Hölder continuous function and the lower-triangular matrix

$$\mathcal{L} = \begin{pmatrix} \ell_{1,1} & \ell_{1,2} \\ \ell_{2,1} & \ell_{2,2} \end{pmatrix} = \begin{pmatrix} 1 & 0 \\ \ell_{2,1} & \ell_{2,2} \end{pmatrix},$$

which defines the leverage effect between the price and the volatility (see Sect. 7.2), is such that $\ell_{2,1}^2 + \ell_{2,2}^2 = 1$. In addition, throughout this chapter we will always assume that

$$\sup_{s\in[0,T]} \int_0^T \mathcal{K}^2(u,s)du < \infty. \tag{8.2}$$

Remark 8.1 Note that we take $\varphi > 0$, so the volatility process Y satisfies

$$0 < \varphi(t) < Y(t) < \psi(t), \quad t \in [0,T] \ \ a.s.$$

In particular, Y is bounded away from zero with probability 1.

Next, let f be a measurable payoff function that satisfies the following assumption.

Assumption 8.1 Function $f\colon \mathbb{R} \to \mathbb{R}$ is locally Riemann integrable function (not necessarily continuous) of polynomial growth, i.e. there exist constants $q > 0$ and $c_f > 0$ such that

$$|f(s)| \le c_f(1 + |s|^q), \quad s > 0.$$

Notation 8.1 We will also denote

$$F(s) := \int_0^s f(u)du, \quad g(x) := f(e^x), \quad G(x) := F(1) + \int_0^x g(v)dv,$$

i.e. $f(S(T)) = g(X(T))$, where $X(t) := \log S(t)$, $t \in [0,T]$.

Let us now introduce another useful parametrization of the SVV model (8.1). Since $\ell_{2,1}^2 + \ell_{2,2}^2 = 1$, observe that the process

$$\widetilde{W}_2(t) = \ell_{2,1}dW_1(t) + \ell_{2,2}dW_2(t), \quad t \in [0,T],$$

is itself a standard Brownian motion, and there exists another Brownian motion $\widetilde{W}_1$ independent of $\widetilde{W}_2$ as well as an upper-triangular matrix

$$\mathcal{U} = \begin{pmatrix} u_{1,1} & u_{1,2} \\ u_{2,1} & u_{2,2} \end{pmatrix} = \begin{pmatrix} u_{1,1} & u_{1,2} \\ 0 & 1 \end{pmatrix}$$

such that

$$\begin{pmatrix} u_{1,1} & u_{1,2} \\ 0 & 1 \end{pmatrix} \begin{pmatrix} \widetilde{W}_1 \\ \widetilde{W}_2 \end{pmatrix} = \begin{pmatrix} 1 & 0 \\ \ell_{2,1} & \ell_{2,2} \end{pmatrix} \begin{pmatrix} W_1 \\ W_2 \end{pmatrix}.$$

Therefore, one can rewrite (8.1) as

$$\begin{aligned} S(t) &= S(0) + \int_0^t \mu(s)S(s)ds + \int_0^t Y(s)S(s)\left(u_{1,1}d\widetilde{W}_1(s) + u_{1,2}d\widetilde{W}_2(s)\right), \\ Y(t) &= Y(0) + \int_0^t b(s, Y(s))ds + \int_0^t \mathcal{K}(t,s)d\widetilde{W}_2(s), \end{aligned} \tag{8.3}$$

with $\widetilde{W} = (\widetilde{W}_1, \widetilde{W}_2)$ being a 2-dimensional Brownian motion.

An attentive reader may have noticed that the parametrization (8.3) was already introduced in Sect. 7.2.2, in the context of discussing the Malliavin differentiability of the price process. This fact hints the nature of the algorithm to be presented in this chapter: we intend to exploit the *Malliavin duality* principle (see Definition A.9) to "integrate by parts" and replace the discontinuous function f with its continuous primitive F inside the expectation.

Recall that, according to Theorem 7.3, the Malliavin derivative

$$\widetilde{D}_v X(t) = (\widetilde{D}_v^1 X(t), \widetilde{D}_v^2 X(t))$$

of the log-price process

$$X(t) := \log S(t), \quad t \in [0, T],$$

with respect to $\widetilde{W}$ turns out to be

$$\begin{aligned} \widetilde{D}_v^1 X(t) &= u_{1,1} Y_1(v) \mathbb{1}_{[0,t]}(v), \\ \widetilde{D}_v^2 X(t) &= \Bigg(- \int_0^t Y(s)\widetilde{D}_v^2 Y(s)ds + u_{1,2}Y(v) \\ &\quad + \int_0^t \widetilde{D}_v^2 Y(s)\left(u_{1,1}d\widetilde{W}_1(s) + u_{1,2}d\widetilde{W}_2(s)\right)\Bigg)\mathbb{1}_{[0,t]}(u). \end{aligned} \tag{8.4}$$

In addition, by Theorem 7.4, $X(t)$ is absolutely continuous w.r.t. the Lebesgue measure. Thus, if G turns out to be globally Lipschitz continuous, we can use the chain rule from Appendix A.3 to claim that

$$\widetilde{D}^1_v G(X(t)) = g(X(t))\widetilde{D}^1_v X(t), \quad t \in (0, T].$$

However, Assumption 8.1 does not guarantee *global* Lipschitz continuity of G, so let us extend the chain rule to match our needs.

Proposition 8.1 *For G defined above and any $t \in (0, T]$, $G(X(t)) \in \widetilde{\mathbb{D}}^{1,2}$ and*

$$\widetilde{D}G(X(t)) = g(X(t))\widetilde{D}X(t), \quad t \in (0, T]. \tag{8.5}$$

Proof For any $n \in \mathbb{N}$, denote

$$g_n(x) := g(x)\mathbb{1}_{[-n,n]}(x)$$

and

$$\begin{aligned} G_n(x) &:= F(1) + \int_0^x g_n(v)dv \\ &= G(x)\mathbb{1}_{[-n,n]}(x) + G(n)\mathbb{1}_{[n,\infty)}(x) + G(-n)\mathbb{1}_{(-\infty,-n]}(x). \end{aligned}$$

It is clear that G_n is globally Lipschitz and thus $G_n(X(t)) \in \widetilde{\mathbb{D}}^{1,2}$ and

$$\widetilde{D}G_n(X(t)) = g_n(X(t))\widetilde{D}X(t).$$

Thus, taking into account the closedness of the operator $\widetilde{D}$, it is sufficient to check that

$$g_n(X(t)) \xrightarrow{L^2(\Omega)} g(X(t)), \quad n \to \infty, \tag{8.6}$$

and

$$g_n(X(t))\widetilde{D}X(t) \xrightarrow{L^2(\Omega; L^2([0,T];\mathbb{R}^2))} g(X(t))\widetilde{D}X(t), \quad n \to \infty. \tag{8.7}$$

Observe that $g_n(X(t)) \to g(X(t))$ a.s., $n \to \infty$, and, moreover, with probability 1,

$$g_n^2(X(t)) \le g^2(X(t)).$$

By Assumption 8.1, f is of polynomial growth and thus, by Theorem 7.2, $\mathbb{E}f^2(S(t)) = \mathbb{E}g^2(X(t)) < \infty$ and (8.6) follows from the dominated convergence theorem.

Concerning (8.7), note that for $i = 1, 2$

$$\mathbb{E}\left[g^2(X(t))\int_0^T(\widetilde{D}^i_u X(t))^2du\right]$$
$$\leq\left(\mathbb{E}\left[g^4(X(t))\right]\right)^{\frac{1}{2}}\left(\mathbb{E}\left[\left(\int_0^T(\widetilde{D}^i_u X(t))^2du\right)^2\right]\right)^{\frac{1}{2}}$$
$$\leq C\left(\mathbb{E}\left[\left(\int_0^T(\widetilde{D}^i_u X(t))^2du\right)^2\right]\right)^{\frac{1}{2}},$$

where we used the polynomial growth of f and Theorem 7.2 to estimate $\left(\mathbb{E}\left[g^4(X(t))\right]\right)^{\frac{1}{2}}$ from above. Moreover, having (8.2) and using the estimates similar to the ones from Corollary 7.2, it is easy to verify that

$$\mathbb{E}\left[\left(\int_0^T(\widetilde{D}^i_u X(t))^2du\right)^2\right]<C<\infty.$$

Therefore, (8.7) also holds due to the dominated convergence theorem. □

Now, having Proposition 8.1, we are ready to proceed with the main result of this subsection.

Theorem 8.1 *Let $S(T)$ be given by* (8.3), *$f:\mathbb{R}\to\mathbb{R}$ satisfy Assumption 8.1, $X(t):=\log S(t)$, $F(s):=\int_0^s f(u)du$, $g(x):=f(e^x)$, $G(x):=F(1)+\int_0^x g(v)dv$. Then*

$$\mathbb{E}f(S(T))=\frac{1}{Tu_{1,1}}\mathbb{E}\left[G(X(T))\int_0^T\frac{1}{Y(u)}d\widetilde{W}_1(u)\right]. \tag{8.8}$$

Alternatively,

$$\mathbb{E}f(S(T))=\mathbb{E}\left[\frac{F(S(T))}{S(T)}\left(1+\frac{1}{Tu_{1,1}}\int_0^T\frac{1}{Y(u)}d\widetilde{W}_1(u)\right)\right]. \tag{8.9}$$

Remark 8.2 By Remark 8.1, Y is bounded away from zero, so the Itô integral

$$\int_0^T\frac{1}{Y(u)}d\widetilde{W}_1(u)$$

is well-defined.

Proof In order to obtain representation (8.8), observe that

$$\mathbb{E}f(S(T)) = \mathbb{E}g(X(T)) = \frac{1}{T}\mathbb{E}\left[\int_0^T g(X(T))\widetilde{D}_u^1 X(T)\frac{1}{\widetilde{D}_u^1 X(T)}du\right]$$
$$= \frac{1}{Tu_{1,1}}\mathbb{E}\left[\int_0^T \widetilde{D}_u^1 G(X(T))\frac{1}{Y(u)}du\right]$$
$$= \frac{1}{Tu_{1,1}}\mathbb{E}\left[G(X(T))\int_0^T \frac{1}{Y(u)}d\widetilde{W}_1(u)\right].$$

Here we used the chain rule from Proposition 8.1 and duality relation (A.18) from Appendix A.3 between the Malliavin derivative and Skorokhod integral (which coincides in our case with the Itô integral).

As for the representation (8.9), note that

$$G(x) = F(1) + \int_0^x f(e^v)e^v\frac{1}{e^v}dv = \frac{F(e^x)}{e^x} + \int_0^x \frac{F(e^v)}{e^v}dv.$$

Moreover, using the same argument as in Proposition 8.1, it is easy to prove that

$$\widetilde{D}^1\int_0^{X(T)} \frac{F(e^v)}{e^v}dv = \frac{F(S(T))}{S(T)}\widetilde{D}^1 X(T),$$

therefore, again by the Malliavin duality, we obtain that

$$\mathbb{E}\left[\int_0^{X(T)} \frac{F(e^v)}{e^v}dv\frac{1}{Tu_{1,1}}\int_0^T \frac{1}{Y(u)}d\widetilde{W}_1(u)\right]$$
$$= \mathbb{E}\left[\widetilde{D}_u^1\int_0^{X(T)} \frac{F(e^v)}{e^v}dv\frac{1}{Tu_{1,1}}\frac{1}{Y(u)}du\right]$$
$$= \mathbb{E}\left[\frac{1}{T}\int_0^T \frac{F(S(T))}{S(T)}\widetilde{D}_u^1 X(T)\frac{1}{\widetilde{D}_u^1 X(T)}du\right]$$
$$= \mathbb{E}\left[\frac{F(S(T))}{S(T)}\right],$$

which implies (8.9). □

Let us reflect on the result established in Theorem 8.1. It is noteworthy that the right-hand sides of both representations (8.8) and (8.8) no longer contain the discontinuous function f, which offers greater promise for the application of Monte Carlo estimation techniques. This optimism will be substantiated by rigorous convergence results presented later in Sect. 8.3.

Remark 8.3 Note that the expectation in Theorem 8.1 is taken with respect to the physical measure, and hence is not an option *price* in the non-arbitrage sense. However, in Sect. 8.2.3, we will show how to generalize this formula with the change of measure in place.

8.2.2 Multidimensional Case

The approach described in Sect. 8.2.1 can also be utilized for basket options in d-dimensional SVV models. To do this, let us first recall the notation for the general SVV model from Chap. 7. Namely, we consider a $2d$-dimensional Brownian motion $W = (W_1, ..., W_{2d})^T$ and a lower triangular matrix with strictly positive diagonal entries $\mathcal{L} = (\ell_{i,j})_{i,j=1}^{2d}$ such that

$$\sum_{j=1}^{2d} \ell_{i,j}^2 = 1, \quad i = 1, ..., 2d.$$

Now, let us define

$$B(t) = (B_1(t), ..., B_{2d}(t))^T := \mathcal{L}W(t),$$

and put

$$\begin{aligned} B^S(t) &= \left(B_1^S(t), ..., B_d^S(t)\right)^T := (B_1(t), ..., B_d(t))^T, \\ B^Y(t) &= \left(B_1^Y(t), ..., B_d^Y(t)\right)^T := (B_{d+1}(t), ..., B_{2d}(t))^T. \end{aligned}$$

For this section, we take a d-dimensional SVV model

$$\begin{aligned} S_i(t) &= S_i(0) + \int_0^t \mu_i(s)S_i(s)ds + \int_0^t Y_i(s)S_i(s)dB_i^S(s), \\ Y_i(t) &= Y_i(0) + \int_0^t b_i(s, Y_i(s))ds + \int_0^t \mathcal{K}_i(t, s)dB_i^Y(s), \end{aligned} \tag{8.10}$$

$i = 1, ..., d$, where each Y_i is a two-sided sandwiched process bounded between H_i-Hölder continuous functions $0 < \varphi_i < \psi_i$, $H_i \in (0, 1)$. The coefficients in (8.10) are assumed to satisfy Assumptions 6.1 and 6.3 with order H_i whereas $\mu_i\colon [0, T] \to \mathbb{R}$, $i = 1, ..., d$, are arbitrary H_i-Hölder continuous functions. In addition, we assume that, for each $i = 1, ..., d$,

$$\sup_{s\in[0,T]} \int_0^T \mathcal{K}_i^2(u, s)du < \infty.$$

As always, we denote the corresponding log-price process $X = (X_1, ..., X_d)$, $X_i = \log S_i$. Finally, in order to use Theorem 7.3, we define an *upper* $2d \times 2d$-triangular matrix with positive values on the diagonal $\mathcal{U} = (u_{i,j})_{i,j=1}^{2d}$ such that

$$\widetilde{W} := \mathcal{U}^{-1} B = \mathcal{U}^{-1} \mathcal{L} W \tag{8.11}$$

is a $2d$-dimensional Brownian motion.

With all these definitions and assumptions in place, consider now a basket option with a payoff $f\left(\sum_{i=1}^d \alpha_i S_i(T)\right)$ with $f\colon \mathbb{R} \to \mathbb{R}$ satisfying Assumption 8.1 and $\alpha_i > 0$, $i = 1, ..., d$. As before, define

$$F(s) := \int_0^s f(u)du,$$

and observe that the function

$$G_1(x_1, ..., x_d) := \frac{F\left(\sum_{i=1}^d \alpha_i e^{x_i}\right)}{\alpha_1 e^{x_1}}, \quad (x_1, ..., x_d) \in \mathbb{R}^d,$$

is locally Lipschitz continuous: indeed, if $\sum_{i=1}^d \alpha_i e^{x_i} > \sum_{i=1}^d \alpha_i e^{x_i'}$, then

$$\begin{aligned}
&|G_1(x_1, ..., x_d) - G_1(x_1', ..., x_d')| \\
&\qquad \le \frac{e^{-x_1}}{\alpha_1} \int_{\sum_{i=1}^d \alpha_i e^{x_i'}}^{\sum_{i=1}^d \alpha_i e^{x_i}} |f(s)|ds \\
&\qquad\quad + \frac{1}{\alpha_1} \int_0^{\sum_{i=1}^d \alpha_i e^{x_i'}} |f(s)|ds |e^{-x_1} - e^{-x_1'}| \\
&\qquad \le c_f \frac{e^{-x_1}}{\alpha_1} \left(1 + \left|\sum_{i=1}^d \alpha_i e^{x_i}\right|^q\right) \sum_{j=1}^d \alpha_j e^{x_j \vee x_j'} |x_j - x_j'|
\end{aligned}$$

$$+ c_f \frac{1}{\alpha_1} \left(\sum_{i=1}^d \alpha_i e^{x_i'} + \frac{1}{q+1} \left| \sum_{i=1}^d \alpha_i e^{x_i'} \right|^{q+1} \right) e^{-(x_1 \wedge x_1')} |x_1 - x_1'|.$$

Note that polynomial growth of f (and hence F) together with Theorem 7.2 imply that random variables $F\left(\sum_{i=1}^d \alpha_i S_i(T)\right)$ and $G_1(X_1(T), ..., X_d(T))$ have moments of all orders. Moreover, Theorem 7.4 implies that the law of X (and hence of S) is absolutely continuous w.r.t. the Lebesgue measure on $\mathbb{R}^d$. Therefore, by approximating $G_1(X(t))$ with Lipschitz functions just like in Proposition 8.1, it is straightforward to get the following chain rule for $G_1(X(t))$.

Proposition 8.2 *For G_1 defined above and any $t \in (0, T]$, $G_1(X(t)) \in \widetilde{\mathbb{D}}^{1,2}$ and*

$$\widetilde{D} G_1(X(t)) = \sum_{i=1}^d \frac{\partial G_1(X(t))}{\partial x_i} \widetilde{D} X_i(t), \quad t \in (0, T]. \tag{8.12}$$

We are now ready to formulate an analogue of Theorem 8.1 for the basket option.

Theorem 8.2 *Let $S = (S_1, ..., S_d)$ be given by (8.10), $X = (X_1, ..., X_d)$ be the corresponding log-price process, $X_i = \log S_i$, $f\colon \mathbb{R} \to \mathbb{R}$ satisfy Assumption 8.1, $F(s) := \int_0^s f(u)du$ and $\alpha_i > 0$, $i = 1, ..., d$. Then*

$$\mathbb{E} f\left(\sum_{i=1}^d \alpha_i S_i(T)\right) = \mathbb{E}\left[\frac{F\left(\sum_{i=1}^d \alpha_i S_i(T)\right)}{\alpha_1 S_1(T)} \left(1 + \frac{1}{T u_{1,1}} \int_0^T \frac{1}{Y_1(u)} d\widetilde{W}_1(u)\right)\right], \tag{8.13}$$

where $u_{1,1}$ is the top left element of the matrix $\mathcal{U}$ in (8.11)

Remark 8.4 Note that we assume $\varphi_i > 0$, $i = 1, ..., d$, i.e. the volatility processes Y_i are bounded away from zero. In particular, the integral

$$\int_0^T \frac{1}{Y_1(u)} d\widetilde{W}_1(u)$$

is well-defined.

Proof Note that

$$\frac{\partial G_1(x_1, ..., x_d)}{\partial x_1} = f\left(\sum_{i=1}^d \alpha_i e^{x_i}\right) - \frac{1}{\alpha_1 e^{x_1}} F\left(\sum_{i=1}^d \alpha_i e^{x_i}\right).$$

Hence, taking into account Proposition 8.2 and Theorem 7.3, we get that

$$\mathbb{E}\left[f\left(\sum_{i=1}^d \alpha_i S_i(T)\right)\right]$$

$$
\begin{aligned}
&= \mathbb{E}\left[\frac{1}{T}\int_0^T \left(f\left(\sum_{i=1}^d \alpha_i e^{X_i(T)}\right) - \frac{1}{\alpha_1 e^{X_1(T)}} F\left(\sum_{i=1}^d \alpha_i e^{X_i(T)}\right)\right) dv\right] \\
&\quad + \mathbb{E}\left[\frac{1}{\alpha_1 e^{X_1(T)}} F\left(\sum_{i=1}^d \alpha_i e^{X_i(T)}\right)\right] \\
&= \mathbb{E}\left[\frac{1}{T}\int_0^T \frac{\partial G_1(X(T))}{\partial x_1} dv\right] + \mathbb{E}\left[G_1(X(T))\right] \\
&= \mathbb{E}\left[\frac{1}{T}\int_0^T \frac{\partial G_1(X(T))}{\partial x_1} \widetilde{D}_v^1 X_1(T) \frac{1}{\widetilde{D}_v^1 X_1(T)} dv\right] + \mathbb{E}\left[G_1(X(T))\right] \\
&= \mathbb{E}\left[\frac{1}{T}\int_0^T \widetilde{D}_v^1 G_1(X(T)) \frac{1}{u_{1,1} Y_1(v)} dv\right] + \mathbb{E}\left[G_1(X(T))\right] \\
&= \mathbb{E}\left[\frac{1}{T u_{1,1}} G_1(X(T)) \int_0^T \frac{1}{Y_1(v)} d\widetilde{W}_1(v)\right] + \mathbb{E}\left[G_1(X(T))\right] \\
&= \mathbb{E}\left[\frac{F\left(\sum_{i=1}^d \alpha_i S_i(T)\right)}{\alpha_1 S_1(T)} \left(1 + \frac{1}{T u_{1,1}} \int_0^T \frac{1}{Y_1(v)} d\widetilde{W}_1(v)\right)\right],
\end{aligned}
$$

as required. □

8.2.3 *Malliavin Smoothing Under the Change of Measure*

Theorems 8.1 and 8.2 concern the expectation w.r.t. the *physical* measure, i.e. the obtained representation of the expected payoff is not the *option price*. In this subsection, we will modify our technique to account for the change of measure.

Let $S = (S_1, ..., S_d)$ be given by (8.10) with all assumptions on the coefficients from Sect. 8.2.2. Consider a numéraire $e^{\int_0^t \nu(s)ds}$, $t \in [0, T]$, where ν: $[0, T] \to \mathbb{R}_+$ denotes a Hölder continuous function of order $\min_{i=1,...,d} H_i$ representing an instantaneous interest rate. In addition, recall the notation $\widetilde{\mu}_i := \mu_i - \nu, i = 1, ..., d$, where μ_i is the drift from the price equation (8.10). Next, recall the lower triangular matrix $\mathcal{L}$ defined in Sect. 8.2.2 and represent it as

$$
\mathcal{L} = \begin{pmatrix} \mathcal{L}_{11} & \mathcal{L}_{12} \\ \mathcal{L}_{21} & \mathcal{L}_{22} \end{pmatrix},
$$

where $\mathcal{L}_{ij}$ we denote the $d \times d$ blocks of matrix $\mathcal{L}$. Since $\mathcal{L}$ is lower-triangular, $\mathcal{L}_{12}$ is a zero matrix whereas $\mathcal{L}_{11}$, $\mathcal{L}_{22}$ are lower triangular matrices with strictly positive elements on the diagonal. In particular, $\mathcal{L}_{11}$ is invertible, so we denote its inverse by $\mathcal{L}_{11}^{-1} = \left(\ell_{j,k}^{(-1)}\right)_{j,k=1}^{d}$.

With this notation in hands, recall that, by Lemma 7.1, all martingale densities on the SVV market have the form

$$M(T) = \mathcal{E}_T \left\{ -\sum_{j=1}^{d} \int_0^{\cdot} \left(\sum_{k=1}^{d} \ell_{j,k}^{(-1)} \frac{\widetilde{\mu}_k(s)}{Y_k(s)} \right) dW_j(s) + \sum_{l=d+1}^{2d} \int_0^{\cdot} \xi_l(s) dW_l(s) \right\},$$

where $\mathcal{E}_T\{\cdot\}$ is the Doléans-Dade exponential and ξ_l, $l = d+1, ..., 2d$, are any predictable processes such that $\int_0^{\cdot} \xi_l(s) dW_l(s)$ are well-defined and

$$\mathbb{E}[M(T)] = 1.$$

In this subsection, we will be interested in one particular density from this class, which we have already seen in the proof of Theorem 7.5:

$$\mathcal{E}_T^{\min} := \mathcal{E}_T \left\{ -\sum_{j=1}^{d} \int_0^{\cdot} \left(\sum_{k=1}^{d} \ell_{j,k}^{(-1)} \frac{\widetilde{\mu}_k(s)}{Y_k(s)} \right) dW_j(s) \right\}. \tag{8.14}$$

Remark 8.5 In the literature, the measure with the density (8.14) is commonly referred to as the *minimal martingale measure*, see e.g. [5]. In this subsection, we will be looking specifically at this measure mainly due to its simplicity compared to other options; however, the idea may also be extended to other martingale measures.

Recall that

$$W = \mathcal{L}^{-1} \mathcal{U} \widetilde{W},$$

so let us introduce the values $\rho_{j,i}$ such that

$$W_j(t) = \sum_{i=1}^{2d} \rho_{j,i} \widetilde{W}_i(t),$$

i.e.

$$\mathcal{E}_T^{\min} = \mathcal{E}_T \left\{ -\sum_{j=1}^{d} \sum_{i=1}^{2d} \int_0^{\cdot} \rho_{j,i} \left(\sum_{k=1}^{d} \ell_{j,k}^{(-1)} \frac{\widetilde{\mu}_k(s)}{Y_k(s)} \right) d\widetilde{W}_i(s) \right\}$$

$$= \mathcal{E}_T \left\{ -\sum_{i=1}^{2d} \int_0^{T} \left(\sum_{j=1}^{d} \rho_{j,i} \left(\sum_{k=1}^{d} \ell_{j,k}^{(-1)} \frac{\widetilde{\mu}_k(s)}{Y_k(s)} \right) \right) d\widetilde{W}_i(s) \right\}$$

$$= \exp \left\{ -\sum_{i=1}^{2d} \int_0^T \left(\sum_{j=1}^{d} \rho_{j,i} \left(\sum_{k=1}^{d} \ell_{j,k}^{(-1)} \frac{\widetilde{\mu}_k(s)}{Y_k(s)} \right) \right) d\widetilde{W}_i(s) \right\} \times$$

$$\times \exp \left\{ -\frac{1}{2} \sum_{i=1}^{2d} \int_0^T \left(\sum_{j=1}^{d} \rho_{j,i} \left(\sum_{k=1}^{d} \ell_{j,k}^{(-1)} \frac{\widetilde{\mu}_k(s)}{Y_k(s)} \right) \right)^2 ds \right\}.$$

Note that each Y_i is bounded away from zero by $\varphi_i^* := \min_{t \in [0,T]} \varphi_i(t)$ and hence $\frac{1}{Y_i(s)}$ can be represented as follows:

$$\frac{1}{Y_i(s)} = h_i(Y_i(s)), \quad h_i(y) := \begin{cases} \frac{1}{y}, & y > \varphi_i^*, \\ -\frac{1}{(\varphi_i^*)^2} y + \frac{2}{\varphi_i^*}, & y \le \varphi_i^*. \end{cases}$$

The function h_i is bounded, continuously differentiable and has bounded derivative. So, by the classical chain rule, $\frac{1}{Y_i(s)} \in \widetilde{\mathbb{D}}^{1,2}$, and

$$\widetilde{D}^1 \left[\frac{1}{Y_i(s)} \right] = 0.$$

In particular, this implies that

$$\widetilde{D}_v^1 \left[-\frac{1}{2} \sum_{i=1}^{2d} \int_0^T \left(\sum_{j=1}^{d} \rho_{j,i} \left(\sum_{k=1}^{d} \ell_{j,k}^{(-1)} \frac{\widetilde{\mu}_k(s)}{Y_k(s)} \right) \right)^2 ds \right] = 0,$$

and

$$\widetilde{D}_v^1 \left[-\sum_{i=1}^{2d} \int_0^T \left(\sum_{j=1}^{d} \rho_{j,i} \left(\sum_{k=1}^{d} \ell_{j,k}^{(-1)} \frac{\widetilde{\mu}_k(s)}{Y_k(s)} \right) \right) d\widetilde{W}_i(s) \right]$$

$$= -\sum_{j=1}^{d} \sum_{k=1}^{d} \rho_{j,1} \ell_{j,k}^{(-1)} \frac{\widetilde{\mu}_k(v)}{Y_k(v)}$$

$$= \sum_{k=1}^{d} \left(-\sum_{j=1}^{d} \rho_{j,1} \ell_{j,k}^{(-1)} \right) \frac{\widetilde{\mu}_k(v)}{Y_k(v)}$$

$$=: \sum_{k=1}^{d} \beta_k \frac{\widetilde{\mu}_k(v)}{Y_k(v)},$$

where $\beta_k := \left(-\sum_{j=1}^{d} \rho_{j,1}\ell^{(-1)}_{j,k}\right)$. Furthermore, using the boundedness of Y and the chain rule from Proposition 7.1, it is easy to establish that $\mathcal{E}^{\min}_T \in \widetilde{\mathbb{D}}^{1,2}$ and

$$\widetilde{D}^1_v \mathcal{E}^{\min}_T = \mathcal{E}^{\min}_T \sum_{k=1}^{d} \beta_k \frac{\widetilde{\mu}_k(v)}{Y_k(v)}.$$

Let now $f\colon \mathbb{R} \to \mathbb{R}$ satisfy Assumption 8.1, $\alpha_i > 0, i = 1, ..., d$. Then the following modification of Theorem 8.2 is true.

Theorem 8.3 *Let $S = (S_1, ..., S_d)$ be given by* (8.10)*, $X = (X_1, ..., X_d)$ be the corresponding log-price process, $X_i = \log S_i$, $f\colon \mathbb{R} \to \mathbb{R}$ satisfy Assumption 8.1, $F(s) := \int_0^s f(u)du$ and $\alpha_i > 0$, $i = 1, ..., d$. Then*

$$\mathbb{E}\left[\mathcal{E}^{\min}_T f\left(\sum_{i=1}^{d} \alpha_i S_i(T)\right)\right] = \mathbb{E}\left[\mathcal{E}^{\min}_T \frac{F\left(\sum_{i=1}^{d} \alpha_i S_i(T)\right)}{\alpha_1 S_1(T)}(1 + \mathcal{I}_T)\right], \tag{8.15}$$

where

$$\mathcal{I}_T := \frac{1}{T u_{1,1}}\left(\int_0^T \frac{1}{Y_1(v)} d\widetilde{W}_1(v) - \sum_{k=1}^{d} \int_0^T \beta_k \frac{\widetilde{\mu}_k(v)}{Y_k(v)Y_1(v)} dv\right).$$

Proof Denote

$$G_1(x_1, ..., x_d) := \frac{F\left(\sum_{i=1}^{d} \alpha_i e^{x_i}\right)}{\alpha_1 e^{x_1}}, \quad (x_1, ..., x_d) \in \mathbb{R}^d.$$

Using polynomial growth of f, boundedness of Y_i, existence of all moments of S_i, Proposition 7.1 and Lemma 6.2, it is easy to prove that

$$\mathcal{E}^{\min}_T G_1(X(T)) \in \widetilde{\mathbb{D}}^{1,2},$$

and

$$\begin{aligned}\widetilde{D}^1_v\left[\mathcal{E}^{\min}_T G_1(X(T))\right] &= \mathcal{E}^{\min}_T \widetilde{D}^1_v G_1(X(T)) + G_1(X(T))\widetilde{D}^1_v \mathcal{E}^{\min}_T \\ &= \mathcal{E}^{\min}_T \widetilde{D}^1_v G_1(X(T)) + \mathcal{E}^{\min}_T G_1(X(T)) \sum_{k=1}^{d} \beta_k \frac{\widetilde{\mu}_k(v)}{Y_k(v)}.\end{aligned}$$

Next, observe that

$$\mathbb{E}\left[\mathcal{E}^{\min}_T f\left(\sum_{i=1}^{d} \alpha_i S_i(T)\right)\right]$$

$$= \mathbb{E}\left[\mathcal{E}_T^{\min}\left(f\left(\sum_{i=1}^{d}\alpha_i e^{X_i(T)}\right) - G_1(X(T))\right)\right] + \mathbb{E}\left[\mathcal{E}_T^{\min} G_1(X(T))\right]$$

and

$$\mathbb{E}\left[\mathcal{E}_T^{\min}\left(f\left(\sum_{i=1}^{d}\alpha_i e^{X_i(T)}\right) - G_1(X(T))\right)\right] = \mathbb{E}\left[\mathcal{E}_T^{\min}\frac{\partial G_1(X(T))}{\partial x_1}\right]$$

$$= \mathbb{E}\left[\frac{1}{T}\int_0^T \mathcal{E}_T^{\min}\frac{\partial G_1(X(T))}{\partial x_1}\widetilde{D}_v^1 X_1(T)\frac{1}{\widetilde{D}_v^1 X_1(T)}dv\right]$$

$$= \mathbb{E}\left[\frac{1}{Tu_{1,1}}\int_0^T \mathcal{E}_T^{\min}\widetilde{D}_v^1 G_1(X(T))\frac{1}{Y_1(v)}dv\right]$$

$$= \frac{1}{Tu_{1,1}}\mathbb{E}\left[\int_0^T \widetilde{D}_v^1\left[\mathcal{E}_T^{\min}G_1(X(T))\right]\frac{1}{Y_1(v)}dv\right]$$

$$- \frac{1}{Tu_{1,1}}\mathbb{E}\left[\int_0^T G_1(X(T))\widetilde{D}_v^1\mathcal{E}_T^{\min}\frac{1}{Y_1(v)}dv\right]$$

$$= \frac{1}{Tu_{1,1}}\mathbb{E}\left[\mathcal{E}_T^{\min}G_1(X(T))\int_0^T \frac{1}{Y_1(v)}d\widetilde{W}_1(v)\right]$$

$$- \frac{1}{Tu_{1,1}}\mathbb{E}\left[G_1(X(T))\mathcal{E}_T^{\min}\int_0^T\sum_{k=1}^{d}\beta_k\frac{\widetilde{\mu}_k(v)}{Y_k(v)Y_1(v)}dv\right]$$

$$= \mathbb{E}\left[\mathcal{E}_T^{\min}\frac{F\left(\sum_{i=1}^{d}\alpha_i S_i(T)\right)}{\alpha_1 S_1(T)}\mathcal{I}_T\right].$$

Taking into account all of the above, we can finally write:

$$\mathbb{E}\left[\mathcal{E}_T^{\min} f\left(\sum_{i=1}^{d}\alpha_i S_i(T)\right)\right] = \mathbb{E}\left[\mathcal{E}_T^{\min}\frac{F\left(\sum_{i=1}^{d}\alpha_i S_i(T)\right)}{\alpha_1 S_1(T)}(1+\mathcal{I}_T)\right]$$

which completes the proof. □

Remark 8.6 Note that the martingale measure should not necessarily be the minimal one: it is possible to obtain a representation of the type (8.15) for other martingale measures provided that the corresponding densities are Malliavin differentiable and one can use the chain rules as in the proof of Theorem 8.3.

8.3 Price Approximation for Options with Discontinuous Payoffs

Having the quadrature formulas established in Theorems 8.1, 8.2 and 8.3, we are now ready to provide some numerical algorithms for computation of

$$\mathbb{E}\left[f\left(\sum_{i=1}^{d}\alpha_i S_i(T)\right)\right].$$

We will consider several variants of the algorithm. In Sect. 8.3.1, we consider a straightforward way of computing $\mathbb{E}\left[f\left(\sum_{i=1}^{d}\alpha_i S_i(T)\right)\right]$ by plugging in $\widehat{Y}$ and $\widehat{S}$ obtained in Sect. 7.4 into (8.8)–(8.9), (8.13) or (8.15). In Sect. 8.3.2, we discuss another algorithm for computation of the expectation for a one-dimensional case: the vector $\left(X(T), \int_0^T \frac{1}{Y(v)}dv\right)^T$ turns out to be conditionally Gaussian w.r.t. an appropriate σ-field and one can use this fact to approximate (8.8) without simulating $\widehat{X}(T)$ at all.

8.3.1 Approximation by Discretizing both Y and S

Notation 8.2 Throughout the rest of the chapter, $\widehat{S}_i$, $\widehat{Y}_i$, $\widehat{X}_i$ will denote the discrete approximations of the processes S_i, Y_i, X_i obtained in Sect. 7.4. When we deal with a 1-dimensional model, we will omit the index i.

Theorem 8.4 *Let $S = (S_1, ..., S_d)$ be given by (8.10), $X = (X_1, ..., X_d)$ be the corresponding log-price process, $X_i = \log S_i$, $f\colon \mathbb{R} \to \mathbb{R}$ satisfy Assumption 8.1, $F(s) := \int_0^s f(u)du$ and $\alpha_i > 0$, $i = 1, ..., d$. Fix a uniform partition $\{0 = t_0 < t_1 < ... < t_N = T\}$, $t_n = \frac{Tn}{N}$, with the mesh satisfying (7.28) and take $\widehat{Y} = (\widehat{Y}_1, ..., \widehat{Y}_d)$, $\widehat{X} = (\widehat{X}_1, ..., \widehat{X}_d)$ and $\widehat{S} = (\widehat{S}_1, ..., \widehat{S}_d)$ as described in Sect. 7.4. Finally, let $\lambda < \min_{i=1,...,d} H_i$ with H_i being the regularity order of the volatility Y_i. Then*

(1) for any $r \geq 1$, there exists a constant $C > 0$ such that

$$\mathbb{E}\left[\left|\frac{F\left(\sum_{i=1}^{d}\alpha_i S_i(T)\right)}{\alpha_1 S_1(T)}\left(1 + \frac{\int_0^T \frac{1}{Y_1(u)}d\widetilde{W}_1(u)}{Tu_{1,1}}\right) - \frac{F\left(\sum_{i=1}^{d}\alpha_i \widehat{S}_i(T)\right)}{\alpha_1 \widehat{S}_1(T)}\left(1 + \frac{\int_0^T \frac{1}{\widehat{Y}_1(u)}d\widetilde{W}_1(u)}{Tu_{1,1}}\right)\right|^r\right] \leq C\Delta_N^{r\lambda},$$

(2) there exists a constant $C > 0$ such that

$$\left|\mathbb{E}\left[f\left(\sum_{i=1}^{d}\alpha_i S_i(T)\right)\right]-\mathbb{E}\left[\frac{F\left(\sum_{i=1}^{d}\alpha_i\widehat{S}_i(T)\right)}{\alpha_1\widehat{S}_1(T)}\left(1+\frac{\int_0^T\frac{1}{\widehat{Y}_1(u)}d\widetilde{W}_1(u)}{Tu_{1,1}}\right)\right]\right|\le C\Delta_N^{\lambda}.$$

Proof By Theorem 8.2, it is enough to prove item 1. We provide the proof only for $d=1$ (in this case we can put $\alpha_1=1$ without any loss in generality). For the general case of $d>1$, the arguments are the same but require a more notation-heavy presentation. For reader's convenience, we split the proof into three steps.

Step 1. We begin with noting that, by the Burkholder-Davis-Gundy inequality and (7.30),

$$\begin{aligned}\mathbb{E}\left[\left|\int_0^T\left(\frac{1}{Y(u)}-\frac{1}{\widehat{Y}(u)}\right)d\widetilde{W}_1(u)\right|^r\right]&\le C\mathbb{E}\left[\left(\int_0^T\left(\frac{1}{Y(u)}-\frac{1}{\widehat{Y}(u)}\right)^2du\right)^{\frac{r}{2}}\right]\\&\le C\Delta_N^{\lambda r}.\end{aligned}$$

Step 2. Next, let us show that for any $r\ge 1$

$$\mathbb{E}\left[\left|\frac{F(S(T))}{S(T)}-\frac{F(\widehat{S}(T))}{\widehat{S}(T)}\right|^r\right]\le C\Delta_N^{r\lambda}. \tag{8.16}$$

It is clear that

$$\begin{aligned}\mathbb{E}\left[\left|\frac{F(S(T))}{S(T)}-\frac{F(\widehat{S}(T))}{\widehat{S}(T)}\right|^r\right]&\le C\mathbb{E}\left[\left|\frac{F(S(T))}{S(T)}-\frac{F(S(T))}{\widehat{S}(T)}\right|^r\right]\\&\quad+C\mathbb{E}\left[\left|\frac{F(S(T))}{\widehat{S}(T)}-\frac{F(\widehat{S}(T))}{\widehat{S}(T)}\right|^r\right].\end{aligned} \tag{8.17}$$

Now we estimate both terms in the right-hand side separately. Observe that F has polynomial growth and thus, by Theorem 7.2, $\mathbb{E}[|F(S(T))|^{2r}]<\infty$. Therefore

$$\begin{aligned}\mathbb{E}\left[|F(S(T))|^r\left|\frac{1}{S(T)}-\frac{1}{\widehat{S}(T)}\right|^r\right]&\le\left(\mathbb{E}\left[|F(S(T))|^{2r}\right]\mathbb{E}\left[\left|\frac{1}{S(T)}-\frac{1}{\widehat{S}(T)}\right|^{2r}\right]\right)^{\frac{1}{2}}\\&\le C\left(\mathbb{E}\left[\left|\frac{1}{S(T)}-\frac{1}{\widehat{S}(T)}\right|^{2r}\right]\right)^{\frac{1}{2}}.\end{aligned}$$

Moreover, observe that, by Theorem 7.2 and Proposition 7.2, there exists constant $C>0$ that does not depend on the partition such that

$$\mathbb{E}\left[e^{-4r(X(T)+\widehat{X}(T))}(e^{X(T)}+e^{\widehat{X}(T)})^{4r}\right]<C,$$

thus, using Theorem 7.6 and the inequality $|e^x - e^y| \le (e^x + e^y)|x-y|$, $x, y \in \mathbb{R}$, we can write

$$\begin{aligned}
\mathbb{E}\left[\left|\frac{1}{S(T)}-\frac{1}{\widehat{S}(T)}\right|^{2r}\right] &= \mathbb{E}\left[e^{-2r(X(T)+\widehat{X}(T))}\left|e^{X(T)}-e^{\widehat{X}(T)}\right|^{2r}\right]\\
&\le \mathbb{E}\left[e^{-2r(X(T)+\widehat{X}(T))}(e^{X(T)}+e^{\widehat{X}(T)})^{2r}\left|X(T)-\widehat{X}(T)\right|^{2r}\right]\\
&\le \left(\mathbb{E}\left[e^{-4r(X(T)+\widehat{X}(T))}(e^{X(T)}+e^{\widehat{X}(T)})^{4r}\right]\right)^{\frac{1}{2}}\left(\mathbb{E}\left[\left|X(T)-\widehat{X}(T)\right|^{4r}\right]\right)^{\frac{1}{2}}\\
&\le C\Delta_N^{2r\lambda},
\end{aligned}$$

where $C > 0$ is again a constant that does not depend on the partition. Therefore

$$\mathbb{E}\left[|F(S(T))|^r\left|\frac{1}{S(T)}-\frac{1}{\widehat{S}(T)}\right|^r\right]\le C\Delta^{r\lambda}. \tag{8.18}$$

Next, observe that for any $s_1, s_2 > 0$

$$|F(s_1)-F(s_2)| = \left|\int_{s_1\wedge s_2}^{s_1\vee s_2} f(u)du\right| \le C(1+s_1^q+s_2^q)|s_1-s_2|,$$

so

$$\begin{aligned}
&\mathbb{E}\left[\frac{1}{\widehat{S}^r(T)}\left|F(S(T))-F(\widehat{S}(T))\right|^r\right]\\
&\le \left(\mathbb{E}\left[\frac{1}{\widehat{S}^{2r}(T)}\right]\right)^{\frac{1}{2}}\left(\mathbb{E}\left[\left|F(S(T))-F(\widehat{S}(T))\right|^{2r}\right]\right)^{\frac{1}{2}}\\
&\le C\left(\mathbb{E}\left[\left|F(S(T))-F(\widehat{S}(T))\right|^{2r}\right]\right)^{\frac{1}{2}}\\
&\le C\left(\mathbb{E}\left[|(1+S^q(T)+\widehat{S}^q(T))|^{2r}\left|S(T)-\widehat{S}(T)\right|^{2r}\right]\right)^{\frac{1}{2}}\\
&\le C\left(\mathbb{E}\left[|(1+S^q(T)+\widehat{S}^q(T))|^{4r}\right]\right)^{\frac{1}{4}}\left(\mathbb{E}\left[\left|S(T)-\widehat{S}(T)\right|^{4r}\right]\right)^{\frac{1}{4}}\\
&\le C\left(\mathbb{E}\left[\left|S(T)-\widehat{S}(T)\right|^{4r}\right]\right)^{\frac{1}{4}},
\end{aligned}$$

where we used Theorem 7.2 together with Proposition 7.2 to estimate $\mathbb{E}\left[\widehat{S}^{-2r}(T)\right]$ together with $\mathbb{E}\left[|(1+S^q(T)+\widehat{S}^q(T))|^{4r}\right]$.

Finally, using the same argument as in the proof of (8.18), we can obtain that

$$\left(\mathbb{E}\left[\left|S(T)-\widehat{S}(T)\right|^{4r}\right]\right)^{\frac{1}{4}} \leq C\Delta_N^{r\lambda},$$

and thus

$$\mathbb{E}\left[\frac{1}{\widehat{S}^r(T)}\left|F(S(T))-F(\widehat{S}(T))\right|^r\right] \leq C\Delta_N^{r\lambda}. \tag{8.19}$$

Now (8.16) follows from (8.17), (8.18) and (8.19).

Step 3. Now we proceed to the claim of the theorem. Since

$$\mathbb{E}\left[\left|\frac{F(S(T))}{S(T)}\right|^{2r}\right] < \infty$$

due to the polynomial growth of the mapping $x \mapsto \frac{F(x)}{x}$, we can deduce by Step 1, Step 2 and Theorem 8.1 that

$$\begin{aligned}
&\mathbb{E}\left[\left|\frac{F(S(T))}{S(T)}\left(1+\frac{\int_0^T \frac{1}{Y(u)}d\widetilde{W}_1(u)}{Tu_{1,1}}\right)-\frac{F(\widehat{S}(T))}{\widehat{S}(T)}\left(1+\frac{\int_0^T \frac{1}{\widehat{Y}(u)}d\widetilde{W}_1(u)}{Tu_{1,1}}\right)\right|^r\right]\\
&\leq C\mathbb{E}\left[\left|\frac{F(S(T))}{S(T)}-\frac{F(\widehat{S}(T))}{\widehat{S}(T)}\right|^r\right]\\
&\quad + C\mathbb{E}\left[\left|\frac{F(S(T))}{S(T)}\int_0^T\left(\frac{1}{Y(u)}-\frac{1}{\widehat{Y}(u)}\right)d\widetilde{W}_1(u)\right|^r\right]\\
&\quad + C\mathbb{E}\left[\left|\int_0^T\frac{1}{\widehat{Y}(u)}d\widetilde{W}_1(u)\left(\frac{F(S(T))}{S(T)}-\frac{F(\widehat{S}(T))}{\widehat{S}(T)}\right)\right|^r\right]\\
&\leq C\Delta_N^{r\lambda}+C\left(\mathbb{E}\left[\left|\frac{F(S(T))}{S(T)}\right|^{2r}\right]\right)^{\frac{1}{2}}\left(\mathbb{E}\left[\left|\int_0^T\left(\frac{1}{Y(u)}-\frac{1}{\widehat{Y}(u)}\right)d\widetilde{W}_1(u)\right|^{2r}\right]\right)^{\frac{1}{2}}\\
&\quad + C\left(\mathbb{E}\left[\left|\int_0^T\frac{1}{\widehat{Y}(u)}d\widetilde{W}_1(u)\right|^{2r}\right]\right)^{\frac{1}{2}}\left(\mathbb{E}\left[\left|\frac{F(S(T))}{S(T)}-\frac{F(\widehat{S}(T))}{\widehat{S}(T)}\right|^{2r}\right]\right)^{\frac{1}{2}}\\
&\leq C\Delta_N^{r\lambda},
\end{aligned}$$

which ends the proof. □

Theorem 8.4 concerns representations (8.9) and (8.13). However, for the one-dimensional case, we also have the representation (8.8) which can also be used to approximate the expected payoff. We give the corresponding result below.

Theorem 8.5 *Let S be given by (8.3), X be the corresponding log-price process, $X = \log S$, $f\colon \mathbb{R} \to \mathbb{R}$ satisfy Assumption 8.1, $F(s) := \int_0^s f(u)du$, $g(x) := f(e^x)$, $G(x) := F(1) + \int_0^x g(v)dv$. Fix a uniform partition $\{0 = t_0 < t_1 < \ldots < t_N = T\}$, $t_n = \frac{Tn}{N}$, with the mesh satisfying (7.28) and take $\widehat{Y}$, $\widehat{X}$ and $\widehat{S}$ as described in Sect. 7.4. Finally, let $\lambda < H$ with H the regularity order of the volatility process Y. Then*

(1) for any $r \geq 1$, there exists a constant $C > 0$ such that

$$\mathbb{E}\left[\left|\frac{1}{Tu_{1,1}}G(X(T))\int_0^T \frac{1}{Y(u)}d\widetilde{W}_1(u) - \frac{1}{Tu_{1,1}}G(\widehat{X}(T))\int_0^T \frac{1}{\widehat{Y}(u)}d\widetilde{W}_1(u)\right|^r\right] \leq C\Delta_N^{r\lambda},$$

(2) there exists a constant $C > 0$ such that

$$\left|\mathbb{E}\left[f(S(T))\right] - \mathbb{E}\left[\frac{1}{Tu_{1,1}}G(\widehat{X}(T))\int_0^T \frac{1}{\widehat{Y}(u)}d\widetilde{W}_1(u)\right]\right| \leq C\Delta_N^{\lambda}.$$

Proof Once again, item 1 implies item 2 by Theorem 8.1 and we sketch the proof of the former omitting the details. First, observe that for any $x_1, x_2 \in \mathbb{R}$

$$\begin{aligned}|G(x_1) - G(x_2)| &= \left|\int_{x_1\wedge x_2}^{x_1\vee x_2} g(v)dv\right| \leq \sup_{v\in[x_1\wedge x_2, x_1\vee x_2]} |g(v)||x_1 - x_2| \\ &= \sup_{v\in[x_1\wedge x_2, x_1\vee x_2]} |f(e^v)||x_1 - x_2| \leq c_f(1 + e^{q(x_1\vee x_2)})|x_1 - x_2| \\ &\leq c_f(1 + e^{qx_1} + e^{qx_2})|x_1 - x_2|\end{aligned}$$

and thus, using the same machinery as in the proof of Theorem 8.4, we can write

$$\begin{aligned}\mathbb{E}\left[\left|G(X(T)) - G(\widehat{X}(T))\right|^r\right] &\leq C\mathbb{E}\left[\left|(1 + S^q(T) + \widehat{S}^q(T))|X(T) - \widehat{X}(T)|\right|^r\right] \\ &\leq C\Delta_N^{r\lambda}.\end{aligned}$$

Due to existence of exponential moments of $X(T)$ as well as polynomial growth of f, it is also clear that for any $r \geq 1$

$$\mathbb{E}\left[|G(X(T))|^r\right] < \infty.$$

Finally, as noted in Step 1 of the proof of Theorem 8.4,

$$\mathbb{E}\left[\left|\int_0^T \left(\frac{1}{Y(u)} - \frac{1}{\widehat{Y}(u)}\right)d\widetilde{W}_1(u)\right|^r\right] \leq C\Delta_N^{r\lambda}.$$

Therefore

$$
\begin{aligned}
&E\left[\left|G(X(T))\int_0^T \frac{1}{Y(u)}d\widetilde{W}_1(u) - G(\widehat{X}(T))\int_0^T \frac{1}{\widehat{Y}(u)}d\widetilde{W}_1(u)\right|^r\right] \\
&\le CE\left[\left|G(X(T))\left(\int_0^T \frac{1}{Y(u)}d\widetilde{W}_1(u) - \int_0^T \frac{1}{\widehat{Y}(u)}d\widetilde{W}_1(u)\right)\right|^r\right] \\
&\quad + CE\left[\left|\left(G(X(T)) - G(\widehat{X}(T))\right)\int_0^T \frac{1}{\widehat{Y}(u)}d\widetilde{W}_1(u)\right|^r\right] \\
&\le C\left(\mathbb{E}\left[|G(X(T))|^{2r}\right]\right)^{\frac{1}{2}}\left(\mathbb{E}\left[\left|\int_0^T\left(\frac{1}{Y(u)} - \frac{1}{\widehat{Y}(u)}\right)d\widetilde{W}_1(u)\right|^{2r}\right]\right)^{\frac{1}{2}} \\
&\quad + C\left(E\left[\left|\int_0^T \frac{1}{\widehat{Y}(u)}d\widetilde{W}_1(u)\right|^{2r}\right]\right)^{\frac{1}{2}}\left(\mathbb{E}\left[\left|G(X(T)) - G(\widehat{X}(T))\right|^{2r}\right]\right)^{\frac{1}{2}} \\
&\le C\Delta_N^{r\lambda}.
\end{aligned}
$$

□

We conclude the subsection with an approximation result in line with Theorem 8.4 involving a change of measure. Our strategy will be the same: replacing S_i, X_i, Y_i with their discretized counterparts $\widehat{S}_i, \widehat{X}_i, \widehat{Y}_i$ – but there is one piece that is missing: a way to approximate the density of the minimal martingale measure $\mathcal{E}_T^{\min}$. So, let us first fill this gap.

Let $\mathcal{E}_T^{\min}$ be the density of the minimal martingale measure defined by (8.14). Fix a uniform partition $\{0 = t_0 < t_1 < ... < t_N = T\}$, $t_n = \frac{Tn}{N}$, with the mesh satisfying (7.28) and put

$$
\begin{aligned}
\widehat{\mathcal{E}}_T^N := \exp&\left\{-\sum_{j=1}^{d}\sum_{n=0}^{N-1}\left(\sum_{k=1}^{d}\ell_{j,k}^{(-1)}\frac{\widetilde{\mu}_k(t_n)}{\widehat{Y}_k(t_n)}\right)(W_j(t_{n+1}) - W_j(t_n))\right\}\times \\
\times \exp&\left\{-\frac{1}{2}\sum_{j=1}^{d}\frac{T}{N}\sum_{n=0}^{N-1}\left(\sum_{k=1}^{d}\ell_{j,k}^{(-1)}\frac{\widetilde{\mu}_k(t_n)}{\widehat{Y}_k(t_n)}\right)^2\right\}.
\end{aligned}
\tag{8.20}
$$

Theorem 8.6 *Let the mesh of the partition Δ_N satisfy (7.28) and take $\lambda < \min_{i=1,\dots,d} H_i$ with H_i being the regularity order of the volatility Y_i. Then, for any $r \ge 1$, there exists a constant $C > 0$ that does not depend on the partition such that*

$$
\mathbb{E}\left[|\mathcal{E}_T^{\min} - \widehat{\mathcal{E}}_T^N|^r\right] \le C\Delta_N^{r\lambda}.
$$

Proof The proof is straightforward, so we will provide an outline omitting all the details.

First, observe that for every $j = 1, ..., d$, using boundedness of each $\frac{1}{\widehat{Y}_k}$, λ-Hölder continuity of each $\widetilde{\mu}_k$, the inequality (7.30) as well as the fact that we set $\widehat{Y}_k(s) = \widehat{Y}_k(t_n)$ for all $s \in [t_n, t_{n+1})$, we can write

$$\mathbb{E}\left[\left|\sum_{n=0}^{N-1}\left(\sum_{k=1}^{d}\ell_{j,k}^{(-1)}\frac{\widetilde{\mu}_k(t_n)}{\widehat{Y}_k(t_n)}\right)(W_j(t_{n+1})-W_j(t_n)) - \int_0^T\left(\sum_{k=1}^{d}\ell_{j,k}^{(-1)}\frac{\widetilde{\mu}_k(s)}{Y_k(s)}\right)dW_j(s)\right|^r\right]$$

$$= \mathbb{E}\left[\left|\sum_{n=0}^{N-1}\int_{t_n}^{t_{n+1}}\left(\sum_{k=1}^{d}\ell_{j,k}^{(-1)}\left(\frac{\widetilde{\mu}_k(t_n)}{\widehat{Y}_k(s)}-\frac{\widetilde{\mu}_k(s)}{Y_k(s)}\right)\right)dW_j(s)\right|^r\right]$$

$$\leq C\mathbb{E}\left[\left(\sum_{n=0}^{N-1}\int_{t_n}^{t_{n+1}}\left(\sum_{k=1}^{d}\ell_{j,k}^{(-1)}\left(\frac{\widetilde{\mu}_k(t_n)}{\widehat{Y}_k(s)}-\frac{\widetilde{\mu}_k(s)}{Y_k(s)}\right)\right)^2 ds\right)^{\frac{r}{2}}\right]$$

$$\leq C\mathbb{E}\left[\left(\sum_{n=0}^{N-1}\int_{t_n}^{t_{n+1}}\sum_{k=1}^{d}\left(\frac{\widetilde{\mu}_k(t_n)}{\widehat{Y}_k(s)}-\frac{\widetilde{\mu}_k(s)}{Y_k(s)}\right)^2 ds\right)^{\frac{r}{2}}\right]$$

$$\leq C\mathbb{E}\left[\left(\sum_{n=0}^{N-1}\int_{t_n}^{t_{n+1}}\sum_{k=1}^{d}\frac{1}{\widehat{Y}_k^2(s)}(\widetilde{\mu}_k(t_n)-\widetilde{\mu}_k(s))^2 ds\right)^{\frac{r}{2}}\right]$$

$$+ C\mathbb{E}\left[\left(\sum_{n=0}^{N-1}\int_{t_n}^{t_{n+1}}\sum_{k=1}^{d}\left(\frac{1}{\widehat{Y}_k(s)}-\frac{1}{Y_k(s)}\right)^2 ds\right)^{\frac{r}{2}}\right]$$

$$\leq C\mathbb{E}\left[\left(\sum_{k=1}^{d}\sum_{n=0}^{N-1}\int_{t_n}^{t_{n+1}}(\widetilde{\mu}_k(t_n)-\widetilde{\mu}_k(s))^2 ds\right)^{\frac{r}{2}}\right]$$

$$+ C\mathbb{E}\left[\left(\sum_{k=1}^{d}\sup_{s\in[0,T]}\left(\frac{1}{\widehat{Y}_k(s)}-\frac{1}{Y_k(s)}\right)^2\right)^{\frac{r}{2}}\right]$$

$$\leq C\mathbb{E}\left[\left(\sum_{k=1}^{d}\sum_{n=0}^{N-1}\int_{t_n}^{t_{n+1}}(t_n-s)^{2\lambda} ds\right)^{\frac{r}{2}}\right] + C\Delta_N^{r\lambda}$$

$$\leq C\Delta_N^{r\lambda},$$

where C is, as always, a constant that does not depend on the partition and which may vary from line to line. Similarly, for each $j = 1, ..., d$,

$$\mathbb{E}\left[\left|\sum_{j=1}^{d} \frac{T}{N} \sum_{n=0}^{N-1} \left(\sum_{k=1}^{d} \ell_{j,k}^{(-1)} \frac{\widetilde{\mu}_k(t_n)}{\widehat{Y}_k(t_n)}\right)^2 - \sum_{j=1}^{d} \int_0^T \left(\sum_{k=1}^{d} \ell_{j,k}^{(-1)} \frac{\widetilde{\mu}_k(s)}{\widehat{Y}_k(s)}\right)^2\right|^r\right] \le C\Delta_N^{r\lambda}.$$

Therefore we conclude that

$$\mathbb{E}\left[\left|\log \mathcal{E}_T^{\min} - \log \widehat{\mathcal{E}}_T^N\right|^r\right] \le C\Delta_N^{r\lambda}.$$

Using the arguments similar to Proposition 7.2, one can see that for all $r \ge 1$ there exists a constant C that does not depend on the partition such that

$$\mathbb{E}\left[(\mathcal{E}_T^{\min})^r\right] + \max_{N\ge 1} \mathbb{E}\left[(\widehat{\mathcal{E}}_T^N)^r\right] < C \tag{8.21}$$

and hence

$$\mathbb{E}\left[|\mathcal{E}_T^{\min} - \widehat{\mathcal{E}}_T^N|^r\right] \le C\mathbb{E}\left[\left((\mathcal{E}_T^{\min})^r + (\widehat{\mathcal{E}}_T^N)^r\right)\left|\log \mathcal{E}_T^{\min} - \log \widehat{\mathcal{E}}_T^N\right|^r\right] \le C\Delta_N^{r\lambda}.$$

□

Now we are ready for the last main result of this subsection. We omit the proof since it is straightforward and utilizes the same machinery as the one of Theorem 8.4 (one just has to additionally use Theorem 8.6 in order to estimate expectations of the type $\mathbb{E}\left[|\mathcal{E}_T^{\min} - \widehat{\mathcal{E}}_T^N|^r\right]$).

Theorem 8.7 *Let $S = (S_1, ..., S_d)$ be given by (8.10), $X = (X_1, ..., X_d)$ be the corresponding log-price process, $X_i = \log S_i$, $f\colon \mathbb{R} \to \mathbb{R}$ satisfy Assumption 8.1, $F(s) := \int_0^s f(u)du$ and $\alpha_i > 0$, $i = 1, ..., d$. Also, let $\mathcal{E}_T^{\min}$ be the density of the minimal martingale measure defined by (8.14). Fix a uniform partition $\{0 = t_0 < t_1 < ... < t_N = T\}$, $t_n = \frac{Tn}{N}$, with the mesh satisfying (7.28) and take the discretized processes $\widehat{Y} = (\widehat{Y}_1, ..., \widehat{Y}_d)$, $\widehat{X} = (\widehat{X}_1, ..., \widehat{X}_d)$, $\widehat{S} = (\widehat{S}_1, ..., \widehat{S}_d)$ with respect to this partition as described in Sect. 7.4. Finally, let $\widehat{\mathcal{E}}_T^N$ given by (8.20) and take $\lambda < \min_{i=1,...,d} H_i$ with H_i being the regularity order of the volatility Y_i. Then*

(1) for any $r \ge 1$, there exists a constant $C > 0$ such that

$$\mathbb{E}\left[\left|\mathcal{E}_T^{\min} \frac{F\left(\sum_{i=1}^d \alpha_i S_i(T)\right)}{\alpha_1 S_1(T)}\left(1 + \frac{\int_0^T \frac{1}{Y_1(v)} d\widetilde{W}_1(v) - \sum_{k=1}^d \int_0^T \beta_k \frac{\widetilde{\mu}_k(v)}{Y_k(v)Y_1(v)} dv}{Tu_{1,1}}\right)\right.\right.$$
$$\left.\left. - \widehat{\mathcal{E}}_T^N \frac{F\left(\sum_{i=1}^d \alpha_i \widehat{S}_i(T)\right)}{\alpha_1 \widehat{S}_1(T)}\left(1 + \frac{\int_0^T \frac{1}{\widehat{Y}_1(v)} d\widetilde{W}_1(v) - \sum_{k=1}^d \int_0^T \beta_k \frac{\widetilde{\mu}_k(v)}{\widehat{Y}_k(v)\widehat{Y}_1(v)} dv}{Tu_{1,1}}\right)\right|^r\right]$$

$$\leq C\Delta_N^{r\lambda},$$

(2) there exists a constant $C > 0$ such that

$$\Bigg|\mathbb{E}\left[\mathcal{E}_T^{\min} f\left(\sum_{i=1}^d \alpha_i S_i(T)\right)\right] - \mathbb{E}\left[\widehat{\mathcal{E}}_T^N \frac{F\left(\sum_{i=1}^d \alpha_i \widehat{S}_i(T)\right)}{\alpha_1 \widehat{S}_1(T)}\left(1 + \frac{\int\limits_0^T \frac{1}{\widehat{Y}_1(v)} d\widetilde{W}_1(v) - \sum_{k=1}^d \int\limits_0^T \beta_k \frac{\widetilde{\mu}_k(v)}{\widehat{Y}_k(v)\widehat{Y}_1(v)} dv}{T u_{1,1}}\right)\right]\Bigg|$$
$$\leq C\Delta_N^{\lambda}.$$

8.3.2 Approximation via Conditional Gaussianity for One-Dimensional Model

The core idea of the approach described in Sect. 8.3.1 was quite straightforward: simply plug in the simulated paths $\widehat{Y}$, $\widehat{S}$ and $\widehat{X}$ into representations (8.8) or (8.9). However, this requires simulation of *both* volatility and the price which inevitably leads to higher variance in Monte Carlo estimations. However, we can do better than that: it is possible to avoid simulation of the price and in this subsection we describe the corresponding approach.

Remark 8.7 For simplicity, throughout this subsection, we assume that $d = 1$ and the price S is given by (8.3), together with all the assumptions going with it.

Our "improved" algorithm is based on the following observation: *the random vector $\left(\widehat{X}(T), \int_0^T \frac{1}{Y(u)} d\widetilde{W}_1(u)\right)^T$ is conditionally Gaussian with respect to the σ-field $\widetilde{\mathcal{F}}_T^2 := \sigma\{\widetilde{W}_2(t),\ t \in [0,T]\}$*. Furthermore, we even know the parameters of the joint conditional distribution of the random vector $\left(\widehat{X}(T), \int_0^T \frac{1}{Y(u)} d\widetilde{W}_1(u)\right)$, as described in the lemma below.

Lemma 8.1 *The random vector $\left(\widehat{X}(T), \int_0^T \frac{1}{Y(u)} d\widetilde{W}_1(u)\right)^T$ is conditionally Gaussian with respect to the σ-field $\widetilde{\mathcal{F}}_T^2 := \sigma\{\widetilde{W}_2(t),\ t \in [0,T]\}$, with conditional mean*

$$\begin{pmatrix} X(0) + \frac{T}{N}\sum_{k=0}^{N-1}\left(\mu(t_k) - \frac{1}{2}\widehat{Y}^2(t_k)\right) + u_{1,2}\sum_{k=0}^{N-1}\widehat{Y}(t_k)(\widetilde{W}_2(t_{k+1}) - \widetilde{W}_2(t_k)) \\ 0 \end{pmatrix}$$
$$=: \begin{pmatrix} m(\widehat{Y}) \\ 0 \end{pmatrix} \tag{8.22}$$

and conditional covariance matrix

$$\widehat{C} = \begin{pmatrix} u_{1,1}^2 \frac{T}{N}\sum_{k=0}^{N-1}\widehat{Y}^2(t_k) & u_{1,1}T \\ u_{1,1}T & \frac{T}{N}\sum_{k=0}^{N-1}\widehat{Y}^{-2}(t_k) \end{pmatrix} =: \begin{pmatrix} u_{1,1}^2\, V_1^2 & u_{1,1}T \\ u_{1,1}T & V_2^2 \end{pmatrix},$$

where $V_1^2 := \frac{T}{N}\sum_{k=0}^{N-1}\widehat{Y}^2(t_k)$ *and* $V_2^2 := \frac{T}{N}\sum_{k=0}^{N-1}\widehat{Y}^{-2}(t_k)$. *Moreover, if, additionally,*

$$\det\widehat{C} = u_{1,1}^2\left(V_1^2V_2^2 - T^2\right) > 0$$

with probability 1, then the random vector $\left(\widehat{X}(T), \int_0^T \frac{1}{\widehat{Y}(u)}d\widetilde{W}_1(u)\right)^T$ *has a conditional density of the form*

$$\phi(x, y) = \frac{1}{2\pi\sqrt{\det\widehat{C}}}\exp\left\{-\frac{V_2^2(x - m(\widehat{Y}))^2 + u_{1,1}^2\, V_1^2 y^2 - 2Tu_{1,1}(x - m(\widehat{Y}))y}{2\det\widehat{C}}\right\}.$$

Proof Recall that the random variable $\widehat{X}(T)$ can be represented in the form

$$\widehat{X}(T) = X(0) + \frac{T}{N}\sum_{k=0}^{N-1}\left(\mu(t_k) - \frac{\widehat{Y}^2(t_k)}{2}\right) + u_{1,1}\sum_{k=0}^{N-1}\widehat{Y}(t_k)\left(\widetilde{W}_1(t_{k+1}) - \widetilde{W}_1(t_k)\right)$$
$$+ u_{1,2}\sum_{k=0}^{N-1}\widehat{Y}(t_k)\left(\widetilde{W}_2(t_{k+1}) - \widetilde{W}_2(t_k)\right).$$

Now the explicit form of conditional mean, conditional covariance matrix and conditional density (provided that the conditional covariance matrix is non-degenerate) can be computed directly. □

Theorem 8.8 *Let* $\det\widehat{C} > 0$ *with probability 1, where* $\widehat{C}$ *is the conditional covariance matrix from Lemma 8.1. Then*

$$\frac{1}{Tu_{1,1}}\mathbb{E}\left[G(\widehat{X}(T))\int_0^T \frac{1}{\widehat{Y}(u)}d\widetilde{W}_1(u)\right]$$
$$= \frac{1}{\sqrt{2\pi}u_{1,1}^2}\int_{\mathbb{R}} G(x)\mathbb{E}\left[\frac{x - m(\widehat{Y})}{V_1^3}\exp\left\{-\frac{(x - m(\widehat{Y}))^2}{2u_{1,1}^2V_1^2}\right\}\right]dx. \tag{8.23}$$

Proof Using Lemma 8.1, we can write

$$
\begin{aligned}
\mathbb{E}\left[G(\widehat{X}(T))\int_0^T \frac{1}{\widehat{Y}(u)}d\widetilde{W}_1(u)\right] &= \mathbb{E}\left[\mathbb{E}\left[G(\widehat{X}(T))\int_0^T \frac{1}{\widehat{Y}(u)}d\widetilde{W}_1(u)\,\middle|\,\widetilde{\mathcal{F}}_T^2\right]\right] \\
&= \mathbb{E}\left[\int_{\mathbb{R}^2} G(x)y\phi(x,y)dxdy\right] \\
&= \int_{\mathbb{R}} G(x)\mathbb{E}\left[\int_{\mathbb{R}} y\phi(x,y)dy\right]dx.
\end{aligned}
$$

The integral $\int_{\mathbb{R}} y\phi(x,y)dy$ can be simplified. Namely, if one denotes $\widetilde{x} := x - m(\widehat{Y})$,

$$
\begin{aligned}
&\int_{\mathbb{R}} y\phi(x,y)dy \\
&= \frac{1}{2\pi\sqrt{\det\widehat{C}}}\int_{\mathbb{R}} y\exp\left\{-\frac{V_2^2\widetilde{x}^2 + u_{1,1}^2 V_1^2 y^2 - 2u_{1,1}T\widetilde{x}y}{2\det\widehat{C}}\right\}dy \\
&= \frac{1}{2\pi\sqrt{\det\widehat{C}}}\int_{\mathbb{R}} y\exp\left\{-\frac{1}{2\det\widehat{C}}\left(\left(u_{1,1}V_1y - \frac{T\widetilde{x}}{V_1}\right)^2 + V_2^2\widetilde{x}^2 - \frac{T^2\widetilde{x}^2}{V_1^2}\right)\right\}dy \\
&= \frac{1}{2\pi\sqrt{\det\widehat{C}}}\exp\left\{-\frac{(V_1^2V_2^2 - T^2)\widetilde{x}^2}{2V_1^2\det\widehat{C}}\right\}\int_{\mathbb{R}} y\exp\left\{-\frac{\left(y - \frac{T\widetilde{x}}{u_{1,1}V_1^2}\right)^2}{2\frac{\det\widehat{C}}{u_{1,1}^2V_1^2}}\right\}dy \\
&= \frac{1}{2\pi\sqrt{\det\widehat{C}}}\exp\left\{-\frac{\widetilde{x}^2}{2u_{1,1}^2V_1^2}\right\}\sqrt{2\pi}\frac{T\widetilde{x}\sqrt{\det\widehat{C}}}{u_{1,1}^2V_1^3} \\
&= \frac{T\widetilde{x}}{\sqrt{2\pi}u_{1,1}^2V_1^3}\exp\left\{-\frac{\widetilde{x}^2}{2u_{1,1}^2V_1^2}\right\}.
\end{aligned}
$$

This yields the result. □

Equation (8.23) from Theorem 8.8 essentially relies on the condition

$$\det\widehat{C} > 0$$

with probability 1. A simple sufficient condition is provided by the following proposition.

Proposition 8.3 *Let $Z(t) = \int_0^t \mathcal{K}(t,s)d\widetilde{W}_2(s)$, $t \in [0,T]$, and $\{0 = t_0 < t_1 < \ldots < t_N = T\}$ be a uniform partition of $[0,T]$ with the mesh $\Delta_N := \frac{T}{N}$ having the following property: $C_b\Delta_N < 1$, where C_b is an upper bound for $\frac{\partial b}{\partial y}$ from Assumption 6.3(b5).*

If the random vector

$$(Z(t_1) - Z(t_0), Z(t_2) - Z(t_1), ..., Z(t_N) - Z(t_{N-1}))^T$$

has absolutely continuous distribution w.r.t. the Lebesgue measure on $\mathbb{R}^N$*, then* $\det \widehat{C} > 0$ *with probability 1 and* (8.23) *holds.*

Proof Note that $\det \widehat{C} \geq 0$ a.s., because $\widehat{C}$ is a conditional covariance matrix, i.e. we need to check that $\det \widehat{C} \neq 0$ with probability 1. Taking into account that

$$\det \widehat{C} = u_{1,1}^2 \left(V_1^2 V_2^2 - T^2\right) = u_{1,1}^2 \left(\left(\frac{T}{N} \sum_{k=0}^{N-1} \widehat{Y}^2(t_k)\right) \left(\frac{T}{N} \sum_{l=0}^{N-1} \widehat{Y}^{-2}(t_l)\right) - T^2\right),$$

where $u_{1,1}^2 > 0$, it is clear that

$$\det \widehat{C} \neq 0 \quad \Longleftrightarrow \quad \left(\frac{1}{N} \sum_{k=0}^{N-1} \widehat{Y}^2(t_k)\right) \left(\frac{1}{N} \sum_{l=0}^{N-1} \widehat{Y}^{-2}(t_l)\right) \neq 1.$$

Note that, by the CauchyâŁ“Schwarz inequality,

$$\left(\frac{1}{N} \sum_{k=0}^{N-1} \widehat{Y}^2(t_k)\right) \left(\frac{1}{N} \sum_{l=0}^{N-1} \widehat{Y}^{-2}(t_l)\right) \geq \left(\frac{1}{N} \sum_{k=0}^{N-1} \widehat{Y}(t_k) \widehat{Y}^{-1}(t_k)\right)^2 = 1,$$

and the equality is possible if and only if

$$\widehat{Y}(t_0) = \widehat{Y}(t_1) = \cdots = \widehat{Y}(t_{N-1}).$$

Therefore, in order to prove that $\det \widehat{C} \neq 0$ a.s., it is sufficient to check that the random vector $(\widehat{Y}(t_1), ..., \widehat{Y}(t_N))^T$ has density with respect to the Lebesgue measure on $\mathbb{R}^N$.

Denote

$$b_n(y) := y - b(t_n, y)\Delta_N, \quad n = 1, ..., N,$$

and observe that for any $n = 1, ..., N$ is a 1-to-1 mapping of the interval $(\varphi(t_n), \psi(t_n))$ onto $\mathbb{R}$ with $b_n' > 0$. By the definition of $\widehat{Y}$, for any $n = 1, ..., N$,

$$\widehat{Y}(t_n) = b_n^{-1} \left(\widehat{Y}(t_{n-1}) + Z(t_n) - Z(t_{n-1})\right)$$

and thus

$$\begin{pmatrix} \widehat{Y}(t_1) \\ \widehat{Y}(t_2) \\ \vdots \\ \widehat{Y}(t_N) \end{pmatrix} = \begin{pmatrix} b_1^{-1}\left(Y(0) + Z(t_1) - Z(t_0)\right) \\ b_2^{-1}\left(b_1^{-1}\left(Y(0) + Z(t_1) - Z(t_0)\right) + Z(t_2) - Z(t_1)\right) \\ \vdots \\ b_N^{-1}\left(b_{N-1}^{-1}(\cdots)\right) + Z(t_N) - Z(t_{N-1})\right) \end{pmatrix}$$
$$=: I(Z(t_1) - Z(t_0), Z(t_2) - Z(t_1), ..., Z(t_N) - Z(t_{N-1})). \tag{8.24}$$

It is straightforward to check that function $I\colon \mathbb{R}^N \to \prod_{n=1}^N (\varphi(t_n), \psi(t_n))$ from (8.24) is a bijection and its Jacobian det $J_I(\mathbf{z})$ is strictly positive for any $\mathbf{z} \in \mathbb{R}^N$. This implies that the random vector $(\widehat{Y}(t_1), ..., \widehat{Y}(t_N))^T$ has density, i.e. is absolutely continuous w.r.t. the Lebesgue measure, which ends the proof. □

8.4 Simulations

Let us illustrate the algorithms described in Sect. 8.3 with numerical simulations. For simplicity, we assume that

- $T = 1, d = 1$;
- an explicit shape of the SVV model is

$$S(t) = 1 + \int_0^t Y(s)S(s)\left(\sqrt{0.75}dB_1(s) - 0.5dB_2(s)\right),$$
$$Y(t) = 0.5 + \int_0^t \left(\frac{0.005}{(Y(s) - 0.05)^5} - \frac{0.005}{(1 - Y(s))^5} + 0.05(0.5 - Y(s))\right) ds$$
$$+ 0.3 \int_0^t \mathcal{K}(t, s)dB_2(s), \tag{8.25}$$

  where B_1, B_2 are independent Brownian motions and

$$\mathcal{K}(t, s) := \frac{1}{\Gamma(H + 1/2)}(t - s)^{H - \frac{1}{2}} \mathbb{1}_{s<t}$$

  with $H = 0.2$;
- the instantaneous interest rate $r(t) \equiv 0$;
- the option payoff function is a discontinuous function given by

$$f(s) = \mathbb{1}_{(0.5,\infty)}(s) + \mathbb{1}_{(1,\infty)}(s) + \mathbb{1}_{(1.5,\infty)}(s).$$

Table 8.1 Computation of $\mathbb{E}[f(S(T))]$: standard Monte Carlo versus Malliavin MC with double discretization versus Malliavin MC via conditional Gaussianity. Columns entitled "Mean" contain the corresponding estimates of $\mathbb{E}[f(S(T))]$ whereas the "Sd" columns present the standard deviations of (8.26) for Std. MC, (8.27) for MMCDD and (8.28) for MMCCG

Partition size, N	Std. MC		MMCDD		MMCCG	
	Mean	Sd	Mean	Sd	Mean	Sd
10	1.5189	0.9220	1.3963	1.3861	1.4110	0.3213
100	1.4554	0.9161	1.4147	1.4149	1.4071	0.3167
1000	1.3960	0.9408	1.4111	1.3842	1.4081	0.3181
10000	1.4038	0.9310	1.4070	1.4008	1.4089	0.3193

Remark 8.8 Comparing (8.25) with the representation (8.3), we have that $B_1 = \widetilde{W}_1$, $B_2 = \widetilde{W}_2$, $u_{1,1} = \sqrt{0.75}$ and $u_{1,2} = -0.5$.

We want to compute the value $\mathbb{E}[f(S(T))]$ and compare the following three numerical methods:

(1) **Standard Monte Carlo (Std. MC)**: we simply simulate 10,000 independent realizations of

$$f(\widehat{S}(T)) \tag{8.26}$$

and average over them;

(2) **Malliavin Monte Carlo with double discretization (MMCDD)**: we simulate 10,000 independent realizations of

$$\frac{F\left(\widehat{S}(T)\right)}{\widehat{S}(T)}\left(1 + \int_0^T \frac{1}{\widehat{Y}(u)} dW_1(u)\right) \tag{8.27}$$

and then average over them;

(3) **Malliavin Monte Carlo via conditional Gaussianity (MMCCG)**: we exploit the method described in Sect. 8.3.2 and average over 10000 independent realizations of

$$\frac{1}{\sqrt{2\pi}} \int_{\mathbb{R}} G(x) \left(\frac{x - m(\widehat{Y})}{V_1^3}\right) \exp\left\{-\frac{(x - m(\widehat{Y}))^2}{2V_1^2}\right\} dx. \tag{8.28}$$

For all three methods, we consider the uniform partitions of sizes $N = 10$, 100, 1000, and 10000. Table 8.1 contains information on means (being also the estimators of $\mathbb{E}[f(S(T))]$) and standard deviations of (8.26), (8.27) and (8.28) for different partition sizes and Hurst indices. Recall also that there are two sources of error in this type of simulation:

(I) the error between $\mathbb{E}[f(S(T))]$ and

(1) $\mathbb{E}[f(\widehat{S}(T))]$ for Std. MC,

(2) $\mathbb{E}\left[\frac{F(\widehat{S}(T))}{\widehat{S}(T)}\left(1+\int\limits_0^T \frac{1}{\widehat{Y}_1(u)} dW_1(u)\right)\right]$ for MMCDD and

(3) $\mathbb{E}\left[\frac{1}{\sqrt{2\pi}}\int_{\mathbb{R}} G(x)\left(\frac{x-m(\widehat{Y})}{V_1^3}\right)\exp\left\{-\frac{(x-m(\widehat{Y}))^2}{2V_1^2}\right\}dx\right]$ for MMCCG;

(II) the error coming from the Monte Carlo estimation of the corresponding expectations.

The error of type (I) is generally controlled by the size of the partition N, and the Malliavin integration-by-parts technique is used specifically to reduce it. The error of type (II) is reduced by increasing the number of Monte Carlo samples. In this context, standard deviations (Sd) of random variables (8.26), (8.27) and (8.28) evaluate how many Monte Carlo simulations are required to achieve a certain level of accuracy of the mean estimation.

As we can see, the estimators of $\mathbb{E}[f(S(T))]$ by the standard Monte Carlo for small N differ substantially from the ones for larger N whereas both MMCDD and MMCCG show better stability with respect to the size of the partitions (thanks to the Malliavin integration-by-parts regularization technique). In addition, the MMCCG method showed by far the lowest standard deviations out of all three algorithms: this happens because there is no additional error from simulating $\widehat{S}(T)$ which results in a far more stable behavior of (8.28) (c.f. [3]).

References

1. Altmayer, M., Neuenkirch, A.: Multilevel Monte Carlo quadrature of discontinuous payoffs in the generalized Heston model using Malliavin integration by parts. SIAM J. Financ. Math. **6**(1), 22–52 (2015). https://doi.org/10.1137/130933629
2. Avikainen, R.: On irregular functionals of SDEs and the Euler scheme. Finance Stochast. **13**(3), 381–401 (2009)
3. Bezborodov, V., Persio, L.D., Mishura, Y.: Option pricing with fractional stochastic volatility and discontinuous payoff function of polynomial growth. Methodol. Comput. Appl. Probab. **21**(1), 331–366 (2018). https://doi.org/10.1007/s11009-018-9650-3
4. Di Nunno, G., Mishura, Y., Yurchenko-Tytarenko, A.: Option pricing in sandwiched Volterra volatility model. SIAM J. Financ. Math. **15**(3), 824–882 (2024). https://doi.org/10.1137/22m1521328
5. Schweizer, M.: On the minimal martingale measure and the Föllmer-Schweizer decomposition. Stoch. Anal. Appl. **13**(5), 573–599 (1995). https://doi.org/10.1080/07362999508809418

Chapter 9
Quadratic Hedging in the SVV Model

Abstract This chapter concludes the study of the Sandwiched Volterra Volatility (SVV) model by addressing the problem of quadratic mean-variance hedging. As perfect replication is not possible within the SVV framework, we focus on minimizing the expected squared hedging error for European-type payoffs. The optimal solution, classically characterized by the Galtchouk–Kunita–Watanabe decomposition, admits an explicit representation via the non-anticipating derivative. To make this representation numerically tractable, we build on previous developments by replacing the original volatility process with its Markovian approximation. Based on this approximation, we propose two numerical algorithms for computing the optimal hedging strategy: Nested Monte Carlo (NMC) and Least Squares Monte Carlo (LSMC).

9.1 Introduction

The primary focus of Chaps. 7 and 8 was the problem of derivative *pricing*: specifically, how well the implied volatility surface generated from SVV option prices replicates observed market volatility, along with a discussion of martingale measures, numerical algorithms, and related topics. However, until now we have overlooked a non-trivial but extremely important topic in the context of stochastic volatility models: *option hedging*.

Let us provide some context. In general, stochastic volatility models yield incomplete markets, and, as demonstrated in Theorem 7.5, the SVV model is no exception. Among other things, it means that the model does not allow to perfectly replicate all derivatives with self-financing strategies, and, as a result, an alternative approach is required to hedge positions as effectively as possible. One possibility here is to perform quadratic hedging in the form of mean-variance hedging (see e.g. [14, 15]) or minimal variance (equivalently, minimum-variance or variance-optimal) hedging (see, e.g. [2, 3, 7]), as we will discuss here after. Given a discounted derivative payoff F and a discounted price process S (that is assumed to be a square integrable martingale w.r.t. a measure $\mathbb{Q}$), one can consider an optimization problem of the form

G. Di Nunno et al., *Volterra Volatility Models*, Springer Finance,
https://doi.org/10.1007/978-3-032-26576-0_9

$$\inf_u \mathbb{E}_{\mathbb{Q}}\left[\left(F - \int_0^T u_s dS_s\right)^2\right], \tag{9.1}$$

where the minimization is performed over all adapted S-integrable strategies u.

We can immediately note that the theory behind such problems is very well studied: in particular, the *existence* of the solution to (9.1) is guaranteed by the celebrated *Galtchouk-Kunita-Watanabe decomposition theorem* (see e.g. [13, 15]). Furthermore, [2] gives the *explicit representation* of the optimal hedging strategy: according to [2, Theorem 2.1], the optimal hedging portfolio u minimizing (9.1) can be written as the *non-anticipating derivative* $\mathfrak{D}F$ defined as the L^2-limit of simple processes

$$\mathfrak{D}F := L^2\text{-}\lim_{|\pi|\to 0} u_\pi, \quad u_\pi := \sum_{k=0}^{n-1} \frac{\mathbb{E}_{\mathbb{Q}}[(S_{t_{k+1}} - S_{t_k})F \mid \mathcal{F}_{t_k}]}{\mathbb{E}_{\mathbb{Q}}[(S_{t_{k+1}} - S_{t_k})^2 \mid \mathcal{F}_{t_k}]} \mathbb{1}_{(t_k, t_{k+1}]}, \tag{9.2}$$

where $|\pi| := \max_k(t_k - t_{k-1})$ is the mesh of the partition

$$\pi = \{0 = t_0 < t_1 < \cdots < t_n = T\}.$$

Note that (9.2) is explicit in the sense that the hedge is written only in terms of the discounted price model, the information flow of reference, and the claim F. Nevertheless, the practical use of (9.2) is limited. If there is a non-trivial dependence of S on the past – a property that we are actually trying to achieve to reproduce the "memory" effect in real-life markets – the analytical expressions for $\mathfrak{D}F$ are usually impossible to obtain. To complicate things even further, the numerical computation of the conditional expectations in (9.2) by e.g. Monte Carlo methods is also very problematic due to the same past-dependence.

The goal of this chapter based on our paper [5] is to explore the numerical aspects of hedging problem (9.1) within the SVV model introduced in Chap. 7. Namely, fix H-Hölder continuous functions $0 < \varphi < \psi$, $H \in (0, 1)$, and consider the SVV model with a risk-free dynamics of the form

$$S(t) = S(0) + \int_0^t Y(s)S(s)\left(\rho dB_1(s) + \sqrt{1-\rho^2}dB_2(s)\right), \tag{9.3}$$

$$Y(t) = Y(0) + \int_0^t b(s, Y(s))ds + Z(t), \tag{9.4}$$

$t \in [0, T]$, where $B = (B_1, B_2)$ is a 2-dimensional Brownian motion, $\rho \in (-1, 1)$, the process Z is defined by

$$Z(t) = \int_0^t \mathcal{K}(t,s)dB_1(s), \quad t \in [0,T],$$

with the kernel $\mathcal{K}$ satisfying Assumption 6.1. We also assume that the coefficients of (9.4) satisfy Assumption 6.3, so that, with probability 1,

$$0 < \varphi(t) < Y(t) < \psi(t), \quad t \in [0,T].$$

For the sake of simplicity, some technicalities have been streamlined in the formulation of equations (9.3)–(9.4).

- First, here and throughout this entire chapter, we assume that the instantaneous interest rate $\nu \equiv 0$ (see Remark 7.2).
- Second, we impose the SVV structure on the *arbitrage-free* dynamics of the asset price, i.e. the model (9.3)–(9.4) is already formulated under a martingale measure $\mathbb{Q}$. Accordingly, in the remainder of this chapter, we will omit the subscript $\mathbb{Q}$ from conditional expectations, i.e. the hedging problem (9.1) reads as

$$\inf_u J(u) := \inf_u \mathbb{E}\left[\left(F - \int_0^T u(s)dS(s)\right)^2\right]. \tag{9.5}$$

- Third, as our filtration, we pick the $\mathbb{Q}$-augmented filtration $\mathbb{F} = (\mathcal{F}_t)_{t\in[0,T]}$ generated by the 2-dimensional Brownian motion (B^1, B^2).
- Finally, we assume that the payoff F is of European type, i.e. $F = f(S(T))$.

Note that the two-sided SVV structure is immediately beneficial in this setting: Theorem 7.2 guarantees that $F = f(S(T)) \in L^2(\mathbb{Q})$ as long as the payoff function is of polynomial growth, i.e. the problem (9.5) is well-posed and has a solution given by the Galtchouk-Kunita-Watanabe decomposition.

Our strategy for finding the solution to (9.5) will be as follows. In Sect. 6.4, we introduced a way to approximate the volatility Y with some multidimensional Markovian process. If we replace Y in (9.3) with this approximation, we get rid of the dependence on the past and may hope that the conditional expectations in (9.2) are easier to compute numerically. Of course, in order for that approach to work, we have to impose some restrictions on the shape of $\mathcal{K}$ and, in addition, there must be a sequence of kernels $\{\mathcal{K}_m,\ m \geq 0\}$ satisfying Assumption 6.5 that converges to $\mathcal{K}$. In other words, throughout this chapter, we will always make the following assumption.

Assumption 9.1 The Volterra kernel $\mathcal{K}$ is a difference kernel, i.e.

$$\mathcal{K}(t,s) = \mathcal{K}(t-s)\mathbb{1}_{s<t}, \quad t,s \in [0,T], \qquad \int_0^T \mathcal{K}^2(x)dx < \infty,$$

and there exists a sequence of difference kernels $\{\mathcal{K}_m,\ m \ge 0\}$,

$$\mathcal{K}_m(t,s) = \mathcal{K}_m(t-s)\mathbb{1}_{s<t}, \quad m \ge 0, \qquad \int_0^T \mathcal{K}_m^2(x)dx < \infty,$$

satisfying Assumption 6.5 such that $\mathcal{K}_m \to \mathcal{K}$, $m \to \infty$, in $L^2([0,T]^2)$.

Assumption 9.1 obviously restricts the choice of the driving noise Z but, as we have seen in Examples 6.3 and 6.4, some of the most popular classes of kernels *do* admit such an approximation: in particular, we cover all Hölder continuous kernels, rough fractional kernels and their linear combinations.

Notation 9.1 Since, by Assumption 9.1, $\mathcal{K}$ is of the difference type, one can write

$$\begin{aligned}
\|\mathcal{K}\|^2_{L^2([0,T]^2)} &:= \int_0^T\int_0^T \mathcal{K}^2(t,s)dsdt = \int_0^T\int_0^t \mathcal{K}^2(t-s)dsdt \\
&= \int_0^T\int_0^t \mathcal{K}^2(x)dxdt \\
&\le T\int_0^T \mathcal{K}^2(x)dx =: T\|\mathcal{K}\|^2_{L^2([0,T])}.
\end{aligned}$$

Here, we slightly abuse the notation: technically speaking, $\mathcal{K} = \mathcal{K}(t,s)$ and $\mathcal{K} = \mathcal{K}(x)$ are two different functions defined on two different sets. However, given the simple connection between them, we will keep denoting both kernels as $\mathcal{K}$ and often characterize the convergence in terms of $\|\cdot\|_{L^2([0,T])}$ rather than $\|\cdot\|_{L^2([0,T]^2)}$.

We organize the chapter as follows. First, we provide more detailed information about the non-anticipating derivatives. Then, we swap the volatility Y inside the non-anticipating derivative with its Markovian approximation and analyze how small is the resulting error. Finally, we provide some numerical techniques to actually compute the "Markovianized" optimal hedge.

9.2 Non-anticipating Derivative and Minimal Variance Hedging in the SVV Model

Let $\xi \in L^2(\mathbb{Q})$ and $\eta = \{\eta(t),\ t \in [0,T]\}$ be a square-integrable martingale w.r.t. some filtration $\mathbb{G} = \{\mathcal{G}_t,\ t \in [0,T]\}$ with quadratic variation process $[\eta] = \{[\eta]_t,\ t \in [0,T]\}$. For an arbitrary partition

$$\pi = \{0 = t_0 < t_1 < \cdots < t_n = T\}$$

of $[0, T]$ with the mesh $|\pi| := \max_k(t_k - t_{k-1})$, denote

$$\begin{aligned} u_\pi &:= \sum_{k=0}^{n-1} u_{\pi,k} \mathbb{1}_{(t_k, t_{k+1}]}, \\ u_{\pi,k} &:= \frac{\mathbb{E}\left[(\eta(t_{k+1}) - \eta(t_k))\xi \mid \mathcal{G}_{t_k}\right]}{\mathbb{E}\left[(\eta(t_{k+1}) - \eta(t_k))^2 \mid \mathcal{G}_{t_k}\right]}, \quad k = 0, 1, ..., n-1. \end{aligned} \tag{9.6}$$

Definition 9.1 Consider a monotone sequence of partitions $\{\pi_M,\ M \geq 1\}$, such that $|\pi_M| \to 0$ as $M \to \infty$. The $L^2(\mathbb{Q} \times [\eta])$-limit

$$\mathfrak{D}\xi := \lim_{|\pi_M| \to 0} u_{\pi_M} \tag{9.7}$$

is called the *non-anticipating derivative* of ξ w.r.t. η.

It turns out that $\mathfrak{D}\xi$ is well-defined and $\int_0^T \mathfrak{D}\xi(s) d\eta(s)$ gives the orthogonal $L^2(\mathbb{Q} \times [\eta])$-projection of ξ on the subspace of stochastic integrals w.r.t. η.

Theorem 9.1 *([2, Theorem 2.1]) The non-anticipating derivative $\mathfrak{D}\xi$ is well-defined, i.e. the limit (9.7) exists and does not depend on the particular choice of the partitions. Moreover, any $\xi \in L^2(\mathbb{Q})$ admits a unique representation of the form*

$$\xi = \xi_0 + \int_0^T \mathfrak{D}\xi(s) d\eta(s), \tag{9.8}$$

where $\xi_0 \in L^2(\mathbb{Q})$ is such that $\mathfrak{D}\xi_0 = 0$ and $\mathbb{E}\left[\xi_0 \int_0^T \mathfrak{D}\xi(s) d\eta(s)\right] = 0$. In other words, the infimum over all $\mathbb{G}$-adapted η-integrable strategies

$$\inf_u \mathbb{E}\left[\left(\xi - \int_0^T u(s) d\eta(s)\right)^2\right]$$

is attained at $u = \mathfrak{D}\xi$.

Notation 9.2 Since the limit (9.7) does not depend on the particular choice of partitions $\{\pi_M,\ M \geq 1\}$, we will use the notation

$$\mathfrak{D}\xi := \lim_{|\pi| \to 0} u_\pi,$$

meaning that the limit in $L^2(\mathbb{Q} \times [\eta])$ is taken along an arbitrary sequence of monotone partitions with the mesh going to zero.

Let us make a couple of comments regarding this result. First, by Theorem 9.1, it is easy to see that the linear operator $\mathfrak{D}$ is actually the dual of the Itô integral. Second, if one interprets ξ as a derivative payoff and the process η as a discounted asset price under the martingale measure $\mathbb{Q}$, the non-anticipating derivative $u = \mathfrak{D}\xi$ of ξ w.r.t. η represents the optimal hedging portfolio, exactly what we are looking for. In fact, if one examines the proof of [2, Theorem 2.1] more thoroughly, it implies that the pre-limit sum u_π in (9.6) is the $L^2(\mathbb{Q} \times [\eta])$-orthogonal projection of ξ onto the subspace generated by stochastic integrals of *simple* processes

$$\sum_{k=0}^{n-1} \zeta_k \mathbb{1}_{(t_k, t_{k+1}]} \tag{9.9}$$

Note that the portfolios of the form (9.9) are very relevant for applications: in real-life situations, trading cannot be "truly" continuous to utilize the limit $\mathfrak{D}\xi$, so simple processes (9.9) cover all realistic portfolios. Finally, a reader familiar with Malliavin calculus would immediately notice striking similarities between (9.8) and the Clark-Haussmann-Ocone (CHO) formula (see e.g. [6, 12]). However, we would like to stress that the non-anticipating derivative has several important advantages. Namely, the CHO formula exploits the Malliavin derivative which is tailored for Gaussian noises and random variables ξ falling in the domain of the corresponding Malliavin operator. In turn, formula (9.8) is available for *all* square integrable claims and all square integrable martingale integrators. For further details, we refer the reader to the survey [4].

9.3 Markovian Approximation of the SVV Price

Let all assumptions outlined in the beginning of this chapter hold and let S be defined by (9.3).

As we announced at the beginning of the chapter, our plan is to utilize the Markovian approximation derived in Sect. 6.4. For compact notation, we denote

$$W(t) := \rho B_1(t) + \sqrt{1-\rho^2} B_2(t), \quad t \in [0, T],$$

where $\rho \in (-1, 1)$ is from (2.21).

First, let us remind the notation associated to the Markovian approximations. Recall that each approximating kernel $\mathcal{K}_m$ from Assumption 9.1 is degenerate, i.e.

$$\mathcal{K}_m(t, s) = \sum_{i=0}^{m} e_{m,i}(t) f_{m,i}(s) \mathbb{1}_{s<t},$$

where $\int_0^T e_{m,i}^2(t)dt < \infty$ and $\int_0^T f_{m,i}^2(t)dt < \infty$, $i = 0, ..., m$, $m \geq 0$. Then, a natural approximation for the model (9.3)–(9.4) reads as

$$S_m(t) = S(0) + \int_0^t Y_m(s)S_m(s)dW(s), \tag{9.10}$$

$$Y_m(t) = Y(0) + \int_0^t b(s, Y_m(s))ds + Z_m(t), \tag{9.11}$$

where Z_m is given by

$$Z_m(t) := \int_0^t \mathcal{K}_m(t, s)dB_1(s) = \sum_{i=0}^m e_{m,i}(t) \int_0^t f_{m,i}(s)dB_1(s), \quad t \in [0, T]. \tag{9.12}$$

Similarly to the original process Y, each Y_m is "sandwiched" between φ and ψ, i.e.

$$0 < \varphi(t) < Y_m(t) < \psi(t), \quad t \in [0, T], \quad m \geq 0. \tag{9.13}$$

Observe that the $(m + 3)$-dimensional process $(S_m, Y_m, U_{m,0}, ..., U_{m,m})$, where

$$U_{m,i}(t) = \int_0^t f_{m,i}(s)dB_1(s), \quad i = 0, 1, ..., m, \tag{9.14}$$

is Markovian w.r.t. the filtration $\mathbb{F} = \{\mathcal{F}_t,\ t \in [0, T]\}$ generated by (B_1, B_2) and, *intuitively*, should approximate the original non-Markovian SVV model if $\mathcal{K}_m$ is close enough to $\mathcal{K}$. We have already performed the first step to back up this intuition by a theoretical foundation with Theorem 6.5 and Corollary 6.3: for any $t \in [0, T]$,

$$|Y(t) - Y_m(t)| \leq C \left(|Z(t) - Z_m(t)| + \xi_m \int_0^t |Z(u) - Z_m(u)|du \right),$$

where $C > 0$ is a constant that does not depend on m and positive random variables ξ_m, $m \geq 1$, are such that

$$\sup_{m \geq 1} \mathbb{E}[\xi_m^r] < \infty \tag{9.15}$$

for any $r > 0$. This, in turn, implies that

$$\mathbb{E}\left[\int_0^T |Y(t) - Y_m(t)|^2 dt\right] \leq C\|\mathcal{K} - \mathcal{K}_m\|^2_{L^2([0,T]^2)},$$

i.e. the "Markovianized" volatility Y_m indeed approximates Y. Our next stop is to prove a similar result for the price process S.

Let us start with a simple auxiliary lemma.

Lemma 9.1 *For any* $r \in \mathbb{R}$

$$\sup_{m\geq 1} \mathbb{E}\left[\sup_{t\in[0,T]} S_m^r(t)\right] < \infty.$$

Proof Using the argument identical to the proof of Theorem 7.2, it is straightforward to show that

$$\mathbb{E}\left[\sup_{t\in[0,T]} S_m^r(t)\right] \leq C\left(1 + C_1\sqrt{T}\max_{s\in[0,T]}\psi(s)\exp\left\{\frac{r^2T}{2}\max_{s\in[0,T]}\psi^2(s)\right\}\right), \tag{9.16}$$

where ψ is the upper bound for the volatility process,

$$C := S^r(0)\exp\left\{\frac{|r|(|r|+1)T}{2}\max_{s\in[0,T]}\psi^2(s)\right\}$$

and the constant C_1 comes from the Burkholder-Davis-Gundy inequality. Then, the result follows from the fact that the right-hand side of (9.16) does not depend on m. □

Theorem 9.2 *Let* $r \geq 2$ *be fixed. Then there exists constant* $C > 0$ *that does not depend on* m *such that*

$$\mathbb{E}\left[\sup_{t\in[0,T]} |S(t) - S_m(t)|^r\right] \leq C\|\mathcal{K} - \mathcal{K}_m\|^r_{L^2([0,T])}.$$

Proof For any $t \in [0, T]$,

$$|S(t) - S_m(t)|^r = \left|\int_0^t (Y(s)S(s) - Y_m(s)S_m(s))dW(s)\right|^r,$$

whence

$$\sup_{u\in[0,t]}|S(u) - S_m(u)|^r \leq \sup_{u\in[0,t]}\left|\int_0^u (Y(s)S(s) - Y_m(s)S_m(s))dW(s)\right|^r.$$

Then, using the Burkholder-Davis-Gundy inequality, we can write

$$\mathbb{E}\left[\sup_{u\in[0,t]}|S(u)-S_m(u)|^r\right] \le C\mathbb{E}\left[\left(\int_0^t \left(Y(s)S(s)-Y_m(s)S_m(s)\right)^2 ds\right)^{\frac{r}{2}}\right]$$
$$\le C\int_0^t \mathbb{E}\left[\left|Y(s)S(s)-Y_m(s)S_m(s)\right|^r\right]ds. \tag{9.17}$$

Using (9.13),

$$\max_{m\ge 1}\max_{t\in[0,T]}|Y_m(t)| \le \max_{t\in[0,T]}|\psi(t)|,$$

so

$$\int_0^t \mathbb{E}\left[|Y(s)S(s)-Y_m(s)S_m(s)|^r\right]ds$$
$$\le C\int_0^t \mathbb{E}\left[Y_m^r(s)|S(s)-S_m(s)|^r\right]ds$$
$$+C\int_0^t \mathbb{E}\left[S^r(s)|Y(s)-Y_m(s)|^r\right]ds \tag{9.18}$$
$$\le C\int_0^t \mathbb{E}\left[\sup_{u\in[0,s]}|S(u)-S_m(u)|^r\right]ds$$
$$+C\int_0^t \mathbb{E}\left[S^r(s)|Y(s)-Y_m(s)|^r\right]ds.$$

Studying the last summand of (9.18), we obtain by Theorem 6.5 that there exist positive random variables ξ_m, $m\ge 1$, such that

$$\int_0^t \mathbb{E}\left[S^r(s)|Y(s)-Y_m(s)|^r\right]ds$$
$$\le C\int_0^t \mathbb{E}\left[S^r(s)|Z(s)-Z_m(s)|^r\right]ds \tag{9.19}$$
$$+C\int_0^t \mathbb{E}\left[S^r(s)\xi_m^r\int_0^s |Z(u)-Z_m(u)|^r du\right]ds.$$

Also, by Theorem 7.2,

$$\mathbb{E}\left[\sup_{t\in[0,T]} S^{2r}(t)\right] < \infty,$$

so the Hölder inequality as well as the Gaussianity of the random variable $Z(u) - Z_m(u)$ yield

$$\begin{aligned}
&\int_0^t \mathbb{E}\left[S^r(s)|Z(s) - Z_m(s)|^r\right] ds \\
&\quad \le \int_0^t \left(\mathbb{E}\left[S^{2r}(s)\right]\right)^{\frac{1}{2}} \left(\mathbb{E}\left[|Z(s) - Z_m(s)|^{2r}\right]\right)^{\frac{1}{2}} ds \\
&\quad \le C \int_0^t \left(\int_0^s (\mathcal{K}(s,u) - \mathcal{K}_m(s,u))^2 du\right)^{\frac{r}{2}} ds \\
&\quad \le C \int_0^t \left(\int_0^s (\mathcal{K}(x) - \mathcal{K}_m(x))^2 dx\right)^{\frac{r}{2}} ds \\
&\quad \le C\|\mathcal{K} - \mathcal{K}_m\|^r_{L^2([0,T])}.
\end{aligned} \tag{9.20}$$

Additionally, (9.15) together with Theorem 7.2 imply that

$$\sup_{m\ge1} \mathbb{E}[\xi_m^{2r} \sup_{t\in[0,T]} S^{2r}(t)] < \infty.$$

Therefore

$$\begin{aligned}
&\int_0^t \mathbb{E}\left[S^r(s)\xi_m^r \int_0^s |Z(u) - Z_m(u)|^r du\right] ds \\
&\quad \le \int_0^t \int_0^s \left(\mathbb{E}\left[S^{2r}(s)\xi_m^{2r}\right]\right)^{\frac{1}{2}} \left(\mathbb{E}\left[|Z(u) - Z_m(u)|^{2r}\right]\right)^{\frac{1}{2}} duds \\
&\quad \le C \int_0^t \int_0^s \left(\mathbb{E}\left[|Z(u) - Z_m(u)|^{2r}\right]\right)^{\frac{1}{2}} duds \\
&\quad \le C \int_0^t \int_0^s \left(\int_0^u (\mathcal{K}(u,v) - \mathcal{K}_m(u,v))^2 dv\right)^{\frac{r}{2}} duds \\
&\quad \le C\|\mathcal{K} - \mathcal{K}_m\|^r_{L^2([0,T])}.
\end{aligned} \tag{9.21}$$

Therefore, taking into account (9.17)–(9.21), we see that there exists a constant $C > 0$ that does not depend on m or the particular choice of $t \in [0, T]$ and such that

$$\mathbb{E}\left[\sup_{u\in[0,t]} |S(u) - S_m(u)|^r\right] \le C \int_0^t \mathbb{E}\left[\sup_{u\in[0,s]} |S(u) - S_m(u)|^r\right] ds + C\|\mathcal{K} - \mathcal{K}_m\|^r_{L^2([0,T])}, \tag{9.22}$$

and the claim of the theorem now follows from the Gronwall inequality. □

Remark 9.1 In principle, we could obtain a result in the spirit of Theorem 9.2 without assuming that $\mathcal{K}$, $\mathcal{K}_m$ are all difference kernels. Indeed, the shape of the kernel plays a role only in two places in the proof: in inequalities (9.20) and (9.21), when we bound from above the integrals

$$\int_0^t \left(\int_0^s (\mathcal{K}(s,u) - \mathcal{K}_m(s,u))^2 du\right)^{\frac{r}{2}} ds,$$

and

$$\int_0^t \int_0^s \left(\int_0^u (\mathcal{K}(u,v) - \mathcal{K}_m(u,v))^2 dv\right)^{\frac{r}{2}} duds,$$

doing this in terms of $\|\mathcal{K} - \mathcal{K}_m\|$. Of course, we could do it differently: for example, if $\mathcal{K}, \mathcal{K}_m \in L^r([0,T]^2)$, it is easy to see that

$$\int_0^t \left(\int_0^s (\mathcal{K}(s,u) - \mathcal{K}_m(s,u))^2 du\right)^{\frac{r}{2}} ds \le C\|\mathcal{K} - \mathcal{K}_m\|^2_{L^r([0,T]^2)}.$$

However, since our main examples are all difference kernels, we do not see much value in imposing some extra assumptions on $\mathcal{K}$ or the sequence $\{\mathcal{K}_m,\ m \ge 0\}$.

9.4 Non-anticipating Derivative and Markovian Approximations

Let us now apply the knowledge from Sect. 9.2 to the specific case of the SVV model. Once again, let all assumptions outlined in the beginning of this chapter hold and let S be defined by (9.3). Consider the mean-square hedging problem (9.5), where the payoff function $f\colon \mathbb{R}_+ \to \mathbb{R}$ satisfies the following assumption.

Assumption 9.2 The payoff function $f\colon \mathbb{R}_+ \to \mathbb{R}$ is globally Lipschitz, i.e. there exists $C > 0$ such that

$$|f(t) - f(s)| \le C|t - s|, \quad s, t \ge 0.$$

In particular, Assumption 9.2 and Theorem 7.2 imply that $F = f(S(T)) \in L^2(\mathbb{Q})$, i.e. the non-anticipating derivative $\mathfrak{D}F$ of F with respect to S exists and Theorem 9.1 holds, so that $\mathfrak{D}F$ is the optimal hedging portfolio. Moreover, the simple processes u_π given by

$$\begin{aligned} u_\pi &:= \sum_{k=0}^{n-1} u_{\pi,k} \mathbb{1}_{(t_k,t_{k+1}]}, \\ u_{\pi,k} &:= \frac{\mathbb{E}\left[(S(t_{k+1}) - S(t_k))F \mid \mathcal{F}_{t_k}\right]}{\mathbb{E}\left[(S(t_{k+1}) - S(t_k))^2 \mid \mathcal{F}_{t_k}\right]}, \quad k = 0, 1, ..., n-1, \end{aligned} \tag{9.23}$$

provide a natural approximation to $\mathfrak{D}F$ on the partition

$$\pi = \{0 = t_0 < t_1 < \cdots < t_n = T\}.$$

However, the *computation* of the conditional expectations in (9.23) is a challenging task that becomes even more complicated since the Volterra noise Z from (9.4) may have memory (and thus, in general, S is not Markovian). Our tactics is to replace the original discounted price S with its approximation S_m given by (9.10), i.e. consider the non-anticipating derivative $\mathfrak{D}f(S_m(T))$ w.r.t. S_m instead of $\mathfrak{D}f(S(T))$ w.r.t. S. In the remainder of this section we aim to implement this idea and prove some convergence of $\mathfrak{D}f(S_m(T))$ to $\mathfrak{D}f(S(T))$.

Denote $F_m := f(S_m(T))$ and observe that, by Lemma 9.1, for any $r \in \mathbb{R}$,

$$\mathbb{E}\left[F^r\right] + \sup_{m\ge 0} \mathbb{E}\left[F_m^r\right] < \infty. \tag{9.24}$$

For a given partition $\pi = \{0 = t_0 < t_1 < \cdots < t_n = T\}$, denote $\Delta S(t_k) := S(t_{k+1}) - S(t_k)$, $\Delta S_m(t_k) := S_m(t_{k+1}) - S_m(t_k)$ so that the pre-limit processes in the non-anticipating derivatives

$$\mathfrak{D}F := \lim_{|\pi|\to 0} u_\pi, \quad \mathfrak{D}F_m := \lim_{|\pi|\to 0} u_\pi^m,$$

admit a more compact notation

$$u_\pi := \sum_{k=0}^{n-1} u_{\pi,k} \mathbb{1}_{(t_k,t_{k+1}]}, \quad u_{\pi,k} := \frac{\mathbb{E}\left[F \Delta S(t_k) \mid \mathcal{F}_{t_k}\right]}{\mathbb{E}\left[(\Delta S(t_k))^2 \mid \mathcal{F}_{t_k}\right]}, \tag{9.25}$$

$$u_\pi^m := \sum_{k=0}^{n-1} u_{\pi,k}^m \mathbb{1}_{(t_k,t_{k+1}]}, \quad u_{\pi,k}^m := \frac{\mathbb{E}\left[F_m \Delta S_m(t_k) \mid \mathcal{F}_{t_k}\right]}{\mathbb{E}\left[(\Delta S_m(t_k))^2 \mid \mathcal{F}_{t_k}\right]}. \tag{9.26}$$

We start from a couple of technical results.

Lemma 9.2 *There exist positive constants $C_1 < C_2$ such that, for any m and pair $0 \le t_1 \le t_2 \le T$,*

$$C_1(t_2 - t_1)S^2(t_1) \le \mathbb{E}\left[(\Delta S)^2 \mid \mathcal{F}_{t_1}\right] \le C_2(t_2 - t_1)S^2(t_1)$$

and

$$C_1(t_2 - t_1)S_m^2(t_1) \le \mathbb{E}\left[(\Delta S_m)^2 \mid \mathcal{F}_{t_1}\right] \le C_2(t_2 - t_1)S_m^2(t_1),$$

with $\Delta S := S(t_2) - S(t_1)$, $\Delta S_m := S_m(t_2) - S_m(t_1)$.

Proof We only give the proof for the process S; the one for S_m is identical. By the definition of S, boundedness of Y and Itô lemma, we have

$$\begin{aligned}\mathbb{E}\left[(\Delta S)^2 \mid \mathcal{F}_{t_1}\right] &= \mathbb{E}\left[2\int_{t_1}^{t_2} (S(s) - S(t_1))Y(s)S(s)dW(s) \,\middle|\, \mathcal{F}_{t_1}\right] \\ &\quad + \mathbb{E}\left[\int_{t_1}^{t_2} Y^2(s)S^2(s)ds \,\middle|\, \mathcal{F}_{t_1}\right] \\ &= \int_{t_1}^{t_2} \mathbb{E}\left[Y^2(s)S^2(s) \mid \mathcal{F}_{t_1}\right]ds.\end{aligned} \tag{9.27}$$

Let now $s > t_1$ be fixed. Note that, with probability 1,

$$\mathbb{E}\left[S^2(s) \mid \mathcal{F}_{t_1}\right] \min_{t\in[0,T]} \varphi(t) \le \mathbb{E}\left[Y^2(s)S^2(s) \mid \mathcal{F}_{t_1}\right] \le \mathbb{E}\left[S^2(s) \mid \mathcal{F}_{t_1}\right] \max_{t\in[0,T]} \psi(t) \tag{9.28}$$

and

$$\begin{aligned}\mathbb{E}\left[S^2(s) \mid \mathcal{F}_{t_1}\right] &= \mathbb{E}\left[\exp\left\{2\int_0^s Y(u)dW(u) - \int_0^s Y^2(u)du\right\} \mid \mathcal{F}_{t_1}\right] \\ &= \mathbb{E}\left[\exp\left\{\int_0^s 2Y(u)dW(u) - \frac{1}{2}\int_0^s (2Y(u))^2du + \int_0^s Y^2(u)du\right\} \mid \mathcal{F}_{t_1}\right],\end{aligned}$$

whence, by the boundedness of Y, we get two-sided bounds

$$\mathbb{E}\left[S^2(s)\mid\mathcal{F}_{t_1}\right]\le e^{T\max_{t\in[0,T]}\psi(t)}\mathbb{E}\left[\exp\left\{\int_0^s 2Y(u)dW(u)-\frac{1}{2}\int_0^s(2Y(u))^2du\right\}\mid\mathcal{F}_{t_1}\right]$$

$$\mathbb{E}\left[S^2(s)\mid\mathcal{F}_{t_1}\right]\ge\mathbb{E}\left[\exp\left\{\int_0^s 2Y(u)dW(u)-\frac{1}{2}\int_0^s(2Y(u))^2du\right\}\mid\mathcal{F}_{t_1}\right].$$

Next, by the Novikov criterion, the process

$$\exp\left\{\int_0^s 2Y(u)dW(u)-\frac{1}{2}\int_0^s(2Y(u))^2du\right\},\quad s\in[0,T],$$

is a martingale, and hence, for $s>t_1$,

$$\begin{aligned}\mathbb{E}&\left[\exp\left\{\int_0^s 2Y(u)dW(u)-\frac{1}{2}\int_0^s(2Y(u))^2du\right\}\mid\mathcal{F}_{t_1}\right]\\&=\exp\left\{\int_0^{t_1} 2Y(u)dW(u)-\frac{1}{2}\int_0^{t_1}(2Y(u))^2du\right\}\\&=S^2(t_1)\exp\left\{-\int_0^{t_1}Y^2(s)ds\right\}.\end{aligned}$$

Therefore, two-sided bounds can be rewritten in the form

$$\mathbb{E}\left[\exp\left\{\int_0^s 2Y(u)dW(u)-\frac{1}{2}\int_0^s(2Y(u))^2du\right\}\mid\mathcal{F}_{t_1}\right]\le S^2(t_1),$$

$$\mathbb{E}\left[\exp\left\{\int_0^s 2Y(u)dW(u)-\frac{1}{2}\int_0^s(2Y(u))^2du\right\}\mid\mathcal{F}_{t_1}\right]\ge S^2(t_1)e^{-T\max_{t\in[0,T]}\psi(t)},$$

and now we can write

$$e^{-T\max_{t\in[0,T]}\psi(t)}S^2(t_1)\le\mathbb{E}\left[S^2(s)\mid\mathcal{F}_{t_1}\right]\le e^{T\max_{t\in[0,T]}\psi(t)}S^2(t_1).\qquad(9.29)$$

Finally, by (9.27)–(9.29),

$$C_1(t_2-t_1)S^2(t_1)\le\mathbb{E}\left[(\Delta S)^2\mid\mathcal{F}_{t_1}\right]\le C_2(t_2-t_1)S^2(t_1),$$

where

$$C_1 := \min_{t\in[0,T]} \varphi(t)e^{-T\max_{t\in[0,T]}\psi(t)}, \quad C_2 := \max_{t\in[0,T]} \psi(t)e^{T\max_{t\in[0,T]}\psi(t)}.$$

□

Lemma 9.3 *For any $r \geq 2$, there exists a constant $C > 0$ such that for any m and $0 \leq t_1 \leq t_2 \leq T$*

$$\mathbb{E}[|\Delta S - \Delta S_m|^r] \leq C(t_2 - t_1)^{\frac{r}{2}} \|\mathcal{K} - \mathcal{K}_m\|^r_{L^2([0,T])}$$

with $\Delta S := S(t_2) - S(t_1)$, $\Delta S_m := S_m(t_2) - S_m(t_1)$.

Proof Using the uniform boundedness of Y and Y_m, the Burkholder-Davis-Gundy and Jensen inequalities as well as Theorem 6.5, one can write:

$$\begin{aligned}
\mathbb{E}\left[|\Delta S - \Delta S_m|^r\right] &= \mathbb{E}\left[\left|\int_{t_1}^{t_2} (Y(s)S(s) - Y_m(s)S_m(s))dW(s)\right|^r\right] \\
&\leq C\mathbb{E}\left[\left|\int_{t_1}^{t_2} (Y(s)S(s) - Y_m(s)S_m(s))^2 ds\right|^{\frac{r}{2}}\right] \\
&\leq C(t_2 - t_1)^{\frac{r}{2}-1} \int_{t_1}^{t_2} \mathbb{E}\left[|Y(s)S(s) - Y_m(s)S_m(s)|^r\right] ds \\
&\leq C(t_2 - t_1)^{\frac{r}{2}-1} \int_{t_1}^{t_2} \mathbb{E}\left[|S(s) - S_m(s)|^r\right] ds \\
&\quad + C(t_2 - t_1)^{\frac{r}{2}-1} \int_{t_1}^{t_2} \mathbb{E}\left[S^r(s)|Y(s) - Y_m(s)|^r\right] ds \\
&\leq C(t_2 - t_1)^{\frac{r}{2}-1} \int_{t_1}^{t_2} \mathbb{E}\left[|S(s) - S_m(s)|^r\right] ds \\
&\quad + C(t_2 - t_1)^{\frac{r}{2}-1} \int_{t_1}^{t_2} \mathbb{E}\left[S^r(s)|Z(s) - Z_m(s)|^r\right] ds \\
&\quad + C(t_2 - t_1)^{\frac{r}{2}-1} \int_{t_1}^{t_2} \mathbb{E}\left[S^r(s)\xi_m^r \int_0^s |Z(u) - Z_m(u)|^r du\right] ds,
\end{aligned}$$

where the random variable ξ_m does not depend on t_1 or t_2. Now, since

$$\mathbb{E}\left[\sup_{s\in[0,T]} |S(s) - S_m(s)|^r\right] \le C\|\mathcal{K} - \mathcal{K}_m\|^r_{L^2([0,T])}$$

by Theorem 9.2, it is clear that

$$\int_{t_1}^{t_2} \mathbb{E}\left[|S(s) - S_m(s)|^r\right] ds \le C(t_2 - t_1)\|\mathcal{K} - \mathcal{K}_m\|^r_{L^2([0,T])}.$$

Next, Theorem 7.2 implies that there exists a constant C that does not depend on t_1 or t_2 such that

$$\int_{t_1}^{t_2} \mathbb{E}\left[S^{2r}(s)\right] ds < C,$$

which yields

$$\begin{aligned}
&\int_{t_1}^{t_2} \mathbb{E}\left[S^r(s)|Z(s) - Z_m(s)|^r\right] ds \\
&\quad \le \left(\int_{t_1}^{t_2} \mathbb{E}\left[S^{2r}(s)\right] ds\right)^{\frac{1}{2}} \left(\int_{t_1}^{t_2} \mathbb{E}\left[|Z(s) - Z_m(s)|^{2r}\right] ds\right)^{\frac{1}{2}} \\
&\quad \le C\left(\int_{t_1}^{t_2} \mathbb{E}\left[|Z(s) - Z_m(s)|^{2r}\right] ds\right)^{\frac{1}{2}} \le C(t_2 - t_1)\|\mathcal{K} - \mathcal{K}_m\|^r_{L^2([0,T])}.
\end{aligned}$$

Similarly, by Theorem 7.2 and (9.15) we can write

$$\begin{aligned}
&\int_{t_1}^{t_2} \mathbb{E}\left[S^r(s)\xi_m^r \int_0^s |Z(u) - Z_m(u)|^r du\right] ds \\
&\quad \le C\int_{t_1}^{t_2}\int_0^s \left(\mathbb{E}\left[|Z(u) - Z_m(u)|^{2r}\right]\right)^{\frac{1}{2}} duds \\
&\quad \le C(t_2 - t_1)\|\mathcal{K} - \mathcal{K}_m\|^r_{L^2([0,T])},
\end{aligned}$$

which finalizes the proof. □

With these technical lemmas in our toolbox, we can now proceed to the main results of the section.

Lemma 9.4 *There exists a constant* $C > 0$ *that does not depend on* m *such that for any partition* $\pi = \{0 = t_0 < t_1 < \cdots < t_n = T\}$

$$\mathbb{E}[|u_{\pi,k} - u^m_{\pi,k}|] \le \frac{C}{\sqrt{t_{k+1} - t_k}} \left(\left(\mathbb{E}\left[(F - F_m)^4\right]\right)^{\frac{1}{4}} + \|\mathcal{K} - \mathcal{K}_m\|_{L^2([0,T])} \right),$$

$k = 0, ..., n-1$, *where* $u_{\pi,k}$ *and* $u^m_{\pi,k}$ *are defined in (9.25) and (9.26).*

Proof Fix $k = 0, ..., n-1$ and denote $\Delta := t_{k+1} - t_k$. It is easy to see that

$$\begin{aligned} &\mathbb{E}[|u_{\pi,k} - u^m_{\pi,k}|] \\ &\quad \le \mathbb{E}\left[\frac{|F - F_m||\Delta S(t_k)|}{\mathbb{E}\left[(\Delta S(t_k))^2 \mid \mathcal{F}_{t_k}\right]} \right] + \mathbb{E}\left[|F_m| \frac{|\Delta S(t_k) - \Delta S_m(t_k)|}{\mathbb{E}\left[(\Delta S(t_k))^2 \mid \mathcal{F}_{t_k}\right]} \right] \\ &\quad + \mathbb{E}\left[|F_m|\,|\Delta S_m(t_k)| \frac{\mathbb{E}\left[|(\Delta S_m(t_k))^2 - (\Delta S(t_k))^2| \mid \mathcal{F}_{t_k}\right]}{\mathbb{E}\left[(\Delta S(t_k))^2 \mid \mathcal{F}_{t_k}\right]\mathbb{E}\left[(\Delta S_m(t_k))^2 \mid \mathcal{F}_{t_k}\right]} \right] \\ &\quad =: I_1 + I_2 + I_3. \end{aligned} \tag{9.30}$$

Now we will deal with each of the above summands separately.
Step 1: upper bound for I_1. By Lemma 9.2, there exist constants C_1, C_2 (not depending on the partition or m) such that

$$\mathbb{E}\left[(\Delta S(t_k))^2 \mid \mathcal{F}_{t_k}\right] \ge \Delta C_1 S^2(t_k)$$

and

$$\mathbb{E}\left[(\Delta S(t_k))^2\right] \le \Delta C_2 \mathbb{E}\left[S^2(t_k)\right].$$

Hence, using Hölder inequality and Theorem 7.2, one can write

$$\begin{aligned} I_1 &:= \mathbb{E}\left[\frac{|F - F_m||\Delta S(t_k)|}{\mathbb{E}\left[(\Delta S(t_k))^2 \mid \mathcal{F}_{t_k}\right]} \right] \le \frac{C}{\Delta} \mathbb{E}\left[|F - F_m||\Delta S(t_k)| S^{-2}(t_k)\right] \\ &\le \frac{C}{\Delta} \left(\mathbb{E}\left[(\Delta S(t_k))^2\right]\right)^{\frac{1}{2}} \left(\mathbb{E}[S^{-8}(t_k)]\right)^{\frac{1}{4}} \left(\mathbb{E}\left[|F - F_m|^4\right]\right)^{\frac{1}{4}} \\ &\le \frac{C}{\sqrt{\Delta}} \left(\mathbb{E}\left[S^2(t_k)\right]\right)^{\frac{1}{2}} \left(\mathbb{E}\left[|F - F_m|^4\right]\right)^{\frac{1}{4}} \le \frac{C}{\sqrt{\Delta}} \left(\mathbb{E}\left[|F - F_m|^4\right]\right)^{\frac{1}{4}}, \end{aligned} \tag{9.31}$$

where $C > 0$ can be chosen so that it does not depend on m or the partition.
Step 2: upper bound for I_2. By Hölder inequality, Lemma 9.2, Lemma 9.3 and (9.24),

$$\begin{aligned} I_2 &:= \mathbb{E}\left[|F_m| \frac{|\Delta S(t_k) - \Delta S_m(t_k)|}{\mathbb{E}\left[(\Delta S(t_k))^2 \mid \mathcal{F}_{t_k}\right]} \right] \\ &\leq \frac{C}{\Delta} \left(\mathbb{E}[|\Delta S(t_k) - \Delta S_m(t_k)|^2]\right)^{\frac{1}{2}} \left(\mathbb{E}\left[F_m^2 S^{-4}(t_k)\right]\right)^{\frac{1}{2}} \\ &\leq \frac{C}{\sqrt{\Delta}} \|\mathcal{K} - \mathcal{K}_m\|^r_{L^2([0,T])}, \end{aligned} \tag{9.32}$$

where C does not depend on m or the partition.

Step 3: upper bound for I_3. Using Hölder's inequality and (9.24), one can verify that

$$\begin{aligned} I_3 &:= \mathbb{E}\left[|F_m|\,|\Delta S_m(t_k)| \frac{\mathbb{E}\left[|(\Delta S_m(t_k))^2 - (\Delta S(t_k))^2| \mid \mathcal{F}_{t_k}\right]}{\mathbb{E}\left[(\Delta S(t_k))^2 \mid \mathcal{F}_{t_k}\right]\mathbb{E}\left[(\Delta S_m(t_k))^2 \mid \mathcal{F}_{t_k}\right]} \right] \\ &\leq \frac{C}{\Delta^2} \left(\mathbb{E}\left[F_m^4\right]\right)^{\frac{1}{4}} \left(\mathbb{E}\left[|\Delta S_m(t_k)|^4\right]\right)^{\frac{1}{4}} \left(\mathbb{E}\left[|(\Delta S_m(t_k))^2 - (\Delta S(t_k))^2|^2\right]\right)^{\frac{1}{2}} \\ &\leq \frac{C}{\Delta^2} \left(\mathbb{E}\left[|\Delta S_m(t_k)|^4\right]\right)^{\frac{1}{4}} \left(\mathbb{E}\left[|(\Delta S_m(t_k))^2 - (\Delta S(t_k))^2|^2\right]\right)^{\frac{1}{2}}. \end{aligned} \tag{9.33}$$

By the Burkholder–Davis–Gundy and Hölder inequalities, uniform boundedness of Y_m and Theorem 7.2,

$$\begin{aligned} \left(\mathbb{E}\left[|\Delta S_m(t_k)|^4\right]\right)^{\frac{1}{4}} &= \left(\mathbb{E}\left[\left|\int_{t_k}^{t_{k+1}} Y_m(s) S_m(s) dW(s)\right|^4\right]\right)^{\frac{1}{4}} \\ &\leq \left(\mathbb{E}\left[\left(\int_{t_k}^{t_{k+1}} Y_m^2(s) S_m^2(s) ds\right)^2\right]\right)^{\frac{1}{4}} \\ &\leq \left(\Delta \int_{t_k}^{t_{k+1}} \mathbb{E}\left[Y_m^4(s) S_m^4(s)\right] ds\right)^{\frac{1}{4}} \\ &\leq C\sqrt{\Delta}\left(\mathbb{E}\left[\sup_{u\in[0,T]} S_m^4(u)\right]\right)^{\frac{1}{4}} \leq C\sqrt{\Delta}. \end{aligned} \tag{9.34}$$

Further, one can verify using Itô formula that

$$
\begin{aligned}
&\mathbb{E}\left[|(\Delta S_m(t_k))^2-(\Delta S(t_k))^2|^2\right]\\
&\le C\mathbb{E}\left[\left(\int_{t_k}^{t_{k+1}}\left(Y_m^2(s)S_m^2(s)-Y^2(s)S^2(s)\right)ds\right)^2\right]\\
&\quad+C\mathbb{E}\left[\left(2\int_{t_k}^{t_{k+1}}\left((S_m(s)-S_m(t_k))Y_m(s)S_m(s)-(S(s)-S(t_k))Y(s)S(s)\right)dW(s)\right)^2\right].
\end{aligned}
\tag{9.35}
$$

We study the last two summands separately. It is clear that

$$
\begin{aligned}
&\mathbb{E}\left[\left(\int_{t_k}^{t_{k+1}}\left(Y_m^2(s)S_m^2(s)-Y^2(s)S^2(s)\right)ds\right)^2\right]\\
&\le \Delta\int_{t_k}^{t_{k+1}}\mathbb{E}\left[\left(Y_m^2(s)S_m^2(s)-Y^2(s)S^2(s)\right)^2\right]ds\\
&\le C\Delta\int_{t_k}^{t_{k+1}}\mathbb{E}\left[\left(S_m^2(s)-S^2(s)\right)^2\right]ds\\
&\quad+C\Delta\int_{t_k}^{t_{k+1}}\mathbb{E}\left[S^2(s)\left(Y_m^2(s)-Y^2(s)\right)^2\right]ds.
\end{aligned}
\tag{9.36}
$$

Taking into account that processes $\{S^2(t),\ t\in[0,T]\}$ and $\{S_m^2(t),\ t\in[0,T]\}$ satisfy the stochastic differential equations

$$
\begin{aligned}
dS^2(t)&=Y^2(t)S^2(t)dt+2Y(t)S^2(t)dW(t),\\
dS_m^2(t)&=Y_m^2(t)S_m^2(t)dt+2Y_m(t)S_m^2(t)dW(t),
\end{aligned}
$$

boundedness of Y, uniform boundedness of $\{Y_m,\ m\ge 1\}$ as well as the same argument as in Theorem 9.2 imply that there exists a constant $C>0$ such that

$$
\mathbb{E}\left[\sup_{s\in[0,T]}\left(S_m^2(s)-S^2(s)\right)^2\right]ds\le C\|\mathcal{K}-\mathcal{K}_m\|^2_{L^2([0,T])},
$$

whereas, again like in the proof of Theorem 9.2, we have

$$
\begin{aligned}
\mathbb{E}\left[S^2(s)\left(Y_m^2(s)-Y^2(s)\right)^2\right]&=\mathbb{E}\left[S^2(s)(Y_m(s)+Y(s))^2(Y_m(s)-Y(s))^2\right]\\
&\le C\mathbb{E}\left[S^2(s)(Y_m(s)-Y(s))^2\right]\le C\|\mathcal{K}-\mathcal{K}_m\|^2_{L^2([0,T])}.
\end{aligned}
$$

Plugging these estimates in (9.36), we obtain that

$$\mathbb{E}\left[\left(\int_{t_k}^{t_{k+1}}\left(Y_m^2(s)S_m^2(s)-Y^2(s)S^2(s)\right)ds\right)^2\right]\leq C\Delta^2\|\mathcal{K}-\mathcal{K}_m\|^2_{L^2([0,T])}.$$

Next, for the second summand of (9.35) we have

$$\begin{aligned}
&\mathbb{E}\Bigg[\Bigg(\int_{t_k}^{t_{k+1}}\Big((S_m(s)-S_m(t_k))Y_m(s)S_m(s)-(S(s)-S(t_k))Y(s)S(s)\Big)dW(s)\Bigg)^2\Bigg]\\
&=\mathbb{E}\Bigg[\int_{t_k}^{t_{k+1}}\Big((S_m(s)-S_m(t_k))Y_m(s)S_m(s)-(S(s)-S(t_k))Y(s)S(s)\Big)^2ds\Bigg]\\
&\leq C\int_{t_k}^{t_{k+1}}\mathbb{E}\left[(S(s)-S(t_k))^2(Y_m(s)S_m(s)-Y(s)S(s))^2\right]ds\\
&\quad+C\int_{t_k}^{t_{k+1}}\mathbb{E}\left[\Big((S_m(s)-S_m(t_k))-(S(s)-S(t_k))\Big)^2Y_m^2(s)S_m^2(s)\right]ds\\
&\leq C\int_{t_k}^{t_{k+1}}\left(\mathbb{E}\left[(S(s)-S(t_k))^4\right]\right)^{\frac12}\left(\mathbb{E}\left[(Y_m(s)S_m(s)-Y(s)S(s))^4\right]\right)^{\frac12}ds\\
&\quad+C\int_{t_k}^{t_{k+1}}\left(\mathbb{E}\left[\Big((S_m(s)-S_m(t_k))-(S(s)-S(t_k))\Big)^4\right]\right)^{\frac12}\left(\mathbb{E}\left[S_m^4(s)\right]\right)^{\frac12}ds.
\end{aligned}\tag{9.37}$$

Using the same argument as in (9.34), one can verify that there exists a constant $C>0$ (not depending on m or the partition) such that, for all $s\in[t_k,t_{k+1}]$,

$$\mathbb{E}\left[(S(s)-S(t_k))^4\right]\leq C(s-t_k)^2.\tag{9.38}$$

Moreover, using Theorem 6.5 and Theorem 7.2, it is easy to show that for any $s\in[0,T]$

$$\begin{aligned}
\mathbb{E}\left[S^4(s)(Y_m(s)-Y(s))^4\right]&\leq\left(\mathbb{E}\left[S^8(s)\right]\right)^{\frac12}\left(\mathbb{E}\left[(Y_m(s)-Y(s))^8\right]\right)^{\frac12}\\
&\leq C\|\mathcal{K}-\mathcal{K}_m\|^4_{L^2([0,T])},
\end{aligned}$$

whence

$$\begin{aligned}&\mathbb{E}\left[(Y_m(s)S_m(s) - Y(s)S(s))^4\right]\\&\le C\mathbb{E}\left[Y_m^4(s)(S_m(s) - S(s))^4\right] + C\mathbb{E}\left[S^4(s)(Y_m(s) - Y(s))^4\right]\\&\le C\mathbb{E}\left[(S_m(s) - S(s))^4\right] + C\mathbb{E}\left[S^4(s)(Y_m(s) - Y(s))^4\right]\\&\le C\|\mathcal{K} - \mathcal{K}_m\|^4_{L^2([0,T])}.\end{aligned} \tag{9.39}$$

Next, by Lemma 9.3, for all $s \in [t_k, t_{k+1}]$,

$$\begin{aligned}&\mathbb{E}\left[\Big((S_m(s) - S_m(t_k)) - (S(s) - S(t_k))\Big)^4\right]\\&\le C(s - t_k)^2\|\mathcal{K} - \mathcal{K}_m\|^4_{L^2([0,T])}.\end{aligned} \tag{9.40}$$

Plugging the estimates (9.38)–(9.40) into (9.37), we finally obtain that there exists a constant $C > 0$ (not depending on m or the partition) such that

$$\begin{aligned}&\mathbb{E}\Bigg[\Bigg(\int_{t_k}^{t_{k+1}}\Big((S_m(s) - S_m(t_k))Y_m(s)S_m(s) - (S(s) - S(t_k))Y(s)S(s)\Big)dW(s)\Bigg)^2\Bigg]\\&\le C\int_{t_k}^{t_{k+1}}(s - t_k)\|\mathcal{K} - \mathcal{K}_m\|^2_{L^2([0,T])}ds\\&\quad + C\int_{t_k}^{t_{k+1}}(s - t_k)\|\mathcal{K} - \mathcal{K}_m\|^2_{L^2([0,T])}\left(\sup_{m\ge 1}\mathbb{E}\left[\sup_{t\in[0,T]} S_m^4(t)\right]\right)^{\frac{1}{2}} ds\\&\le C\Delta^2\|\mathcal{K} - \mathcal{K}_m\|^2_{L^2([0,T])}.\end{aligned}$$

Taking into account all of the above, we can finally write

$$\mathbb{E}\left[|(\Delta S_m(t_k))^2 - (\Delta S(t_k))^2|^2\right] \le C\Delta^2\|\mathcal{K} - \mathcal{K}_m\|^2_{L^2([0,T])}.$$

So, we receive the upper bound

$$\begin{aligned}I_3 &:= \mathbb{E}\left[|F_m|\,|\Delta S_m(t_k)|\,\frac{\mathbb{E}\left[|(\Delta S_m(t_k))^2 - (\Delta S(t_k))^2|\mid \mathcal{F}_{t_k}\right]}{\mathbb{E}\left[(\Delta S(t_k))^2 \mid \mathcal{F}_{t_k}\right]\mathbb{E}\left[(\Delta S_m(t_k))^2 \mid \mathcal{F}_{t_k}\right]}\right]\\&\le \frac{C}{\Delta^2}\left(\mathbb{E}\left[|\Delta S_m(t_k)|^4\right]\right)^{\frac{1}{4}}\left(\mathbb{E}\left[|(\Delta S_m(t_k))^2 - (\Delta S(t_k))^2|^2\right]\right)^{\frac{1}{2}}\\&\le \frac{C}{\Delta^2}\sqrt{\Delta}\Delta\|\mathcal{K} - \mathcal{K}_m\|_{L^2([0,T])} \le \frac{C}{\sqrt{\Delta}}\|\mathcal{K} - \mathcal{K}_m\|_{L^2([0,T])},\end{aligned}$$

which completes the proof. □

Theorem 9.3 *(1) There exists a constant $C > 0$ that does not depend on m such that, for any partition $\pi = \{0 = t_0 < t_1 < \cdots < t_n = T\}$ with the mesh $|\pi| := \max_k |t_{k+1} - t_k|$,*

$$\mathbb{E}\left[\int_0^T |u_\pi(s) - u_\pi^m(s)|ds\right] \leq Cn\sqrt{|\pi|}\|\mathcal{K} - \mathcal{K}_m\|_{L^2([0,T])}, \tag{9.41}$$

where u_π and u_π^m are defined by (9.25) and (9.26). In particular, if the partition π is uniform, i.e. $t_k = \frac{kT}{n}$ and $|\pi| = \frac{T}{n}$,

$$\mathbb{E}\left[\int_0^T |u_\pi(s) - u_\pi^m(s)|ds\right] \leq C\sqrt{n}\|\mathcal{K} - \mathcal{K}_m\|_{L^2([0,T])}. \tag{9.42}$$

(2) If $\{m_n,\ n \geq 1\}$ is such that

$$n\sqrt{|\pi|}\|\mathcal{K} - \mathcal{K}_{m_n}\|_{L^2([0,T])} \to 0, \quad n \to \infty,$$

then

$$\mathbb{E}\left[\int_0^T |\mathfrak{D}F(s) - u_\pi^{m_n}(s)|ds\right] \to 0, \quad n \to \infty.$$

Proof By Assumption 9.2,

$$\begin{aligned}\mathbb{E}[|F - F_m|^4] &= \mathbb{E}[|f(S(T)) - f(S_m(T))|^4] \\ &\leq C\mathbb{E}[|S(T) - S_m(T)|^4] \\ &\leq C\|\mathcal{K} - \mathcal{K}_m\|^4_{L^2([0,T])},\end{aligned}$$

where the constant C does not depend on m or the partition. Furthermore, by Lemma 9.4,

$$\begin{aligned}\mathbb{E}\left[\int_0^T |u_\pi(s) - u_\pi^m(s)|ds\right] &= \sum_{k=0}^{n-1}\mathbb{E}\left[|u_{\pi,k} - u_{\pi,k}^m|\right](t_{k+1} - t_k) \\ &\leq C\left(\sum_{k=0}^{n-1}\sqrt{t_{k+1} - t_k}\right)\left(\left(\mathbb{E}[|F - F_m|^4]\right)^{\frac{1}{4}} + \|\mathcal{K} - \mathcal{K}_m\|_{L^2([0,T])}\right) \\ &\leq Cn\sqrt{|\pi|}\|\mathcal{K} - \mathcal{K}_m\|_{L^2([0,T])},\end{aligned}$$

which gives item 1).

To prove item 2), first note that

$$\mathbb{E}\left[\int_0^T |\mathfrak{D}F(s) - u_\pi(s)|\, ds\right] = \mathbb{E}\left[\int_0^T |\mathfrak{D}F(s) - u_\pi(s)|\, Y(s)S(s)\frac{1}{Y(s)S(s)} ds\right]$$
$$\leq \left(\mathbb{E}\left[\int_0^T |\mathfrak{D}F(s) - u_\pi(s)|^2\, Y^2(s)S^2(s) ds\right]\right)^{\frac{1}{2}} \left(\mathbb{E}\left[\int_0^T \frac{1}{Y^2(s)S^2(s)} ds\right]\right)^{\frac{1}{2}}$$
$$\leq C\|\mathfrak{D}F - u_\pi\|_{L^2(\mathbb{Q}\times[S])} \to 0$$

as $|\pi| \to 0$. Therefore,

$$\mathbb{E}\left[\int_0^T |\mathfrak{D}F(s) - u_\pi^{m_n}(s)| ds\right]$$
$$\leq \mathbb{E}\left[\int_0^T |\mathfrak{D}F(s) - u_\pi(s)|\, ds\right] + \mathbb{E}\left[\int_0^T \left|u_\pi(s) - u_\pi^{m_n}(s)\right| ds\right]$$
$$\leq \mathbb{E}\left[\int_0^T |\mathfrak{D}F(s) - u_\pi(s)|\, ds\right] + Cn\sqrt{|\pi|}\|\mathcal{K} - \mathcal{K}_{m_n}\|_{L^2([0,T])} \to 0$$

as $n \to \infty$. This completes the proof. □

Example 9.1 Let S be the SVV discounted price (9.3) with the kernel $\mathcal{K}$ being as in Example 6.3, i.e. $\mathcal{K}(t, s) = \mathcal{K}(t - s)\mathbb{1}_{s<t}$, $\mathcal{K}(0) = 0$ and $\mathcal{K} \in C^H([0, T])$ for some $H \in (0, 1)$. In this situation, we may take $\mathcal{K}_m$ defined by the Bernstein polynomial approximation of $\mathcal{K}$ given by (6.51): as explained in Example 6.3, this setup satisfies Assumption 9.1. Then, taking into account (6.52) and Theorem 9.2, for $m_n := n^\alpha$ for $\alpha > \frac{1}{H}$ and any $r \geq 2$,

$$\mathbb{E}\left[\int_0^T |u_\pi(s) - u_\pi^{m_n}(s)| ds\right] \leq Cn^{-\frac{\alpha H}{2}+\frac{1}{2}} \to 0, \quad n \to \infty. \tag{9.43}$$

Example 9.2 Let the kernel $\mathcal{K}$ be the rough fractional kernel from Example 6.4 and $\mathcal{K}_m$ be the approximation of $\mathcal{K}$ with exponentials given by (6.55) with $\tau_{m,i}$, $i = 0, ..., m$, $m \geq 1$, given by (6.56). As discussed in Example 6.4, Assumption 9.1 is satisfied and, by (6.57) and Theorem 9.2, for any $r \geq 2$,

$$\mathbb{E}\left[\sup_{t\in[0,T]} |S(t) - S_m(t)|^r\right] \leq Cm^{-\frac{4rH}{5}}.$$

Moreover, by Theorem 9.3, for a uniform partition $\pi = \{0 = t_0 < t_1 < \cdots < t_n = T\}$, $t_k = \frac{kT}{n}$, and $m_n := n^\alpha$ for $\alpha > \frac{5}{8H}$,

$$\mathbb{E}\left[\int_0^T |u_\pi(s) - u_\pi^{m_n}(s)| ds\right] \le C n^{-\frac{4\alpha H}{5} + \frac{1}{2}} \to 0, \quad n \to \infty. \tag{9.44}$$

9.5 Monte Carlo Computation of the Optimal Hedge

In Sect. 9.4, we approximated the optimal hedging portfolio $\mathfrak{D} f(S(T))$ with $\mathfrak{D} f(S_m(T))$, where S_m is a coordinate of some multidimensional Markov process. However, the main question remains unanswered: how can we leverage the Markovianity to estimate the conditional expectations in (9.26) numerically?

One way to do it is as follows: observe that

$$\begin{aligned}
&\mathbb{E}\left[f(S_m(T))(S_m(t_{k+1}) - S_m(t_k)) \mid \mathcal{F}_{t_k}\right] \\
&\quad = \mathbb{E}\left[f(S_m(T))(S_m(t_{k+1}) - S_m(t_k)) \mid S_m(t_k),\ Y_m(t_k),\ U_{m,0}(t_k), \ldots, U_{m,m}(t_k)\right] \\
&\quad =: \Phi_1\big(t_k, S_m(t_k), Y_m(t_k), U_{m,0}(t_k), \ldots, U_{m,m}(t_k)\big),
\end{aligned}$$

and

$$\begin{aligned}
&\mathbb{E}\left[(S_m(t_{k+1}) - S_m(t_k))^2 \mid \mathcal{F}_{t_k}\right] \\
&\quad = \mathbb{E}\left[(S_m(t_{k+1}) - S_m(t_k))^2 \mid S_m(t_k),\ Y_m(t_k),\ U_{m,0}(t_k), \ldots, U_{m,m}(t_k)\right] \\
&\quad =: \Phi_2\big(t_k, S_m(t_k), Y_m(t_k), U_{m,0}(t_k), \ldots, U_{m,m}(t_k)\big).
\end{aligned}$$

In other words, our computational problem boils down to learning the shape of functions Φ_1 and Φ_2 with fixed and finite list of arguments; since we can simulate the processes S_m, Y_m, $U_{m,0}$, ..., $U_{m,m}$ without any limitations, we can estimate Φ_1 and Φ_2 via Monte Carlo methods.

In this section, we propose two algorithms for estimating Φ_1 and Φ_2: *Nested Monte Carlo* (NMC) and *Least-Squares Monte Carlo* (LSMC) algorithms. In all examples, F will be a payoff of a standard call option. All simulations are performed using R programming language on the system with Intel Core i9-9900K CPU and 64 Gb RAM.

9.5.1 *Nested Monte Carlo Method*

The first (and more straightforward) way to compute Φ_1 and Φ_2 is the *Nested Monte Carlo* (NMC) approach that can be summarized as follows (see also Fig. 9.1):

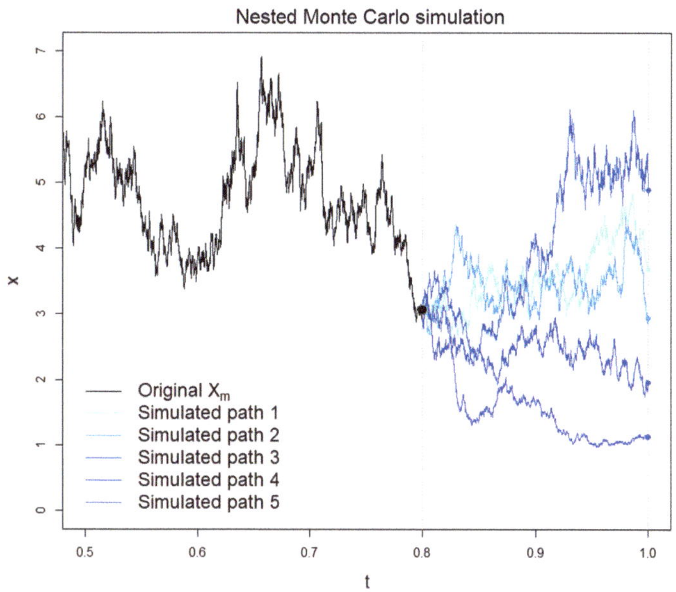

Fig. 9.1 Nested Monte Carlo approach, $t_k = 0.8$, $T = 1$. Given the values $S_m(t_k)$, $Y_m(t_k)$, $U_{m,1}(t_k)$, ..., $U_{m,m}(t_k)$, we simulate a number of trajectories $S_m^{(i)}$ on $(t_k, T]$ (blue), use each of those to compute $f(S_m^{(i)}(T))(S_m^{(i)}(t_{k+1}) - S_m^{(i)}(t_k))$ and $(S_m^{(i)}(t_{k+1}) - S_m^{(i)}(t_k))^2$. The means of the latter values are then used as approximations of the required conditional expectations

(1) given $S_m(t_k) = x$, $Y_m(t_k) = y$, $U_{m,0}(t_k) = u_0$, ..., $U_{m,m}(t_k) = u_m$, simulate N independent trajectories $\{S_m^{(i)}(t),\ t \in (t_k, T]\}$, $i = 1, ..., N$;
(2) for each trajectory, compute $f(S_m^{(i)}(T))(S_m^{(i)}(t_{k+1}) - S_m^{(i)}(t_k))$ and $(S_m^{(i)}(t_{k+1}) - S_m^{(i)}(t_k))^2$, $i = 1, ..., N$;
(3) put

$$\widehat{\Phi}_1(t_k, x, y, u_0, ..., u_m) := \frac{1}{N}\sum_{i=1}^{N} f(S_m^{(i)}(T))(S_m^{(i)}(t_{k+1}) - S_m^{(i)}(t_k)),$$

$$\widehat{\Phi}_2(t_k, x, y, u_0, ..., u_m) := \frac{1}{N}\sum_{i=1}^{N}(S_m^{(i)}(t_{k+1}) - S_m^{(i)}(t_k))^2.$$

Example 9.3 *(Hölder continuous kernel)* Consider the SVV model

$$S(t) = S(0) + \int_0^t Y(s)S(s)\left(\rho dB_1(s) + \sqrt{1-\rho^2}dB_2(s)\right),$$

$$Y(t) = Y(0) + \int_0^t b(s, Y(s))ds + \int_0^t \mathcal{K}(t, s)dB_1(s),$$

with $S(0) = 5, Y(0) = 1, \rho = 0.5, b(t, y) = \frac{1}{(y-0.01)^4} - \frac{1}{(5-y)^4}$ and $\mathcal{K}(t, s) = \mathcal{K}(t - s) = (t - s)^{0.4}$. Consider also the European call payoff function $f(x) := \max\{x - 4, 0\}$. The approximation $(S_m, Y_m, U_{m,0}, U_{m,1}, ..., U_{m,m})$ is constructed using the Bernstein polynomial approximation $\mathcal{K}_m$ from Example 6.3.

Figure 9.2 illustrates the approximation of the SVV model; original kernel and generated sample paths of the corresponding processes are depicted in black whereas approximations using the Bernstein polynomials are depicted in red (of order $m = 10$) and blue (of order $m = 30$). Note that the path of S simulated using the original noise (black) is not visible on Fig. 9.2d due to the high degree of overlapping with the approximations (red and blue trajectories). Figure 9.3 contains a path of the process

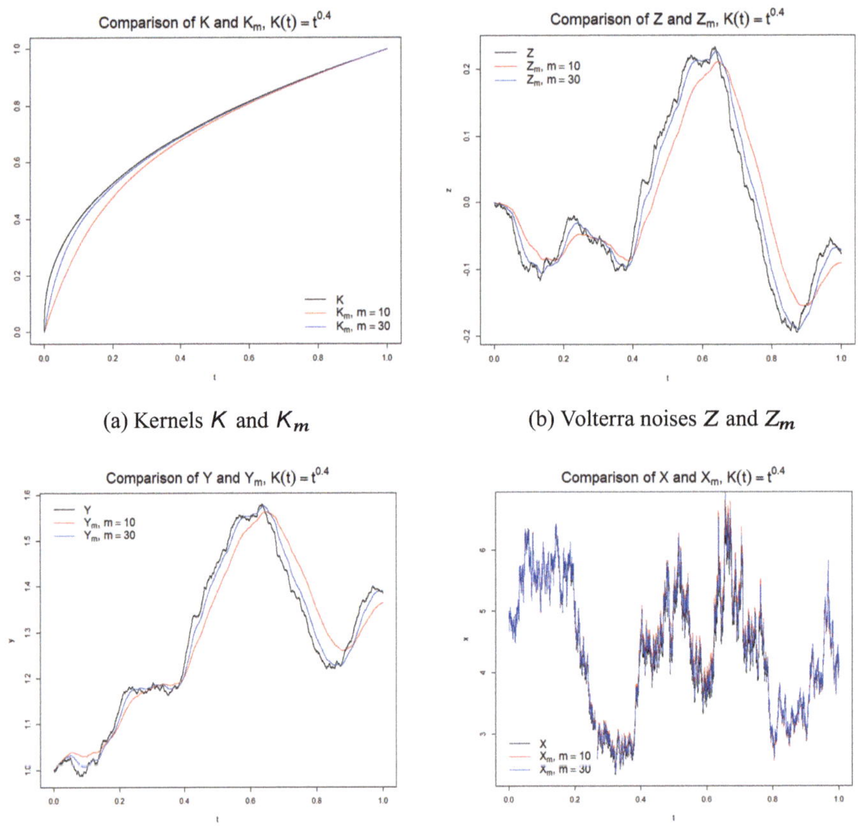

(a) Kernels K and K_m

(b) Volterra noises Z and Z_m

(c) Volatility processes Y and Y_m

(d) Price processes S and S_m

Fig. 9.2 Approximation of the SVV model with Hölder continuous kernel

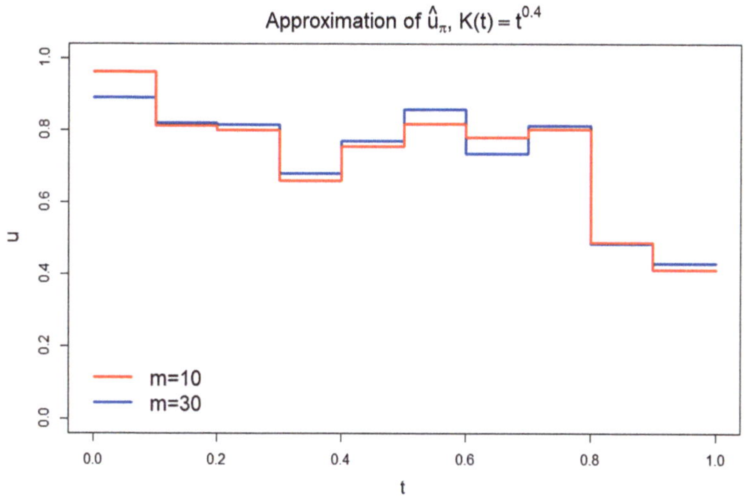

Fig. 9.3 Hedging strategy estimated for the path from Fig. 9.2 for $m = 10$ (red) and $m = 30$ (blue). The corresponding partition is $k/10$, $k = 0, 1, ..., 10$. Simulating the red line took 63231 s whereas the blue line took 73410 s

$$\widehat{u}_\pi(t) := \sum_{k=0}^{9} \frac{\widehat{\Phi}_1\left(\frac{k}{10}, S_m\left(\frac{k}{10}\right), Y_m\left(\frac{k}{10}\right), U_{m,0}\left(\frac{k}{10}\right), ..., U_{m,m}\left(\frac{k}{10}\right)\right)}{\widehat{\Phi}_2\left(\frac{k}{10}, S_m\left(\frac{k}{10}\right), Y_m\left(\frac{k}{10}\right), U_{m,0}\left(\frac{k}{10}\right), ..., U_{m,m}\left(\frac{k}{10}\right)\right)} \mathbb{1}_{\left(\frac{k}{10}, \frac{k+1}{10}\right]}(t), \tag{9.45}$$

with the values $S_m\left(\frac{k}{10}\right)$, $Y_m\left(\frac{k}{10}\right)$, $U_{m,0}\left(\frac{k}{10}\right)$, ..., $U_{m,m}\left(\frac{k}{10}\right)$ coming exactly from the trajectory depicted on Fig. 9.2. In order to estimate

$$\widehat{\Phi}_1\left(\frac{k}{10}, S_m\left(\frac{k}{10}\right), Y_m\left(\frac{k}{10}\right), U_{m,0}\left(\frac{k}{10}\right), ..., U_{m,m}\left(\frac{k}{10}\right)\right),$$
$$\widehat{\Phi}_2\left(\frac{k}{10}, S_m\left(\frac{k}{10}\right), Y_m\left(\frac{k}{10}\right), U_{m,0}\left(\frac{k}{10}\right), ..., U_{m,m}\left(\frac{k}{10}\right)\right),$$

100000 simulations were used for each $k = 0, 1, ..., 9$.

Example 9.4 *(Rough kernel)* Consider the SVV model

$$S(t) = S(0) + \int_0^t Y(s)S(s)\left(\rho dB_1(s) + \sqrt{1-\rho^2}dB_2(s)\right),$$

$$Y(t) = Y(0) + \int_0^t b(s, Y(s))ds + \int_0^t \mathcal{K}(t,s)dB_1(s),$$

with $S(0) = 5, Y(0) = 1, \rho = 0.5, b(t, y) = \frac{1}{(y-0.01)^4} - \frac{1}{(5-y)^4}$ and $\mathcal{K}(t, s) = \mathcal{K}(t - s) = \frac{1}{\Gamma(0.8)(t-s)^{0.2}}$. The approximation $(S_m, Y_m, U_{m,1}, ..., U_{m,m})$ is constructed by approximating the Volterra noise $Z(t) := \frac{1}{\Gamma(0.8)} \int_0^t \frac{1}{(t-s)^{0.2}} dB_1(s)$ by a linear combination of standard Ornstein-Uhlenbeck processes as described in Example 6.4 and Remark 6.8. Consider also the payoff function $f(x) := \max\{x - 4, 0\}$.

Just as in Example 9.3, Fig. 9.4 illustrates approximation of the SVV model; original kernel and generated sample paths of the corresponding processes are depicted in black whereas approximations are depicted in red ($m = 10$ summands), green ($m = 100$ summands) and blue ($m = 1000$ summands). Note that rough volatility requires much higher values of m in order to ensure a decent level of approximation in comparison to Example 9.3. Figure 9.5 contains a path of the process

$$\widehat{u}_\pi(t) := \sum_{k=0}^{9} \frac{\widehat{\Phi}_1\left(\frac{k}{10}, S_m\left(\frac{k}{10}\right), Y_m\left(\frac{k}{10}\right), U_{m,0}\left(\frac{k}{10}\right), ..., U_{m,m}\left(\frac{k}{10}\right)\right)}{\widehat{\Phi}_2\left(\frac{k}{10}, S_m\left(\frac{k}{10}\right), Y_m\left(\frac{k}{10}\right), U_{m,0}\left(\frac{k}{10}\right), ..., U_{m,m}\left(\frac{k}{10}\right)\right)} \mathbb{1}_{\left(\frac{k}{10}, \frac{k+1}{10}\right]}(t), \tag{9.46}$$

$t \in [0, 1]$, where the values $S_m\left(\frac{k}{10}\right)$, $Y_m\left(\frac{k}{10}\right)$, $U_{m,1}\left(\frac{k}{10}\right)$, ..., $U_{m,m}\left(\frac{k}{10}\right)$ are exactly the ones from the trajectory depicted on Fig. 9.4. In order to estimate

$$\widehat{\Phi}_1\left(\frac{k}{10}, S_m\left(\frac{k}{10}\right), Y_m\left(\frac{k}{10}\right), U_{m,0}\left(\frac{k}{10}\right), ..., U_{m,m}\left(\frac{k}{10}\right)\right),$$
$$\widehat{\Phi}_2\left(\frac{k}{10}, S_m\left(\frac{k}{10}\right), Y_m\left(\frac{k}{10}\right), U_{m,0}\left(\frac{k}{10}\right), ..., U_{m,m}\left(\frac{k}{10}\right)\right),$$

100000 simulations were used for each $k = 0, 1, ..., 9$.

9.5.2 *Least Squares Monte Carlo Method*

Despite its simplicity and clear theoretical justification, the nested Monte Carlo approach has a substantial disadvantage: it requires heavy time-consuming computations for each step of the strategy and thus demands powerful computational resources in order to be used in practice. In order to overcome this issue, one can use the following *Least Squares Monte Carlo* (LSMC) method instead, that can be regarded as a simplified version of the approach suggested in [1, 11, 16] for pricing American options or in [8–10] for modeling of life insurance companies.

(1) Simulate N independent realizations

$$(S_m^{(i)}, Y_m^{(i)}, U_{m,0}^{(i)}, ..., U_{m,m}^{(i)}) = \left\{\left(S_m^{(i)}(t), Y_m^{(i)}(t), U_{m,0}^{(i)}(t), ..., U_{m,m}^{(i)}(t)\right),\ t \in [0, T]\right\},$$

$i = 1, ..., N;$

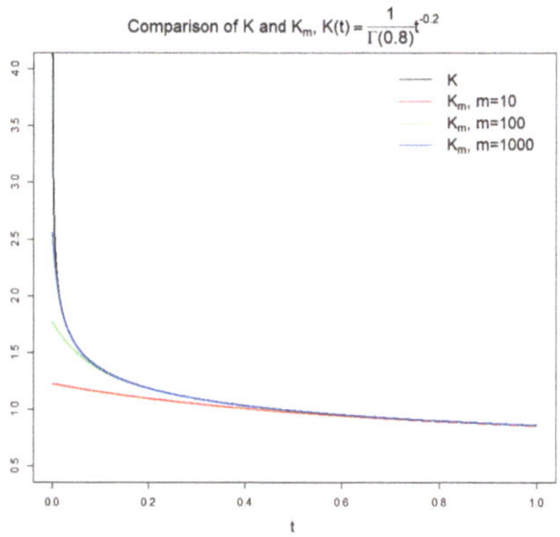

(a) Kernels K and K_m

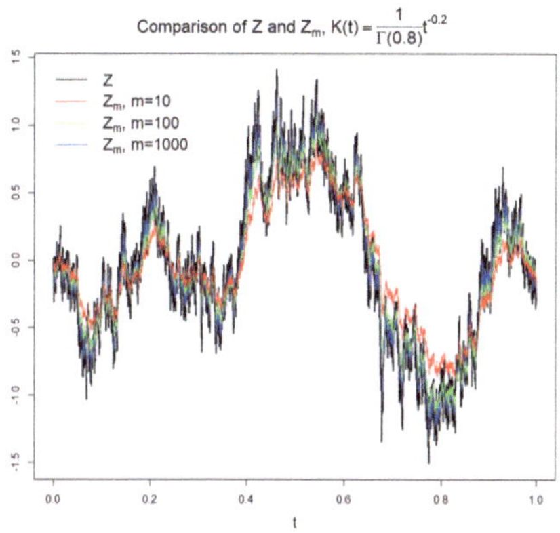

(b) Volterra noises Z and Z_m

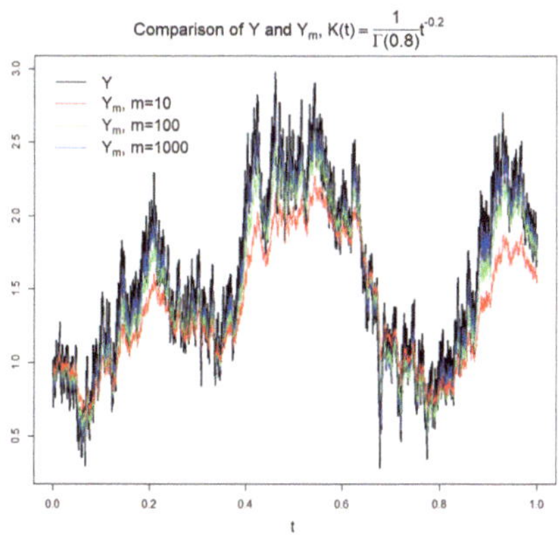

(c) Volatility processes Y and Y_m

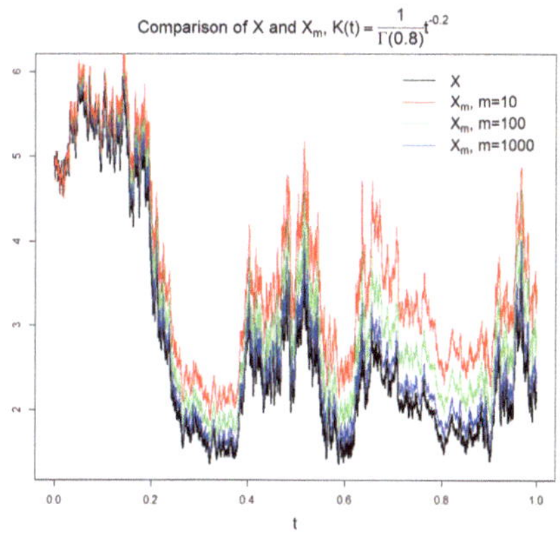

(d) Price processes S and S_m

Fig. 9.4 Approximation of the SVV model with rough fractional kernel

(2) for each path $i = 1, ..., N$, evaluate

$$\begin{gathered}\left(S_m^{(i)}(t_k), Y_m^{(i)}(t_k), U_{m,0}^{(i)}(t_k), ..., U_{m,m}^{(i)}(t_k)\right),\\ f(S_m^{(i)}(T))(S_m^{(i)}(t_{k+1}) - S_m^{(i)}(t_k)),\\ (S_m^{(i)}(t_{k+1}) - S_m^{(i)}(t_k))^2;\end{gathered}$$

(3) apply an appropriate non-parametric regression (with a mean squared error loss function) to the generated "dataset" treating

$$\left(S_m^{(i)}(t_k), Y_m^{(i)}(t_k), U_{m,0}^{(i)}(t_k), ..., U_{m,m}^{(i)}(t_k)\right), \quad i = 1, ..., N,$$

as "inputs" and

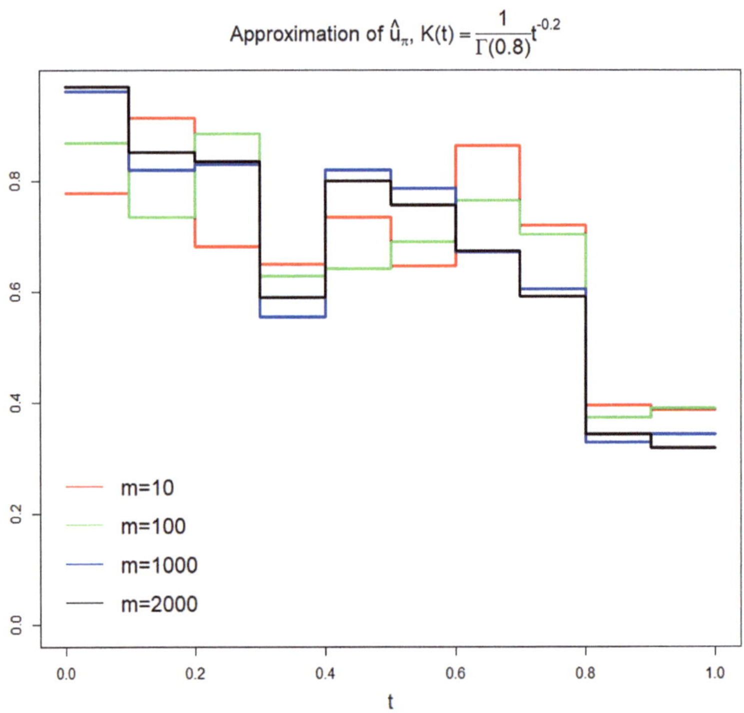

Fig. 9.5 Hedging strategy estimated for the path from Fig. 9.4 for $m = 10$ (red), $m = 100$ (green), $m = 1000$ (blue) and $m = 2000$ (black). The corresponding partition is $k/10$, $k = 0, 1, ..., 10$. The figure also illustrates slower rate of convergence in comparison to the Hölder kernel case: the black and blue lines are close to each other but the red and green lines (corresponding to relatively low values of m) differ substantially. Computation time: 24162 s for $m = 10$, 25348 s for $m = 100$, 38530 s for $m = 1000$ and 42431 s for $m = 2000$

$$f(S_m^{(i)}(T))(S_m^{(i)}(t_{k+1}) - S_m^{(i)}(t_k)), \quad (S_m^{(i)}(t_{k+1}) - S_m^{(i)}(t_k))^2, \quad i = 1, ..., N,$$

as "output" variables. The resulting estimates $\widehat{\Phi}_1(t_k, \cdot)$, $\widehat{\Phi}_2(t_k, \cdot)$ of the regression functions are then used to calculate $\widehat{u}_\pi$ on the interval $(t_k, t_{k+1}]$.

With this method, time consuming computations have to be performed only *once*: that is, when generating the artificial sample $(S_m^{(i)}, Y_m^{(i)}, U_{m,0}^{(i)}, ..., U_{m,m}^{(i)})$, $i = 1, ..., N$, and training the machine learning model. Afterwards, the calculation of the next step of the optimal hedge is almost instantaneous.

Example 9.5 Figure 9.6 contains approximations of the optimal hedging strategy constructed for the path of S_m from Example 9.3 that corresponds to $m = 10$. In order to obtain the dark green line, we simulated the "dataset" containing the "input" variables

$$\left(S_m^{(i)}(t_k), Y_m^{(i)}(t_k), U_{m,0}^{(i)}(t_k), ..., U_{m,m}^{(i)}(t_k)\right), \quad i = 1, ..., N,$$

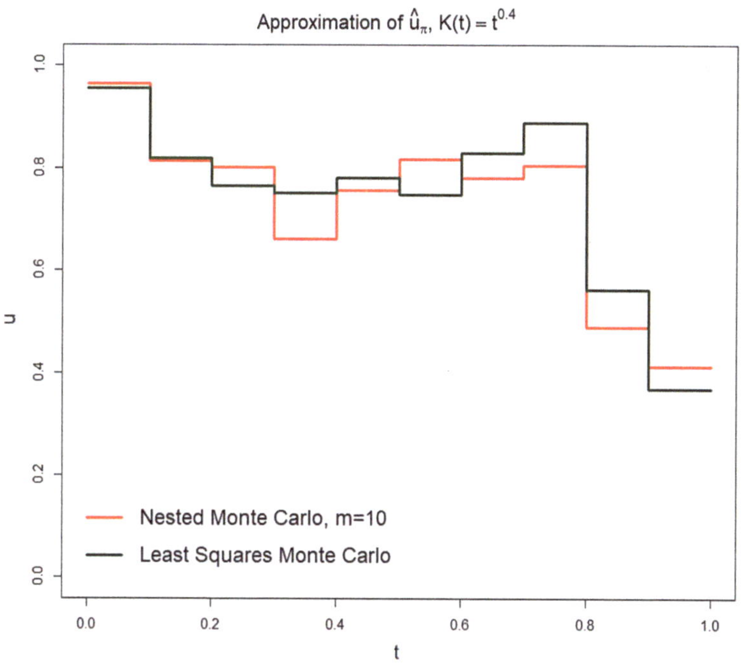

Fig. 9.6 Approximation of the optimal hedge using the Nested Monte Carlo method for $m = 10$ from Fig. 9.3 (red) and using the Least Squares Monte Carlo method (dark green)

together with the corresponding "output" variables

$$f(S_m^{(i)}(T))(S_m^{(i)}(t_{k+1}) - S_m^{(i)}(t_k)), \quad (S_m^{(i)}(t_{k+1}) - S_m^{(i)}(t_k))^2, \quad i = 1, ..., N,$$

with $N = 1000000$ for each $k = 0, 1, ..., 9$. In order to compute $\widehat{\Phi}_1(t_k, \cdot)$, $\widehat{\Phi}_2(t_k, \cdot)$, we used the idea from [9] and utilized a neural network with 3 hidden layers and 20 nods in each layer. The red line is exactly the path of $\hat{u}_\pi$ for $m = 10$ from Fig. 9.3 and is treated as a reference.

Note that the Nested approach described in Example 9.3 required 63231 seconds to simulate a path of $\hat{u}_\pi$ corresponding to a *single* realization of

$$(S_m, Y_m, U_{m,0}, ... U_{m,m})$$

whereas the Least Squares modification took roughly 6 days to simulate the initial dataset and about 5 h to fit the neural network. Once ready, the actual computations were conducted almost instantly.

References

1. Carriere, J.F.: Valuation of the early-exercise price for options using simulations and non-parametric regression. Insur. Math. Econ. **19**(1), 19–30 (1996). https://doi.org/10.1016/s0167-6687(96)00004-2
2. Di Nunno, G.: Stochastic integral representations, stochastic derivatives and minimal variance hedging. Stoch. Stoch. Rep. **73**(1–2), 181–198 (2002). https://doi.org/10.1080/10451120212869
3. Di Nunno, G., Eide, I.B.: Minimal-variance hedging in large financial markets: random fields approach. Stoch. Anal. Appl. **28**(1), 54–85 (2009). https://doi.org/10.1080/07362990903417979
4. Di Nunno, G., Rozanov, Y.A.: Stochastic integrals and adjoint derivatives. In: Stochastic Analysis and Applications, pp. 265–307. Springer Berlin Heidelberg, Berlin, Heidelberg (2007)
5. Di Nunno, G., Yurchenko-Tytarenko, A.: Sandwiched Volterra volatility model: Markovian approximations and hedging. Finance Stochast. **30**(1), 277–325 (2026). https://doi.org/10.1007/s00780-025-00584-2
6. Di Nunno, G., Øksendal, B., Proske, F.: Malliavin calculus for Lévy processes with applications to finance. Springer, Berlin, Heidelberg (2009)
7. Föllmer, H., Sondermann, D.: Hedging of non-redundant contingent claims. In: Contributions to Mathematical Economics, pp. 205–223. North-Holland, Amsterdam (1986)
8. Krah, A.S., Nikolić, Z., Korn, R.: A least-squares Monte Carlo framework in proxy modeling of life insurance companies. Risks **6**(2), 62 (2018). https://doi.org/10.3390/risks6020062
9. Krah, A.S., Nikolić, Z., Korn, R.: Least-squares Monte Carlo for proxy modeling in life insurance: neural networks. Risks **8**(4), 116 (2020). https://doi.org/10.3390/risks8040116
10. Krah, A.S., Nikolić, Z., Korn, R.: Machine learning in least-squares Monte Carlo proxy modeling of life insurance companies. Risks **8**(1), 21 (2020). https://doi.org/10.3390/risks8010021
11. Longstaff, F.A., Schwartz, E.S.: Valuing American options by simulation: a simple least-squares approach. Rev. Financ. Stud. **14**(1), 113–147 (2001). https://doi.org/10.1093/rfs/14.1.113
12. Nualart, D.: The Malliavin Calculus and Related Topics. Springer-Verlag, Berlin/Heidelberg (2006)
13. Pham, H.: On quadratic hedging in continuous time. Math. Methods Oper. Res. **51**(2), 315–339 (2000). https://doi.org/10.1007/s001860050091
14. Rheinländer, T., Schweizer, M.: On L2-projections on a space of stochastic integrals. Ann. Probab.**25**(4), 1810–1831 (1997). http://www.jstor.org/stable/2959516
15. Schweizer, M.: A guided tour through quadratic hedging approaches. In: E. Jouini, J. Cvitanic, Musiela (eds.) Handbooks in Mathematical Finance: Option Pricing, Interest Rates and Risk Management, pp. 538–574. Cambridge University Press (2001). https://doi.org/10.1017/CBO9780511569708.016
16. Tsitsiklis, J.N., Van Roy, B.: Regression methods for pricing complex American-style options. IEEE Trans. Neural Networks **12**(4), 694–703 (2001). https://doi.org/10.1109/72.935083

Appendix
Selected Results from Stochastic Analysis

Regularity Properties of Stochastic Processes

Let $(\Omega, \mathcal{F}, \mathbb{P})$ be a complete probability space and let S be a set equipped with some σ-algebra $\mathcal{S}$. Consider an S-valued random process $X = \{X(t),\ t \in [0, T]\}$, $T < \infty$, defined on $(\Omega, \mathcal{F}, \mathbb{P})$, i.e. a collection of $\mathcal{F}$-$\mathcal{S}$-measurable random variables $X(t, \cdot) : \Omega \to S$ indexed by $t \in [0, T]$.

Definition A.1 Given an S-valued random process $X = \{X(t),\ t \in [0, T]\}$ and any point $\omega \in \Omega$, the mapping

$$X(\cdot, \omega) : [0, T] \to S$$

is called a *realization*, a *trajectory* or a *sample path* of X, and the measurable space $(S, \mathcal{S})$ is called the *state space* of X.

Recall that two S-valued random variables ξ, η: $\Omega \to S$ are said to be equivalent if they coincide with probability 1, i.e.

$$\mathbb{P}(\xi = \eta) := \mathbb{P}(\{\omega \in \Omega : \xi(\omega) = \eta(\omega)\}) = 1.$$

For random processes, the situation is a bit more complex. In fact, there are several notions of equivalence, as summarised in the definition below.

Definition A.2 Let $X = \{X(t),\ t \in [0, T]\}$ and $Y = \{Y(t),\ t \in [0, T]\}$ be stochastic processes defined on the same probability space $(\Omega, \mathcal{F}, \mathbb{P})$ with the same state space $(S, \mathcal{S})$.

(1) Y is said to be a *modification* or a *version* of X, if

$$\mathbb{P}(X(t) = Y(t)) = 1, \quad t \in [0, T].$$

G. Di Nunno et al., *Volterra Volatility Models*, Springer Finance,
https://doi.org/10.1007/978-3-032-26576-0

(2) X and Y are said to be *indistinguishable* or *equivalent up to evanescence*, if

$$\bigcap_{t\in[0,T]} \{X(t) = Y(t)\} \in \mathcal{F}$$

and

$$\mathbb{P}\left(\bigcap_{t\in[0,T]} \{X(t) = Y(t)\}\right) = 1.$$

Clearly, indistinguishability is a stronger property: if X and Y are indistinguishable, Y is a modification of X (and X is a modification of Y). However, the converse is not true, as shown in the example below.

Example A.1 Let $(\Omega, \mathcal{F}, \mathbb{P}) = ([0, 1], \mathcal{B}([0, 1]), \lambda)$, where λ is the Lebesgue measure, $(S, \mathcal{S}) = (\mathbb{R}, \mathcal{B}(\mathbb{R}))$ and

$$Y(t, \omega) = 0, \quad X(t, \omega) = \mathbb{1}_{\{t\}}(\omega), \quad \omega \in [0, 1], \quad t \in [0, 1].$$

Then, for all $t \in [0, 1]$,

$$\mathbb{P}(X(t) = Y(t)) = \mathbb{P}(X(t) = 0) = \lambda([0, 1] \setminus \{t\}) = 1,$$

but

$$\mathbb{P}\left(\bigcap_{t\in[0,T]} \{X(t) = Y(t)\}\right) = \lambda\left(\bigcap_{t\in[0,T]} ([0, 1] \setminus \{t\})\right) = \lambda(\varnothing) = 0.$$

In many applications, in particular, if one is interested only in finite-dimensional distributions of a stochastic process, worrying about indistinguishability may be an overkill, and one may treat modifications as equivalent. In such cases, it is common to select a "*good*" modification with desirable properties and work exclusively with it. For example, if S is a complete separable metric space equipped with a metric μ, we may be interested in a modification such that the mapping $[0, T] \ni t \mapsto X(t, \omega) \in S$ is continuous with respect to t for (almost) all $\omega \in \Omega$. A further encouraging result supports this approach: if a continuous modification exists, it is unique up to indistinguishability. This claim is formalized in the theorem below.

Theorem A.1 *Let S be a complete separable metric space and $X, Y\colon [0, T] \times \Omega \to S$ be continuous processes such that*

$$\mathbb{P}(X(t) = Y(t)) = 1, \quad t \in [0, T].$$

Then X and Y are indistinguishable.

Our next task is to understand when a given process has a continuous modification. One of the most well-known and useful sufficient conditions is given by the celebrated Kolmogorov-Chentsov continuity theorem.

Theorem A.2 (Kolmogorov-Chentsov) *Let S be a complete separable metric space equipped with a metric μ and $\widetilde{X}$: $[0, T] \times \Omega \to S$ be a stochastic process. Assume that there exist positive constants α, β, $C > 0$ such that, for all $s, t \in [0, T]$,*

$$\mathbb{E}\left[\mu(\widetilde{X}(t), \widetilde{X}(s))^\alpha\right] \le C|t-s|^{1+\beta}. \tag{A.1}$$

Then there exists a modification X of $\widetilde{X}$ such that the paths of X are λ-Hölder continuous for every $\lambda \in \left(0, \frac{\beta}{\alpha}\right)$, i.e., for any $\lambda \in \left(0, \frac{\beta}{\alpha}\right)$ and $s, t \in [0, T]$,

$$|X(t) - X(s)| \le \Lambda_{\lambda,T}|t-s|^\lambda, \tag{A.2}$$

where $\Lambda_{\lambda,T}$ is a positive random variable that depends on λ and T.

Example A.2 Let $B = \{B(t),\ t \in [0, T]\}$ be a standard Brownian motion, i.e. a stochastic process satisfying the following conditions:

(1) $B(0) = 0$;
(2) B has independent increments;
(3) $B(t) - B(s) \sim \mathcal{N}(0, t-s)$ for all $0 \le s \le t \le T$.

Observe that, by the well-known formula for the moments of normal random variables,

$$\mathbb{E}\left[|B(t) - B(s)|^\alpha\right] = 2^{\frac{\alpha}{2}} \frac{\Gamma\left(\frac{\alpha+1}{2}\right)}{\sqrt{\pi}} |t-s|^{\frac{\alpha}{2}} =: C_\alpha |t-s|^{\frac{\alpha}{2}}$$

for any $\alpha > 0$. In particular, taking $\alpha := 2 + 2\beta$ for an arbitrary $\beta > 0$, we obtain

$$\mathbb{E}\left[|B(t) - B(s)|^{2+2\beta}\right] = C_{2+\beta}|t-s|^{1+\beta},$$

which shows that condition (A.1) is satisfied. Therefore, Kolmogorov-Chentsov theorem guarantees the existence of a modification that is λ-Hölder continuous for every $\lambda \in \left(0, \frac{\beta}{2+2\beta}\right)$. Moreover, since $\beta > 0$ is arbitrary, we conclude that Brownian motion admits a modification that is Hölder continuous of all orders up to (but not including!) $\frac{1}{2}$.

As a final remark, note that some textbooks (see e.g. [5, p. 19]) explicitly select this continuous modification in the definition of Brownian motion.

Kolmogorov-Chentsov theorem becomes especially convenient in the Gaussian case. Indeed, if $X : [0, T] \times \Omega \to \mathbb{R}$ is a centered Gaussian process with

$$\mathbb{E}[|X(t) - X(s)|^2] = \sigma^2(t, s), \quad 0 \le s \le t \le T,$$

one can express any absolute moment of order $\alpha > 0$ as

$$\mathbb{E}[|X(t) - X(s)|^\alpha] = 2^{\frac{\alpha}{2}} \frac{\Gamma\left(\frac{\alpha+1}{2}\right)}{\sqrt{\pi}} \sigma^\alpha(t, s).$$

Therefore, if

$$\mathbb{E}[|X(t) - X(s)|^2] \le C_2 |t - s|^{2H}, \quad 0 \le s \le t \le T,$$

for some $H \in (0, 1)$ and a positive constant C_2, it automatically holds that

$$\mathbb{E}[|X(t) - X(s)|^\alpha] \le C_\alpha |t - s|^{\alpha H}, \quad 0 \le s \le t \le T, \tag{A.3}$$

for some positive constant C_α. This inequality guarantees the existence of a modification that is Hölder continuous up to the order $\frac{\alpha H - 1}{\alpha} = H - \frac{1}{\alpha}$ for an arbitrary $\alpha > \frac{1}{H}$. This, in turn, implies the existence of a modification that is λ-Hölder continuous for every $\lambda \in (0, H)$.

However, while Kolmogorov-Chentsov theorem gives us information about the Hölder order λ in (A.2), it says nothing about the properties of the Hölder constant $\Lambda_{\lambda,T}$. For the analysis of the latter, it is convenient to use another result, known as the *Garsia-Rodemich-Rumsey inequality*.

Let Ψ be a non-negative even function on $\mathbb{R}$ and p be a non-negative even continuous function on $\mathbb{R}$ such that

- both Ψ and p are non-decreasing on $[0, \infty)$;
- $p(0) = 0$, $\Psi(\infty) = \infty$.

For $u \ge \Psi(0)$, denote

$$\Psi^{-1}(u) := \sup\{v : \Psi(v) \le u\}.$$

Theorem A.3 [3, Lemma 1.1] *Let f be a continuous real function on $[0, T]$ such that*

$$\int_0^T \int_0^T \Psi\left(\frac{f(x) - f(y)}{p(x - y)}\right) dx dy \le B < \infty.$$

Then, for all $s, t \in [0, T]$,

$$|f(s) - f(t)| \le 8 \int_0^{|s-t|} \Psi^{-1}\left(\frac{4B}{u^2}\right) dp(u).$$

If one chooses $\Psi(u) := |u|^\beta$ and $p(u) := |u|^{\alpha + \frac{1}{\beta}}$, where $\beta \ge 1$ and $\alpha > \frac{1}{\beta}$, Theorem (A.3) yields the following corollary.

Corollary A.1 *Let $f\colon [0, T] \to \mathbb{R}$ be a continuous function, $\beta \geq 1$ and $\alpha > \frac{1}{\beta}$. Then, for all $t, s \in [0, T]$,*

$$|f(t) - f(s)| \leq A_{\alpha,\beta}|t - s|^{\alpha - \frac{1}{\beta}} \left(\int_0^T \int_0^T \frac{|f(x) - f(y)|^\beta}{|x - y|^{\alpha\beta+1}} dxdy \right)^{\frac{1}{\beta}},$$

with the convention $0/0 = 0$, where

$$A_{\alpha,\beta} = 2^{3+\frac{2}{\beta}} \left(\frac{\alpha\beta + 1}{\alpha\beta - 1} \right).$$

Remark A.1 Note that Corollary A.1 holds even if

$$\int_0^T \int_0^T \frac{|f(x) - f(y)|^\beta}{|x - y|^{\alpha\beta+1}} dxdy = \infty.$$

Let $\widetilde{X} = \{\widetilde{X}(t),\ t \in [0, T]\}$ be a centered Gaussian process such that

$$\mathbb{E}[|\widetilde{X}(t) - \widetilde{X}(s)|^2] \leq C|t - s|^{2H}, \quad 0 \leq s \leq t \leq T,$$

for some $H \in (0, 1)$ and a positive constant $C > 0$. As noted above, it has a continuous modification which we denote $X = \{X(t),\ t \in [0, T]\}$. Fix an arbitrary continuous path $t \mapsto X(t, \omega)$, take $\beta > \frac{2}{H}$ and observe that, by Corollary A.1 with $\alpha = H - \frac{1}{\beta}$,

$$|X(t, \omega) - X(s, \omega)| \leq A_{H-\frac{1}{\beta},\beta}|t - s|^{H - \frac{2}{\beta}} \left(\int_0^T \int_0^T \frac{|X(x, \omega) - X(y, \omega)|^\beta}{|x - y|^{H\beta}} dxdy \right)^{\frac{1}{\beta}}. \tag{A.4}$$

If we can prove that

$$\int_0^T \int_0^T \frac{|X(x) - X(y)|^\beta}{|x - y|^{H\beta}} dxdy < \infty \quad a.s., \tag{A.5}$$

the inequality (A.4) can be viewed as the Hölder condition of order $H - \frac{2}{\beta}$ with the explicit (random) constant

$$\Lambda_{H-\frac{2}{\beta},T} = A_{H-\frac{1}{\beta},\beta} \left(\int_0^T \int_0^T \frac{|X(x) - X(y)|^\beta}{|x - y|^{H\beta}} dxdy \right)^{\frac{1}{\beta}} \quad a.s. \tag{A.6}$$

And, indeed, for any $r \geq 1$, Minkowski's integral inequality as well as the moment estimate (A.3) for X yield

$$\left(\mathbb{E}\left[\left(\int_0^T\int_0^T \frac{|X(x)-X(y)|^\beta}{|x-y|^{H\beta}}dxdy\right)^r\right]\right)^{\frac{1}{r}}$$
$$\leq \int_0^T\int_0^T \left(\frac{\mathbb{E}\left[|X(x)-X(y)|^{r\beta}\right]}{|x-y|^{Hr\beta}}\right)^{\frac{1}{r}} dxdy$$
$$\leq \int_0^T\int_0^T \left(\frac{C_{r\beta}|x-y|^{Hr\beta}}{|x-y|^{Hr\beta}}\right)^{\frac{1}{r}} dxdy$$
$$\leq T^2 C_{r\beta}^{\frac{1}{r}} < \infty,$$

which implies (A.5).

Observe that the last bound immediately yields that the random Hölder constant $\Lambda_{H-\frac{2}{\beta},T}$ given by (A.6) has moments of all orders. In summary, the following important result holds.

Theorem A.4 *Let* $\widetilde{X} = \{\widetilde{X}(t),\ t \in [0,T]\}$ *be a centered Gaussian process such that*

$$\mathbb{E}[|\widetilde{X}(t)-\widetilde{X}(s)|^2] \leq C|t-s|^{2H}, \quad 0 \leq s \leq t \leq T, \tag{A.7}$$

for some $H \in (0,1)$ *and a positive constant* $C > 0$. *Then* $\widetilde{X}$ *has a modification* X *that is* λ*-Hölder continuous for every* $\lambda \in (0,H)$, *i.e.*

$$|X(t)-X(s)| \leq \Lambda_{\lambda,T}|t-s|^\lambda, \quad 0 \leq s \leq t \leq T,$$

where $\Lambda_{\lambda,T}$ *is a positive random variable that depends on* λ *and* T. *Moreover,* $\Lambda_{\lambda,T}$ *can be chosen in such a way that, for any* $r \in \mathbb{R}$,

$$\mathbb{E}[\Lambda_{\lambda,T}^r] < \infty.$$

In fact, it is possible to prove an even stronger result: it turns out that a condition quite similar to (A.7) is not only *sufficient*, but also *necessary* for a Gaussian process to have a Hölder continuous modification. Namely, the following theorem holds.

Theorem A.5 [1, Theorem 1] *Fix* $H \in (0,1)$. *A centered Gaussian process* $\widetilde{X} = \{\widetilde{X}(t),\ t \in [0,T]\}$ *has a* λ*-Hölder continuous modification* X *for any* $\lambda \in (0,H)$, *i.e.*

$$|X(t)-X(s)| \leq \Lambda_{\lambda,T}|t-s|^\lambda, \quad 0 \leq s \leq t \leq T,$$

if and only if, for any $\lambda \in (0,H)$, *there exists a constant* $C_\lambda > 0$ *such that*

$$\mathbb{E}[|\widetilde{X}(t) - \widetilde{X}(s)|^2] \le C|t-s|^{2\lambda}, \quad 0 \le s \le t \le T.$$

Furthermore, the random variable $\Lambda_{\lambda,T} > 0$ can be chosen in such a way that

$$\mathbb{E}\left[\exp\left(a\Lambda_{\lambda,T}^{\alpha}\right)\right] < \infty$$

for all $a \in \mathbb{R}$ and $\alpha < 2$. In particular, the moments of all orders of $\Lambda_{\lambda,T}$ are finite.

Occupation Measures and Local Times

One of the key questions in the study of stochastic processes is how to characterize the "amount of time" a given process spends in a particular set. This appendix briefly outlines some essential concepts related to a powerful tool for addressing such problems: occupation measures and local times. It is an extremely broad and interesting topic, but we have to limit ourselves to the definitions and results that are explicitly used in the main part of the book. For more details on local times, we refer the reader to [6] as well as to the summary in [7, Section 10.1] and the excellent notes [2] which can give a good intuitive idea on this issue.

Let $X = \{X(t),\ t \ge 0\}$ be a real-valued measurable stochastic process.

Definition A.3 The *occupation measure* of the process X is the σ-finite (random) measure μ_X on $[0,\infty) \times \mathbb{R}$ defined by

$$\mu_X(A) := \lambda\left(\{t \ge 0 :\ (t, X(t)) \in A\}\right), \quad A \in \mathcal{B}([0,\infty) \times \mathbb{R}),$$

where λ denotes the Lebesgue measure on $[0,\infty)$.

The measure μ_X is a very natural instrument for quantifying the amount of time the process X spends in a given set. Indeed, if $A \in \mathcal{B}([0,\infty) \times \mathbb{R})$ is of the form $A = [0,t] \times B$ for some $B \in \mathcal{B}(\mathbb{R})$, then $\mu_X([0,t] \times B)$ represents the "total time" that X spends in the set B over the interval $[0,t]$. Moreover, by the definition of μ_X, for any $t \ge 0$ and $B \in \mathcal{B}(\mathbb{R})$,

$$\mu_{X,t}(B) := \mu_X([0,t] \times B) = \int_0^t \mathbb{1}_B(X(s))ds,$$

where the ds-integral is understood as the integral with respect to the Lebesgue measure λ. Since $\mu_{X,t}$ is itself a measure on $\mathbb{R}$, this implies that, for every bounded measurable function $f\colon \mathbb{R} \to \mathbb{R}$,

$$\int_0^t f(X(s))ds = \int_{\mathbb{R}} f(x)\mu_{X,t}(dx). \tag{A.8}$$

Next, assume that, with probability 1, the measure $\mu_{X,t}$ is absolutely continuous with respect to the Lebesgue measure λ with the Radon-Nikodym derivative

$$L^X(t, x) := \frac{d\mu_{X,t}}{d\lambda}(x), \quad x \in \mathbb{R}. \tag{A.9}$$

Definition A.4 If the Radon-Nikodym derivative L^X defined by (A.9) exists with probability 1, we say that the process X has a *local time* or *occupation density* L^X over $[0, t]$.

Remark A.2 By definition, if X has local time over $[0, t]$, (A.8) can be re-written as

$$\int_0^t f(X(s))ds = \int_{\mathbb{R}} f(x)L^X(t, x)dx. \tag{A.10}$$

This expression is known as the *occupation density* or *occupation time* formula.

Remark A.3 Note that, if $L^X(t, \cdot)$ exists, the Lebesgue differentiation theorem implies that, for almost all $x \in \mathbb{R}$,

$$L^X(t, x) = \lim_{\varepsilon \downarrow 0} \frac{1}{2\varepsilon} \int_0^t \mathbb{1}_{[x-\varepsilon, x+\varepsilon]}(X(s))ds. \tag{A.11}$$

Remark A.4 The representation (A.11) immediately yields some insights on several interesting properties of L^X under assumption that X has local time over $[0, t]$ for all $t \in [0, T]$, where $T < \infty$.

- One can choose a version of the local time such that the mapping

$$[0, T] \times \mathbb{R} \times \Omega \ni (t, x, \omega) \mapsto L^X(t, x, \omega)$$

  is jointly measurable w.r.t. all three variables. In particular, we can treat $L^X(t, x)$ as a well-defined random variable for every $t \in [0, T]$ and $x \in \mathbb{R}$.
- For any fixed $x \in \mathbb{R}$, $L^X(t, x)$ is a.s. monotonically non-decreasing w.r.t. $t \in [0, T]$.

Having the necessary definitions, let's now move on to a particular example. In the theorem below, we summarize without proof some of the well-known results related to the local times of standard Brownian motion.

Theorem A.6 *Let $B = \{B(t),\ t \ge 0\}$ be a standard Brownian motion. Then*

(1) the corresponding local time L^B is well-defined for all $t \ge 0$;
(2) for any $x \in \mathbb{R}$ and almost all $\omega \in \Omega$, the process $t \mapsto L^B(t, x)$ is non-negative, non-decreasing and increases only on the set $\{t \ge 0 :\ B(t, \omega) = x\}$;

(3) *L^B admits a modification that is jointly continuous w.r.t. (t, x). Moreover, this modification is a.s. jointly α-Hölder continuous of all orders $\alpha \in \left(0, \frac{1}{2}\right)$ w.r.t. t and x on all bounded intervals for t.*

Remark A.5 Theorem A.6 highlights that the paths of L are quite hard to imagine. On the one hand, for any fixed x, $t \mapsto L^B(t, x)$ is continuous and non-decreasing w.r.t. t a.s. On the other hand, $t \mapsto L^B(t, x)$ can increase only on the set $\{t \geq 0 : B(t) = x\}$, which is known to have Lebesgue measure 0 with probability 1!

Let us finalize this section by outlining another very important angle of view on local times.

In Definitions A.3 and A.4 above, we did not impose any specific assumptions on the process X. However, the concept of local times reveals its full richness when X is a continuous semimartingale. To conclude this section, let us briefly present an alternative and very important perspective on local times, starting with a fundamental result known as the *Tanaka formula*.

Theorem A.7 (Tanaka formula, [6, IV.43]) *Let $X = \{X(t),\ t \in [0, T]\}$ be a real continuous semimartingale. Then, for any fixed $x \in \mathbb{R}$, there exists a continuous non-decreasing adapted process $\{L^X_{sem}(t, x),\ t \in [0, T]\}$ such that*

$$|X(t) - x| - |X(0) - x| = \int_0^t \operatorname{sign}(X(s) - x)dX(s) + L^X_{sem}(t, x),$$

where $\operatorname{sign}(x) := -\mathbb{1}_{(-\infty,0]}(x) + \mathbb{1}_{(0,\infty)}(x)$. *The random mapping $t \mapsto L^X_{sem}(t, x)$ is called the* semimartingale local time *of X at x.*

Remark A.6 It is possible to prove that the semimartingale local time L^X_{sem} can be viewed as a two-parameter stochastic process $L^X_{\text{sem}} = \{L^X_{\text{sem}}(t, x),\ (t, x) \in [0, T] \times \mathbb{R}\}$. In fact, by [6, Theorem IV.44.2], L^X_{sem} has a modification that is jointly continuous in t and right-continuous with left limits in x. Moreover,

$$L(t, x) - L(t, x-) = 2 \int_0^t \mathbb{1}_{\{X(s)=x\}} dA(s),$$

where A is the bounded variation component in the canonical decomposition of X.

An obvious question arises: how does the semimartingale local time defined in Theorem A.7 relate to the occupation density introduced in Definition A.4? The connection between these two notions is established by the following result, which presents a "semimartingale variant" of the occupation density formula (A.10).

Theorem A.8 [6, Theorem IV.45.1] *Let $X = \{X(t),\ t \in [0, T]\}$ be a real continuous semimartingale with semimartingale local time process $L^X_{sem} =$*

$\{L^X_{sem}(t,x),\ (t,x)\in[0,T]\times\mathbb{R}\}$ *assumed to be continuous in t and right-continuous with left limits in x. Then, for every bounded measurable function* $f:\mathbb{R}\to\mathbb{R}$,

$$\int_0^t f(X(s))d[X]_s = \int_{\mathbb{R}} f(x)L^X_{sem}(t,x)dx. \tag{A.12}$$

Example A.3 Let $B=\{B(t),\ t\geq 0\}$ be a standard Brownian motion. Then, by (A.10) and (A.12), for any bounded measurable function f,

$$\int_0^t f(B(s))ds = \int_{\mathbb{R}} f(x)L^B(t,x)dx = \int_{\mathbb{R}} f(x)L^B_{\text{sem}}(t,x)dx,$$

which implies that $L^B = L^B_{\text{sem}}$.

Example A.4 Let $U=\{U(t),\ t\geq 0\}$ be an Ornstein-Uhlenbeck process defined by

$$U(t) = U(0) - a\int_0^t U(s)ds + \sigma B(t), \quad t\in[0,T],$$

where $B=\{B(t),\ t\geq 0\}$ is a standard Brownian motion and $U(0), a\in\mathbb{R}$, $\sigma>0$. Then, by (A.12), for any bounded measurable function f,

$$\sigma^2\int_0^t f(U(s))ds = \int_{\mathbb{R}} f(x)L^U_{\text{sem}}(t,x)dx.$$

Using this relation, it is straightforward to prove that L^U exists and, for any $t\in[0,T]$, $x\in\mathbb{R}$,

$$L^U(t,x) = \frac{1}{\sigma^2}L^U_{\text{sem}}(t,x).$$

Finally, by computations similar to the ones in [6, Section IV.44], we remark that the process $x\mapsto L^U(t,x)$ is Hölder continuous w.r.t. x up to order $\frac{1}{2}$ for all fixed $t\in[0,T]$.

Example A.5 Let $B=\{B(t),\ t\geq 0\}$ be a standard Brownian motion and $X(t)=|B(t)|$, $t\in[0,T]$. By (A.12), for any bounded measurable function f,

$$\int_0^t f(X(s))d[X]_s = \int_0^t f(X(s))ds = \int_{\mathbb{R}} f(x)L^X_{\text{sem}}(t,x)dx,$$

so it is easy to see that L^X exists and $L^X = L^X_{\text{sem}}$. Moreover, using (A.11) as well as the calculations in [6, IV.44, p. 101], one can prove that

$$L^X(t,x) = \begin{cases} L^B(t,x) + L^B(t,-x), & x \geq 0, \\ 0, & x < 0. \end{cases}$$

In particular, Theorem A.6 implies that, for any fixed $t \in [0, T]$, the mapping

$$[0, \infty) \ni x \mapsto L^X(t,x)$$

has a modification that is α-Hölder continuous of all orders $\alpha \in \left(0, \frac{1}{2}\right)$ w.r.t. x.

Remark A.7 In Example A.5, we restrict attention to the domain $x \in [0, \infty)$ when discussing Hölder continuity. This is because we cannot claim that the process $x \mapsto L^X(t,x)$ defined on $x \in \mathbb{R}$ is continuous at $x = 0$ since

$$L^X(t,0) - L^X(t,0-) = 2L^B(t,0),$$

see, e.g., the discussion in [6, IV.44, p. 101–102].

Malliavin Calculus

In this section, we recall several main definitions and results related to Malliavin calculus. For more details, we refer to [4]. Let $(\Omega, \mathcal{F}, \mathbb{P})$ be a complete probability space and suppose that $\mathcal{H}$ is a real separable Hilbert space with scalar product $\langle \cdot, \cdot \rangle_{\mathcal{H}}$.

Definition A.5 A stochastic process $W = \{W(h),\ h \in \mathcal{H}\}$ is called an *isonormal Gaussian process* on $\mathcal{H}$ if

- W is a centered Gaussian family of random variables;
- for any $h_1, h_2 \in \mathcal{H}$,
$$\mathbb{E}[W(h_1)W(h_2)] = \langle h_1, h_2 \rangle_{\mathcal{H}}.$$

Example A.6 Let $\mathcal{H} = L^2([0,T];\mathbb{R}^d)$ and W be a d-dimensional Brownian motion on $[0, T]$. For any $h \in L^2([0,T];\mathbb{R}^d)$, define

$$W(h) := \int_0^T h(s)dW(s).$$

Then $W = \{W(h),\ h \in \mathcal{H}\}$ is an isonormal Gaussian process on $L^2([0,T];\mathbb{R}^d)$.

Let now $W = \{W(h),\ h \in \mathcal{H}\}$ be an isonormal Gaussian process on $\mathcal{H}$. Denote by $C_p^\infty(\mathbb{R}^n)$ the set of all infinitely differentiable functions with the derivatives of at most polynomial growth at infinity.

Definition A.6 Random variables ξ of the form

$$\xi = f(W(h_1), \dots, W(h_n)), \tag{A.13}$$

where $f \in C_p^\infty(\mathbb{R}^n)$, $h_1, \dots, h_n \in \mathcal{H}$, $n \geq 1$, are called smooth.

Definition A.7 The *stochastic* or *Malliavin derivative* of a smooth random variable ξ of the form (A.13) is the $\mathcal{H}$-valued random variable given by

$$D\xi = \sum_{i=1}^{n} \frac{\partial f}{\partial x_i}(W(h_1), \dots, W(h_n))h_i. \tag{A.14}$$

According to Proposition 1.2.1 from [4], the operator D is closable from $L^p(\Omega)$ to $L^p(\Omega; \mathcal{H})$ for any $p \geq 1$. We shall use the same notation D for the closure and, for any $p \geq 1$, we will denote the domain of D in $L^p(\Omega)$ by $\mathbb{D}^{1,p}$. In other words, $\mathbb{D}^{1,p}$ is the closure of the class of smooth random variables with respect to the norm

$$\|\xi\|_{1,p} := \left(\mathbb{E}|\xi|^p + \mathbb{E}\|D\xi\|_{\mathcal{H}}^p\right)^{1/p}.$$

Remark A.8 For $p = 2$, the space $\mathbb{D}^{1,2}$ is a Hilbert space with respect to the inner product

$$\langle \xi, \eta \rangle_{1,2} = \mathbb{E}\xi\eta + \mathbb{E}\left[\langle D\xi, D\eta \rangle_{\mathcal{H}}\right]. \tag{A.15}$$

We can also define the iteration of the operator D: for any $k \geq 1$ and a smooth random variable ξ, the iterated derivative $D^k\xi$ is a random variable with values in $\mathcal{H}^{\otimes k}$. For any $p \geq 1$, the operator D^k is also closable from the set of smooth random variables into $L^p(\Omega; \mathcal{H}^{\otimes k})$. We will denote by $\mathbb{D}^{k,p}$ the domain of the closure, i.e. $\mathbb{D}^{k,p}$ is the completion of the family of smooth random variables with respect to the norm

$$\|\xi\|_{k,p} := \left(\mathbb{E}|\xi|^p + \sum_{i=1}^{k} \mathbb{E}\left[\|D^i\xi\|_{\mathcal{H}^{\otimes i}}^p\right]\right)^{1/p}.$$

Definition A.8 We say that a random variable ξ is Malliavin differentiable if $\xi \in \mathbb{D}^{1,p}$ for some $p \geq 1$.

Next, let us provide two "chain rule" results that are very useful for establishing Malliavin differentiability.

Theorem A.9 ([4], Proposition 1.2.3) *Let $F: \mathbb{R}^m \to \mathbb{R}$ be a continuously differentiable function with bounded partial derivatives, and fix $p \geq 1$. Suppose that $\xi = (\xi_1, \dots, \xi_m)$ is a random vector whose components belong to the space $\mathbb{D}^{1,p}$. Then $F(\xi) \in \mathbb{D}^{1,p}$ and*

$$DF(\xi) = \sum_{i=1}^{m} \frac{\partial F(\xi)}{\partial x_i} D\xi_i. \tag{A.16}$$

Theorem A.10 ([4], Proposition 1.2.4) *Let $F : \mathbb{R}^m \to \mathbb{R}$ be Lipschitz, i.e. there exists a constant $C > 0$ such that, for all $x_1, x_2 \in \mathbb{R}^m$,*

$$|F(x_1) - F(x_2)| \leq C|x_1 - x_2|.$$

Suppose that $\xi = (\xi_1, \ldots, \xi_m)$ is a random vector whose components belong to the space $\mathbb{D}^{1,2}$ and assume that the law of ξ is absolutely continuous with respect to the Lebesgue measure on $\mathbb{R}^m$. Then $F(\xi) \in \mathbb{D}^{1,2}$ and

$$DF(\xi) = \sum_{i=1}^{m} \frac{\partial F(\xi)}{\partial x_i} D\xi_i,$$

where the derivatives $\frac{\partial F(\xi)}{\partial x_i}$ are understood in the weak sense.

Next, let us move to the notion of the Skorokhod operator.

Definition A.9 The *divergence* or *Skorokhod operator* δ is the adjoint of the operator D, i.e. an unbounded operator on $L^2(\Omega; \mathcal{H})$ with values in $L^2(\Omega)$ such that:

(i) the domain of δ, denoted by $\mathsf{Dom}\,\delta$, is the set of $\mathcal{H}$-valued square integrable random variables $\zeta \in L^2(\Omega; \mathcal{H})$ such that for all $\xi \in \mathbb{D}^{1,2}$:

$$|\mathbb{E}\left[\langle D\xi, \zeta \rangle_{\mathcal{H}}\right]| \leq C_\zeta (\mathbb{E}\xi^2)^{1/2}, \tag{A.17}$$

where C_ζ is some constant depending on ζ;
(ii) if ζ belongs to $\mathsf{Dom}\,\delta$, then $\delta(\zeta)$ is the element of $L^2(\Omega)$ characterized by

$$\mathbb{E}[\xi\delta(\zeta)] = \mathbb{E}\left[\langle D\xi, \zeta \rangle_{\mathcal{H}}\right] \quad \text{for any } \xi \in \mathbb{D}^{1,2}. \tag{A.18}$$

Remark A.9 The Skorokhod operator δ is closed.

Let W be a standard d-dimensional Brownian motion on $[0, T]$, $\mathcal{H} = L^2([0, T]; \mathbb{R}^d)$ be the associated Hilbert space (see Example A.6) and δ be the corresponding divergence operator. In this case, the divergence $\delta(\zeta)$ is called the *Skorokhod stochastic integral* of the process ζ with respect to W and is denoted as follows:

$$\delta(\zeta) = \int_0^T \zeta(t)\delta W(t). \tag{A.19}$$

In fact, according to [4, Section 1.3.3], if the process ζ is Itô integrable w.r.t. W, the Skorokhod integral (A.19) is well-defined and coincides with the Itô integral.

References

1. Azmoodeh, E., Sottinen, T., Viitasaari, L., Yazigi, A.: Necessary and sufficient conditions for Hölder continuity of Gaussian processes. Statist. Probab. Lett. **94**, 230–235 (2014). https://doi.org/10.1016/j.spl.2014.07.030
2. Björk, T.: The pedestrian's guide to local time. In: Barrieu, P. (ed.) Risk and Stochastics, pp. 43–67. World Scientific (2019)
3. Garsia, A., Rodemich, E., Rumsey, H.: A real variable lemma and the continuity of paths of some Gaussian processes. Indiana Univ. Math. J. **20**, 565–578 (1970)
4. Nualart, D.: The Malliavin Calculus and Related Topics. Springer, Berlin/Heidelberg (2006)
5. Revuz, D., Yor, M.: Continuous Martingales and Brownian Motion. Springer, Berlin, Heidelberg (1999)
6. Rogers, L.C.G., Williams, D.: Diffusions. Cambridge University Press, Markov Processes and Martingales (2000)
7. Samorodnitsky, G.: Stochastic Processes and Long Range Dependence. Springer, Cham (2016)

Index

G. Di Nunno et al., *Volterra Volatility Models*, Springer Finance,
https://doi.org/10.1007/978-3-032-26576-0

MIX
Papier aus verantwortungsvollen Quellen
Paper from responsible sources
FSC® C105338

If you have any concerns about our products,
you can contact us on
ProductSafety@springernature.com

In case Publisher is established outside the EU,
the EU authorized representative is:
Springer Nature Customer Service Center GmbH
Europaplatz 3, 69115 Heidelberg, Germany

Printed by Libri Plureos GmbH
in Hamburg, Germany